"As an archipelago, islands are better able to flex and stre... influence. Their political strength lies in their commonali... ferences; although – when presented as complementary assets – differences may be of immense value in attracting a wide range of types of tourism, while encouraging multi-island hopping. These are some of the forces and discontinuities that make archipelagos so fascinating to examine, and why this volume is so significant."

Richard Butler, *Emeritus Professor of Tourism, University of Strathclyde, Glasgow, Scotland, United Kingdom.*

"Has island tourism shifted since the COVID-19 pandemic? Is the industry intent to 'build back better' or has it become even more unsustainable? Whichever the case, archipelagos offer illuminating insights to such questions. A complex interplay of size, remoteness, urbanization, connectivity and governance issues impacts island and archipelago tourism in specific ways. This book builds on Baldacchino's 2015 work, proposing a unique lens to better understand the many dynamics affecting island tourism. Worth a read!"

Rachel Dodds, *Ted Rogers School of Hospitality and Tourism Management, Toronto Metropolitan University, Canada.*

"Anyone interested in untangling the conundrum of how tourism-dependent islands in archipelagos can move towards more sustainable futures is in for a treat. This edited book significantly updates Baldacchino's earlier volume on the subject, responding to impacts of the COVID-19 pandemic and combining fascinating contextual material with in-depth analysis of multi-island locations from around the world."

Regina Scheyvens, *Professor of International Development, Massey University, New Zealand.*

"*Archipelago Tourism Revisited* is a must read and a welcome addition to island tourism scholarship. This collection probes the tensions implicit in the imperatives of cooperation and competition that exist in and between islands as post-pandemic tourism destinations. Assumptions of 'island homogeneity' are upended by the contents of this fine collection. Instead, here are vibrant examples of how islands are navigating core-periphery relations in search of a more fulfilling post-pandemic future."

Sherma Roberts, *University of the West Indies, Cave Hill Campus, Barbados.*

"Clearly, some islands are more equal than others: this book uses a 'core-periphery' framework to deepen our understanding of the nuances of island tourism, post-COVID-19. Twelve case studies highlight the challenges of tourism within archipelagos, including transport infrastructure at airports and seaports, supply-side constraints, marginalization of outer islands, and quality of life issues. Archipelagic tourism requires new perspectives: Baldacchino drives this 'turn' in island and tourism studies forward."

Michelle McLeod, *University of the West Indies, Mona Campus, Jamaica.*

Archipelago Tourism Revisited

This timely and innovative book explores the dynamics of inter-island/island-island tourism – also known as archipelago tourism – on the cusp of the post-pandemic epoch.

Embellished with illustrative maps and diagrams, the volume examines what novel approaches have been developed, if at all, so as not to repeat past mistakes, and nurture a more sustainable, 'island tourism' business model. It looks at how the political-economic relationship between main and outer islands changed during the pandemic and, if so, whether this shift has had a bearing on current tourism policy. The book also explores how these and other changes are reflected in how: islands are branded; island destinations are marketed; and island transport logistics play out. An array of archipelagos of varying sizes and locations is explored, assuring a global perspective. The book furthers our understanding of core-periphery dynamics in archipelago tourism.

The volume will be of interest to students, researchers, policy makers and academics in the fields of tourism policy and planning, sustainability, island studies and development studies.

Godfrey Baldacchino, PhD (Warwick, UK), is Professor of Sociology at the University of Malta and Malta Ambassador-at-Large for Islands and Small States. He is founding Editor of *Island Studies Journal* (2006–2016), former Canada Research Chair and UNESCO Co-Chair in Island Studies at the University of Prince Edward Island (2003–2013; 2016–2020), and former President of the International Small Islands Studies Association (ISISA) (2014–2022). His tourism-related books include *Global Tourism and Informal Labour Relations: The Small Scale Syndrome at Work* (1997), *Extreme Tourism: Lessons from the World's Cold-Water Islands* (2006), and *Archipelago Tourism: Practices and Policies* (2015).

New Directions in Tourism Analysis

Edited by **Dimitri Ioannides**, *E-TOUR, Mid Sweden University, Sweden*

Although tourism is becoming increasingly popular both as a taught subject and an area for empirical investigation, the theoretical underpinnings of many approaches have tended to be eclectic and somewhat underdeveloped. However, recent developments indicate that the field of tourism studies is beginning to develop in a more theoretically informed manner, but this has not yet been matched by current publications.

The aim of this series is to fill this gap with high quality monographs or edited collections that seek to develop tourism analysis at both theoretical and substantive levels using approaches which are broadly derived from allied social science disciplines such as Sociology, Social Anthropology, Human and Social Geography, and Cultural Studies. As tourism studies covers a wide range of activities and sub fields, certain areas such as Hospitality Management and Business, which are already well provided for, would be excluded. The series will therefore fill a gap in the current overall pattern of publication.

Suggested themes to be covered by the series, either singly or in combination, include: consumption; cultural change; development; gender; globalisation; political economy; social theory; and sustainability.

For more information about this series, please visit www.routledge.com/New-Directions-in-Tourism-Analysis/book-series/ASHSER1207

Archipelago Tourism Revisited

Core-Periphery Dynamics after the Pandemic

Edited by
Godfrey Baldacchino

Routledge
Taylor & Francis Group

LONDON AND NEW YORK

First published 2025
by Routledge
4 Park Square, Milton Park, Abingdon, Oxon OX14 4RN

and by Routledge
605 Third Avenue, New York, NY 10158

Routledge is an imprint of the Taylor & Francis Group, an informa business

British Library Cataloguing-in-Publication Data
A catalogue record for this book is available from the British Library

Library of Congress Cataloging-in-Publication Data
Names: Baldacchino, Godfrey, editor.
Title: Archipelago tourism revisited : core-periphery dynamics after the pandemic / Edited by Godfrey Baldacchino.
Description: Abingdon, Oxon ; New York, NY : Routledge, 2025. |
Series: New directions in tourism analysis | Includes bibliographical references and index. |
Identifiers: LCCN 2024030969 (print) | LCCN 2024030970 (ebook) |
ISBN 9781032586779 (hardback) | ISBN 9781032586786 (paperback) |
ISBN 9781003451037 (ebook)
Subjects: LCSH: Tourism--Management--Case studies. | Archipelagoes--Case studies. | COVID-19 Pandemic, 2020---Influence.
Classification: LCC G155.A1 A638 2025 (print) | LCC G155.A1 (ebook) |
DDC 910.68--dc23/eng/20240719
LC record available at https://lccn.loc.gov/2024030969
LC ebook record available at https://lccn.loc.gov/2024030970

ISBN: 978-1-032-58677-9 (hbk)
ISBN: 978-1-032-58678-6 (pbk)
ISBN: 978-1-003-45103-7 (ebk)

DOI: 10.4324/9781003451037

Typeset in Times New Roman
by Taylor & Francis Books

Contents

Figures

Tables

Acknowledgements

It was a southern hemisphere autumn morning in March 2011 when I walked into the Office of Dr Elaine Stratford, at the University of Tasmania campus in Hobart, Tasmania, Australia, with the germs of an idea. She had asked me the previous week to suggest a theme that, in my opinion, was crying out loud for research within the general field of island studies. After our two-hour meeting that morning, when we also brainstormed on a large sheet of newsprint, we had convinced each other that archipelagos could be that theme.

We can probably admit that 'archipelago studies' took off that day. There was a cascade and flurry of more conversations, and other colleagues at UTAS – notably Andrew Harwood and alumnus Carol Farbotko – came on board, as did Elizabeth McMahon from the University of New South Wales, still in Australia, and then, later, Jonathan Pugh at the University of Newcastle, United Kingdom. There was also an outreach and sharing of thoughts and conceptualisations that found their way onto the research agendas of others elsewhere. These include the series of postgraduate seminars at Rutgers State University of New Jersey, USA, with Yolanda Martínez-San Miguel, over 2015–2016; with Brian Russell Roberts and Michelle Ann Stephens, who edited *Archipelagic American studies* in 2017 (with Elaine's Stratford chapter in that book); as well as *Contemporary archipelagic thinking: Towards new comparative methodologies and disciplinary formations* (2020) edited by Yolanda and Michelle Ann (and to which I contributed a chapter). Beatriz Llenin-Figueroa (2022) renders Puerto Rico as a core member of the abundant Caribbean community of islands in *Affect, archive, archipelago*. Most recently, Alicia Albinia weaves a portrait of Great Britain with and through its islands in *The Britannias: An archipelago's tale* (2024). I can hazard to say that the 'archipelago effect' is now recognised as a fundamental variable in island studies: along with the likes of land area, population, altitude, location, boundedness, distance from mainland, and the powerful role of the engirdling sea.

I was also looking for a policy and industrial application to the concept, and that is how tourism came naturally into the picture. A collaboration with Eduardo Costa Duarte Ferreira, then at the University of the Azores, Portugal, led to a fascinating paper on archipelago tourism there. There was clearly enough material and excitement for a proper edited book on archipelago tourism, nurtured with a workshop at an international conference held in Maspalomas, Gran Canaria, in the Canary Islands, Spain (another archipelago). *Archipelago tourism* appeared in 2015, and this book is a sequel to that volume.

This text departs from and builds on that volume in three ways. First, it proposes an analysis of archipelago tourism dynamics from an explicit 'core-periphery' perspective: all archipelagos are unequal, in spite of the glossy and colourful brochures suggesting one big happy family of different islands and constituent islanders. The role of regional and national capital cities cannot be underestimated in decisions that, for example, determine what transport infrastructure happens and where. Second, it examines the manner in which archipelagic churns and twists were halted during Covid-19 and then restated as the world started emerging from the pandemic. (And, as we read in this book, tourism dynamics did not change much in the immediate post-Covid-19 from pre-Covid-19.) Third, is the organisation of material in geo-political terms: with sub-national, national, and supra-national clusters of islands involved in review. The political and administrative instruments at policy makers' disposal can change when one's capital city is on another island or even on a mainland many kilometres away; or when the main regional airport and/or seaport is located in another island jurisdiction altogether.

I thank my publisher Routledge / Taylor and Francis for their confidence and trust. Jennifer Hicks ably dealt with edits at the proofs stage; and Emma Travis and then Harriet Cunningham from their tourism editorial office helped me to steer the project forward as smoothly as can be imagined.

I am grateful to John Connell for his perspicacious comments on an earlier draft of the editorial introduction. And a special thank you to a patient Mr Andrea Pace for preparing all the maps that accompany the text.

Additional acknowledgements

For *Chapter 2*, the authors thank Yoshitaka Kuninaka, President of the Karimata Neighbourhood Association, for providing important insights into the topic of this chapter.

For *Chapter 3*, the lyrics of the song 'El turismo eres tú' are kindly reproduced with the permission of the author of the lyrics, Rubén Antonio García Sancho, who uses the artistic name Rubén Memories. Aina Gomis has been funded in her work by the Conselleria d'Educació, Universitat i Recerca del Govern Balear through the Direcció General de Política Universitària i Recerca and by the Fondo Social Europeo from July 2022 onwards (Grant No. FPI /029/2021).

For *Chapter 5*, the author thanks Associate Professor Kristoffer Esquejo for his valuable materials and validation, Professor Vicente Villan for providing additional information, Miss Myla Villegas of Romblon Investment and Tourisms Promotion Office for sharing relevant documentation, and to anonymous respondents for their support.

For *Chapter 8*, Godfrey Baldacchino thanks the University of Seychelles for hosting him in March 2022, and the University of Malta for the granting of sabbatical leave. Hervé Atayi thanks Jemima Doudee from the Seychelles

Ministry of Agriculture, Climate Change and Environment for providing pertinent data.

For *Chapter 11*, Arjen Alberts thanks Thierry Beltrand, director of the Guadeloupe, Saint-Martin and Saint-Barthélemy department of the Institut d'Émission des Départements d'Outre-Mer (IEDOM) and Veronica Jansen-Webster, head of the Statistical Department of the Government of Sint Maarten (STAT).

The usual disclaimers apply.

Godfrey Baldacchino
Marsaskala, Malta, May 2024

List of abbreviations

A	*The Australian* (newspaper)
ACAT	Atlantic Canada Agreement on Tourism
ACIF	Associação Comercial e Industrial do Funchal (Madeira Chamber of Commerce)
ACOA	Atlantic Canada Opportunities Agency
ACP	Africa, Caribbean, Pacific
AIHRIM	Association des hôteliers et restaurateurs Ile Maurice
AIMS	Asian Institute of Maritime Studies, The Philippines (in Chapter 5)
AIMS	Atlantic Institute of Market Studies, Canada (in Chapter 6)
ANA	All Nippon Airlines
ANSA	Agenzia Nazionale Stampa Associata (Italian News Agency)
ATR	Association du Tourisme Réunie (in Chapter 12)
ATR	Avions de Transport Régional (in Chapter 5)
CARICOM	Caribbean Community
CEO	Chief Executive Officer
CLIA	Cruise Line Industry Association
Covid-19	Coronavirus disease 2019, caused by the virus SARS-CoV-2
DAR	Dominion Atlantic Railway, Canada
DICT	Department of Information and Communications Technology, The Philippines
DILG	Department of Interior and Local Government, The Philippines
DoT	Department of Tourism, The Philippines
DREM	Direção Regional de Estatística da Madeira
DTCAR	Department of Tourism, Culture, Arts and Recreation
EEZ	Exclusive Economic Zone
ERDF	European Regional Development Funds
EU	European Union
FHTA	Fiji Hotel & Tourism Association
FM Radio	Frequency modulation radio broadcasting
FY	Fiscal Year
GDP	Gross Domestic Product

GTA	Gozo Tourism Association (in Chapter 10)
GTA	Grenada Tourism Authority (in Chapter 7)
GVA	Gross Value Added
IATA	International Air Transport Association
IBESTAT	Statistical Agency of the Balearic Islands
IDC	Islands Development Company, Seychelles
IEDOM	Institut d'Emission des Départements d'Outre-Mer
IMF	International Monetary Fund
INSEE	Institut National de la Statistique et des Etudes Economiques
IUCN	International Union for the Conservation of Nature
JTA	Japan Transocean Air
JTB	Japan Tourist Board
KCA	Karimata Community Association
KCU	Karimata Cooperative Union
LDS	Linyon Demokratik Seselwa
LGU	Local Government Unit
LP	Lonely Planet
MIMAROPA	The Southwestern Tagalog Administrative Region, The Philippines
MIRAB	Migration, remittances, aid and bureaucracy
MP	Member of Parliament
MPA	Marine Protected Area
MTPA	Mauritius Tourism Promotion Authority
NB	New Brunswick, Canada
NDC	National Democratic Congress, Grenada
NGO	non-governmental organisation
NL	Newfoundland & Labrador, Canada
NNP	New National Party
NRPB	National Recovery Program Bureau
NRRP	National Recovery and Resilience Plan
NS	Nova Scotia, Canada
NSO	National Statistical Office
NZ	New Zealand
OCT	Overseas Countries and Territories
OECS	Organisation of Eastern Caribbean States
OI	Outer islands
OMT	Outermost Territory
PAL	Provincial Airlines, Canada
PEI	Prince Edward Island, Canada
PESD	Plan for Economic and Social Development, Region of Madeira
PJIA	Princess Juliana International Airport
PLDT	Philippine Long Distance Telephone Company
PROFIT	People, resources, overseas management, finance and transport
PS	Porto Santo

PSO	Public Service Obligation
RAC	Ryukyu Air Commuter
RAM	Região Autónoma da Madeira (Autonomous Region of Madeira)
RO-RO	roll-on, roll off ferries
SBS	Seychelles Bureau of Statistics
S-H	*The Sun-Herald* (newspaper)
SIDS	Small island developing state(s)
SIF	Seychelles Islands Foundation
SITE	Small Island Tourist Economy
SMH	*Sydney Morning Herald* (newspaper)
SNIJ	Sub-national island jurisdiction
SONA	State of the Nation Address, The Philippines
ST	*Sunday Telegraph* (newspaper)
TALC	Tourism Area Life Cycle
TouRAB	Tourism, remittances, aid and bureaucracy
TPI	Tourism Penetration Index
UK	United Kingdom
UKOT	United Kingdom Overseas Territory
UN	United Nations
UNCLOS	United Nations Convention on the Law of the Sea
UNDP	United Nations Development Programme
UNESCO	United Nations Educational, Scientific and Cultural Organization
UNWTO	United Nations World Tourism Organization
UREC	University Research Ethics Committee, University of Malta
US	United States (of America)
WHC	World Heritage Convention (UNESCO)
WHO	World Health Organization
WWII	World War II

Contributors

Karl Agius PhD (Malta) is Visiting Lecturer in the Department of Tourism Management and at the Islands and Small States Institute, both at the University of Malta, Malta, where he lectures on island tourism, research methods in island studies, as well as sustainability of the tourism and hospitality sectors. His research interests include ecotourism and archipelagos. His doctoral research studied the marine ecotourism potential of the central Mediterranean Islands. He is the author of various papers and book chapters on core-periphery relationships, domestic tourism, protected area tourism, tourism seasonality, destination recovery and resilience, climate change, island connectivity, the cost of insularity, and the promotion of island destinations. He is inspired to delve deeper into the field of island studies.

Arend Jan (Arjen) Alberts PhD (Amsterdam) is Project Leader for the Netherlands' Ministry of the Interior and Kingdom Relations, working on programs to promote the resilience of the autonomous countries in the Caribbean that are part of the Dutch Kingdom. For 25 years, he was a researcher, educator, and policy advisor on social and economic issues in Aruba, Sint Maarten, and Curaçao. He has published academic articles on aspects of the tourism area lifecycle, and the socioeconomic development and resilience of small island tourism economies.

António Manuel Martins de Almeida is Associate Professor in the Department of Management and Economics at the University of Madeira, Portugal. His research interests cover regional development, island economies, and tourism economics. He is the author of some 60 papers and book chapters, with a focus on the economy and tourism development of Madeira Island. He contributed to several consultancy and EU-funded projects in the fields of circular economy, air transport, and local development. His teaching interests lie in the areas of regional science, microeconomics, and tourism development.

Hervé Atayi PhD (Leicester, UK) is the Marketing Communications Manager at Cable & Wireless Seychelles. He is also a Chartered Manager and member of the Chartered Management Institute (CMI) and student member of the British Psychological Society (BPS). His research interests include Old Testament exegesis, marketing consumption, management and leadership, tourism studies, and organisational psychology. Hervé is also Associate Lecturer at the University of Seychelles.

Laurie Brinklow PhD (Tasmania, Australia) is Assistant Professor and Co-ordinator of the Master of Arts in Island Studies Program and Chair of the Institute of Island Studies, both at the University of Prince Edward Island in Charlottetown, Canada. There she facilitates and supports research on sustainable communities on islands around the world, as well as knowledge mobilisation and public engagement activities. She also teaches 'islandness' in the Master of Arts in Island Studies program and supervises graduate students. A passionate Islander, she is President of the International Small Islands Studies Association and is Iceland's Honorary Consul to Prince Edward Island.

Richard W. Butler PhD (Glasgow, UK) is a geographer, and Emeritus Professor at Strathclyde University, Glasgow, Scotland, UK. He taught at the Universities of Western Ontario, Surrey, and Strathclyde, and has held visiting positions in Italy, the Netherlands, and Hong Kong. He has consulted government agencies and the private sector. He has published numerous journal articles and book chapters, as well as 26 books on tourism, including several on islands and tourism. His principal research interests are the development of tourist destinations and the resultant impacts of tourism (particularly in island and peripheral areas), sustainable development, and Indigenous tourism. He is former president of the International Academy for the Study of Tourism, and in 2016 he was awarded the UNWTO Ulysses medal 'for excellence in the creation and dissemination of knowledge'.

Louise Campbell MA (UPEI, Canada) is a consultant, researcher, and writer who embraces every opportunity to exercise her passion for anything tourism related. She has presented at a number of conferences, including the inaugural Island States/Island Territories Conference in Aruba (2019) and the Sustainable Island Communities Conference in St. John's, Newfoundland, Canada (2019). Her paper on island tourism brand identities appeared in the *Journal of Marine and Island Studies* (2021).

John Connell is Professor of Geography at the University of Sydney, Australia. His main research focus has centred on development issues in island states, especially migration and health, and mainly in the Pacific. His recent books include *Islands at risk* (2013), *Change and continuity in the Pacific* (2018), *Covid in the islands: A comparative perspective on the Caribbean and the Pacific* (2021), and *Pacific Island guestworkers in Australia* (2023).

Anica Čuka is Professor of Geography at the University of Zadar, Croatia. Her research deals mainly with the socioeconomic and demographic development, as well as landscape and land use changes, on Croatian islands. Since 2022, she has been an Executive Committee member of the International Small Islands Studies Association (ISISA). She collaborates in several scientific and professional projects including those dealing with the island research, and has been active in organising and participating in the scientific boards of various island conferences.

Brian Garrod is Professor of Marketing at Swansea University, Wales, United Kingdom. He is founding co-editor of the *Journal of Destination Marketing & Management* and founding Editor of *Tourism and Hospitality*. He has written over 50 refereed journal papers and eight books, including *Managing visitor attractions* (2022), now in its third edition. He has consulted such clients as the United Nations World Tourism Organization (UNWTO) and the Organisation for Economic Cooperation and Development (OECD).

Ayano Ginoza PhD (Washington State, USA) is Associate Professor of Gender and Postcolonial Studies in the Research Institute for Islands and Sustainability at the University of the Ryukyus, Okinawa, Japan. Her research interests include transoceanic feminist demilitarisation movements, Okinawan indigeneity, and women's peace education. Her work has appeared in *American Quarterly, Critical Ethnic Studies Journal*, and *Hiroshima Peace Research Journal*. She is Editor-in-Chief of the *Okinawan Journal of Island Studies*, and a member of the Executive Board of the International Small Islands Studies Association (ISISA).

Aina Gomis is Predoctoral Researcher at the University of the Balearic Islands. She has a degree in Catalan Language and Literature. She is currently working on her doctoral thesis on the poetry of Minorca and the cultural conception of the island territory, linked to the research group *Literatura contemporània: estudis teòrics i comparatius* (LiCETC).

Andrew Halliday MA (UPEI, Canada) is Adjunct Professor in Island Studies, as well as a Sessional Lecturer at the Department of Political Science at the University of Prince Edward Island, Canada. He is a member of the International Small Islands Studies Association (ISISA) and the Canadian Political Science Association (CPSA). He is a PhD candidate at the University of New Brunswick, Canada, researching 'Covid-islands' and 'Covid-archipelagos', with a specific interest in the 'Atlantic Bubble'.

Željka Kordej-De Villa is Senior Research Fellow at the Institute of Economics, Zagreb, Croatia. She is an experienced researcher in the field of environmental economics and environmental policy. Her additional research interests include the economics of natural resources, local economic development, regional and urban economics, and public policies. She is primarily engaged in applied research in the field of environmental, regional, and urban economics, and in policy-oriented studies and research-based consulting projects. Her recent scientific articles are related to island policy, and environmental and energy issues on the local level.

Isabel MacDougall is a graduate student in the Island Studies program at the University of Prince Edward Island, Canada. Her research interests are in island tourism seasonality and, specifically, year-round tourism on Prince Edward Island (PEI). She has 10 years of experience working in marketing and media relations for Tourism PEI and currently works for the Government of Canada, developing social programs.

Andrea Pace BA (Hons) Geography, MSc Sustainable Development is a Chartered Geographer, Fellow of the Royal Geographical Society and PhD candidate at the University of Malta, Malta. He is a planning consultant and has experience dealing with geographic information systems, land use surveys, and traffic data collection and studies.

Hélène Pébarthe-Désiré is Lecturer in Geography at the University of Angers and Deputy Director of the research laboratory CNRS ESO, 'Spaces and societies', in Angers, France. Her latest publications deal with the maturation and diversification of mobility practices, and destination governance strategies. Responsible for university exchanges with Mauritius and Réunion Island, she has been coordinating degree programs in tourism studies for the University of Angers in partnership with the Mauritius Chamber of Commerce and Industry (MCCI) Business School in Mauritius.

Mercè Picornell is Senior Lecturer in Literary Theory at the University of the Balearic Islands. Her research interests lie in the cultural effects of tourism on the local island populations and the relationship between subaltern and affective studies in Catalan culture with a specific focus on Mallorcan culture. She has worked on testimonial literature (*Discursos testimonials en la literatura catalana recent*, 2002), transitional cultures (*Continuïtats i desviacions. Debats crítics sobre la cultura catalana en el vèrtex 1960/1970*, 2013), and the politics of ruins in contemporary Catalan culture (*Sumar les restes: Ruïnes i mals endreços en la cultura catalana postfranquista*, 2020). She is a member of the LiCETC research group.

Joefe B. Santarita is Professor and former Dean of the University of the Philippines' Asian Center, Diliman, The Philippines. He is the current President of the ASEAN Studies Association of the Philippines. His research interests focus on India-ASEAN relations, data diplomacy, maritime history, and island studies. His books include: *Plying the Straits: Batel mobilities in Central Philippines* (2020), *Toyota in the Philippines: 30 years and beyond in nation-building* (2020), *and 2018 Philippine development report* (co-edited, 2019).

Nenad Starc is Professor Emeritus at the Institute of Economics, Zagreb, Croatia. His activities include research, mentoring, preparation, and evaluation of local and regional development programmes, as well as postgraduate teaching in Croatia. In the 1990s, he advised the Croatian Ministry of Development and Reconstruction and coordinated the preparation of the National Island Development Programme (1997) and the drafting of the Island Act (1999). He is an active member of the International Small Islands Studies Association (ISISA) and one of the founders of the Croatian NGO Anatomija Otoka [*Anatomy of Islands*].

John N. Telesford is Lecturer at the T. A. Marryshow Community College, St George's, Grenada; and Research Associate with the Institute of Island Studies, University of Prince Edward Island, Canada. He is a Deputy Editor-in-Chief for the *Journal of Island and Marine Studies* and served as Editor for both the *Global Islands Report* (2022) and for the *Annual Report on Islands and Sustainable Development: Case Studies* (2022). His research interests include the examination of regional transportation issues in the Caribbean, island metabolism, and island industrial ecology.

Atsushi Toriyama is Professor of Modern and Contemporary History of Okinawa in the Research Institute for Islands and Sustainability at the University of the Ryukyus, Okinawa, Japan. His publications include the book *Okinawa: Origin and conflicts of the military base society between 1945 and 1956* (2013), and the article 'Overcoming the past: Concerning the war experiences on Kumejima' in *Okinawan Journal of Island Studies* (2022). He has served on the editorial board for the *Anthologies of Okinawan Prefectural History, Nago City History, Ginowan City History*, and *Tomigusuku City History*. He is also engaged in the preservation and utilisation of the Ahagon Shoko archives as the Chair for research projects.

Foreword – Richard W. Butler

The study of islands remains fascinating to those of us to whom 'islandness' is a puzzling but tangible emotion, and examining islands in the context of archipelagos and what that means widens the realm of interesting knowledge. *Archipelago tourism revisited: Core-periphery relations after the pandemic* builds on the ground-breaking efforts of *Archipelago tourism* (2015) – the editor's first book on this topic – thus taking further the examination of tourism as it affects, and is affected by, groups of islands. He is correct in commenting that archipelago tourism is not yet regarded as a genre; indeed, to some, island tourism is hardly thought of in that manner also. Yet there is no doubt that islands are different, and that different emotions are often felt when arriving on an island compared to visiting a mainland destination. Why that is the case remains a conundrum and is not the focus of this volume, which examines a range of relatively unexplored aspects of islands and their sense of togetherness and/or solitude.

Since the publication of the first archipelago tourism volume, the number and range of trends and impacts that have affected islands at all scales and in all locations is quite remarkable. Few islands and their neighbours did not experience a continued rise in visitation until the end of the second decade of this century; the effects of this increased popularity brought with it a number of problems previously mostly confined to mainland destinations. One of course is overtourism, with the popular media commenting on islands 'sinking' under the weight of their tourist visitors. In the context of archipelago studies, this aspect raises a number of fascinating areas of study. Is such overtourism uniformly spread, or is it confined to one island or a few? Does it take a common form with similar results in all the islands in a group? Have there been common responses or has each island produced individual and idiosyncratic 'solutions'? Perhaps, more importantly, how has this problem related to and affected any drive towards a more sustainable future for archipelagos in question? The challenge of sustainable development has invariably been a factor of island life in almost every case, if only because such an approach is often the only path to continued survival. Where unsustainable approaches have been adopted or forced upon islands, the results can be devastating, as witnessed in the case of Rapa Nui; while in an archipelago context the sharing of problems and potential solutions may be present a viable pathway to recovery if such positive links can be evoked and maintained. Given the way in which bottlenecks are quickly reached and dynamics are 'articulated by compression' in small island contexts, successful initiatives can be proudly showcased as examples of good practice; but the dire consequences of initiatives that go horribly wrong are similarly impossible to dismiss or discount.

One can find in some cases that the growth of tourism spreads through an archipelago from one island to another in a similar fashion, while in other cases the movement from one island to a neighbour may take a very different form in scale and type of tourism and tourist. Is there a common pattern in how such flows occur, or is each archipelago a truly unique case? One can imagine a range of options, as many as there are different forms of island groupings, as discussed in this volume. While tourism also has many forms, large-scale tourism, whether to islands or mainland destinations, tends to bring the same forms of development and the same issues wherever it unfolds. The issue then becomes how such a phenomenon is managed and controlled, and whether such decisions can be made effectively within an archipelago setting, which in turn may depend upon the level of self-government for and within the grouping. The different archipelagos discussed in this volume illustrate these issues and difficulties very clearly.

Whether we deal with warm-water or cold-water tourism archipelagos, the problems are not that different. Is tourism to be viewed as yet another natural resource-based monocrop economic activity, just like bananas, phosphate, or sugar; or can it be converted to something more permanent and more widely beneficial? Can it be promoted, managed, and made use of equally throughout an archipelago; or will it inevitably be focused on one or two islands with associated benefits and costs? If so, what do the resulting inequalities imply for an island group? It may be that the inter-island issues that emerge are not very different from those that emerge on single islands, where residents of some parts of an island desire continued development and others in different parts do not. Do archipelagos simply represent one set of problems multiplied by the number of islands within that group; or is each archipelago uniquely different to such a degree that every story is different? These and similar types of questions are apparent throughout the wide variety of examples in this volume, which is one aspect that makes it particularly important.

One feature that all archipelagos and islands have faced in recent years, of course, has been the impact of Covid-19. One or two archipelagos and islands managed to avoid the pandemic, but still suffered the economic consequences of a global shut-down of tourism travel, and restrictions imposed by countries of origin of tourists. Other archipelagos encouraged alternative forms of tourism: one being nomad tourism, relying on good communications to attract semi-permanent visitors to stay and work from those locations. Another was 'staycationing', encouraging locals to patronise their own local hospitality establishments. But this was not a solution for the missing millions that would normally have been present. Covid-19 proved devastating to many individual enterprises at all scales; but it did serve to convince any agencies doubting the importance of tourism that the industry has enormous economic value in terms of income and employment, even if it brings with it environmental, social, and economic problems. The calls from some quarters that the coronavirus pandemic should prove an opportunity to radically change tourism and the way it is delivered are not without merit. But, as is almost always the case, and as this volume clearly illustrates, the economic argument proves too powerful to be ignored. In most cases, including in islands and archipelagos, the emphasis has been placed on recovery; and, as discussed, rejigging and some relaunching of tourism, with relatively little reflection or reconsideration. Most archipelagos were too small in population for domestic tourism anyway, even where this was allowed, to significantly replace international visitors. Thus they remain, so to speak, at the mercy of air carriers, cruise ship companies and other elements of the tourism industry to restart and revive their visitor business. If anything, perhaps, the situation in terms of the degree of dependency of islands and archipelagos on mainlands has become greater following Covid-19 than it was before, and this is one area that archipelagos are perhaps better able to combat than single island destinations. But such a response depends greatly on the degree of inter-island cooperation and integration, and the specific power dynamics that invariably obtain.

Hence the relevance of the sub-title to this new and exciting volume. Any political island unit, from the smallest dyad (Malta-Gozo; Antigua-Barbuda) to the most extensive and complex (Indonesia; The Philippines) will be subject to power-based, 'core-periphery' dynamics that direct, siphon, and channel resources and investments in particular ways. Tourism is not exempt from these circulations. Political-economic elites would be well entrenched in the tourism sector, and an implicit hierarchy of islands as targets of tourism infrastructure, branding, and marketing and policy generally would manifest itself. 'Togetherness' and 'solitude' take on added significance when there are relative winners and losers; and when livelihoods may depend on whether an island is deemed 'in' or 'out' in an environment where, as always, there is a keen competition for attention and scarce resources.

What might we expect for archipelagos and tourism in the future? Tourism is likely to continue to grow in most areas that have experienced it before Covid-19, and it is likely that, without major efforts, little will change in terms of arrangements, scale, and type of tourism delivered. Much closer cooperation and integration between islands within archipelagos will be needed if change is desired and to be achieved. Issues at an individual island level within an archipelago are miniscule compared to issues between the group and the dominant forces on the mainland or in the tourist origin countries. Thus, only a combined, integrated, and universally supported viewpoint is likely to be successful. It may be that change is not desired or required, but inevitably change will take place. Inertia is a powerful force but, over time, dynamic elements normally result in change. It is much better on most occasions to ensure that such change, when it occurs, is met with consistently strong and unified responses rather than weak individual complaints.

Not all islands in all archipelagos are happy in their position and with their limited powers and ability to bring about change. Things can look quite bleak when standing on the beach of a small, forgotten island, at multiple degrees of insularity removed from the centre of decision-making. But, as an archipelago, islands are better able to flex and strengthen their agency and influence; they would almost always be better poised to have some sway over external forces. The political strength of archipelagos lies in their commonality rather than their differences, although such differences – presented as complementary assets – may be of immense value in attracting a wide range of types of tourism, while encouraging multi-island hopping.

These are some of the forces and discontinuities that make archipelagos so fascinating to examine and why this volume is so significant.

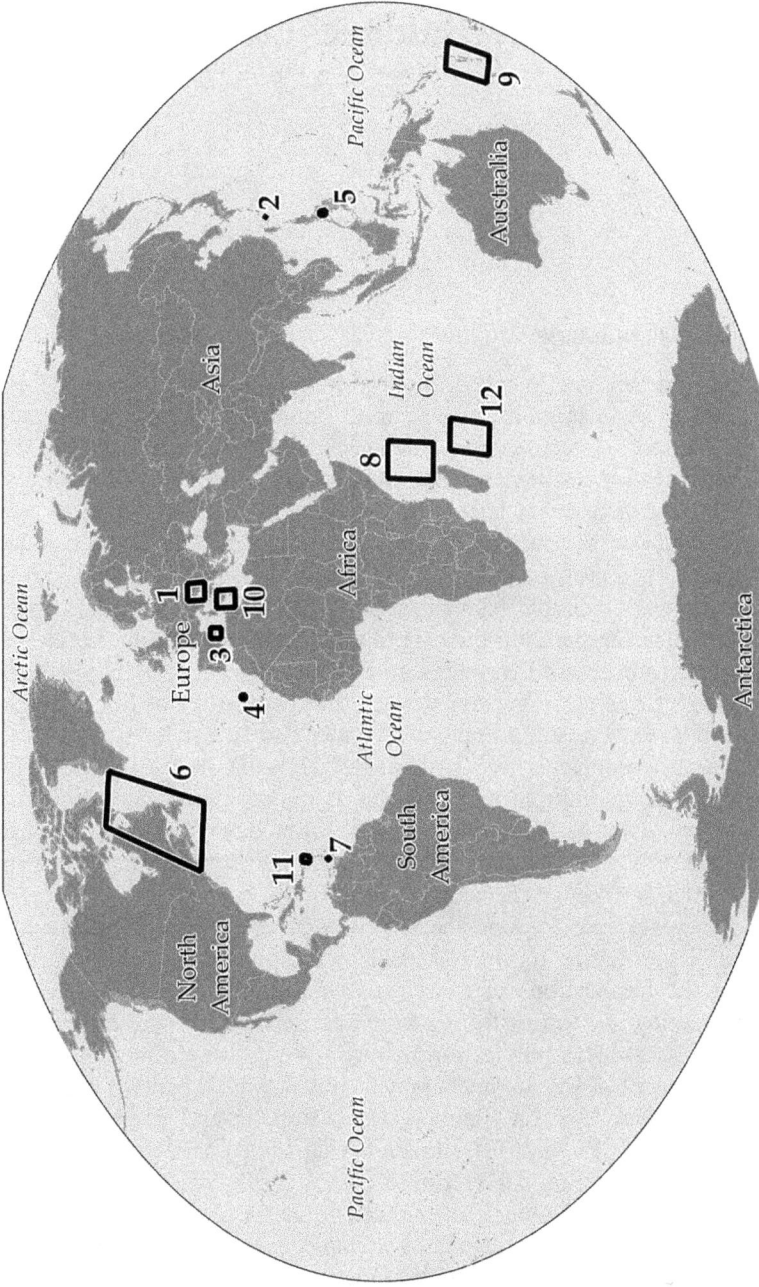

Figure 0.1 Archipelago tourism: Locations of the case studies (by chapter number)

Editorial: The archipelagic turn in island tourism – Godfrey Baldacchino

Introduction and rationale

International tourism was knocked out by Covid-19 but has bounced back. This is especially so in small islands. So much talk of alarming and creeping 'over-tourism', and a return to concerns around 'carrying capacity' by local residents and engaged academics, have been swiftly and deftly reversed and replaced – hardly three years later – by an urgency driven by industry and governments to grow the tourism economy to pre-pandemic levels and as fast as possible, and then even some more. But there is also now a keener awareness of the need for suitable 'climate action', green/blue growth and more planet-friendly development, on the part of major tourism industry players as well as savvy consumers and travellers, in a process that is probably iterative and virtuously cyclical. It may very well be that "advocates of [the tourism] industry's rapid recovery stand opposed to wider efforts to reform tourism to be more ethical, responsible and sustainable" (Higgins-Desbiolles, 2021, p. 551). How do these multiple and complex issues play out in multi-island locations, where there are always tensions between different islands, central and peripheral, urban and underdeveloped, larger and smaller, more and less accessible, better and less connected?

Almost ten years ago, I edited the text *Archipelago tourism: Policies and practices*: a first attempt to offer a deliberate, systematic, and concerted appraisal of the implications of a *multi-island* tourism destination, from a logistics, marketing, as well as from a branding perspective, across the globe (Baldacchino, 2015). This publication followed in the wake of some seminal work on the concept of the archipelago, which is now a recognised sub-genre within island studies (e.g. La Flamme, 1983; Roberts and Stephens, 2017; Stratford et al., 2011; Pugh 2013); though perhaps not yet in tourism studies (Baldacchino and Ferreira, 2013; Couto et al., 2020; Siegel et al., 2013). Apart from conceptual chapters as bookends, this text had pioneering case studies of islands and island regions *as archipelagos*, and therefore as both multi-island assemblages with a common sea, as well as places that may sit uncomfortably within core-periphery relations, whether 'island-mainland' or 'island-island'. Under scrutiny were regions like the Caribbean and the

DOI: 10.4324/9781003451037-1

Aegean; island states like the Bahamas, Cape Verde, Fiji, Maldives, Malta, and Mauritius; and sub-national island units like the self-governing territory of Azores (within Portugal) and the province of Sardinia (within Italy). This continues a specific genre, pioneered by DeLoughrey (2001, p. 23), who appealed for an 'archipelagraphy' that "considers chains of islands in a fluctuating relationship to their surrounding seas, islands and continents [or mainlands]".

Since then, I have been considering a sequel to this book. So much has happened in the interim. Tourism is by nature an expanding industry, and smaller or peripheral islands have been keen to jump on the proverbial bandwagon and benefit from being identified as suitable tourism destinations, and thus share in the value added that the industry generates across so many sectors of the economy and society. Tourists, meanwhile, are encouraged by tourism operators, travel agencies and state-led tourism authorities to explore 'new' and 'undiscovered' island paradises, with their representations of enticing and temptingly empty sandy beaches. The evidence of environmental damage and degradation is irrefutable, leading specific island destinations to 'lock down' even before Covid-19 struck: Proclamation No. 475 in 2018 shut down Boracay, in The Philippines, for six months, declaring "a state of calamity" (Government of The Philippines, 2018). Two decades before, Trousdale (1999) had somewhat presciently predicted, with reference to Boracay that, unless a major calamity – such as an epidemic – scares the tourists away, it will continue to be 'business as usual'. Elsewhere, in the Faroe Islands, a 'Closed for maintenance, Open for voluntourism' project was launched in 2019, meant to improve visitor infrastructure while minimising the tourist impact on the local, fragile, natural environment (Ecott, 2019). The project has been so successful that it has been extended (Visit Faroe Islands, 2023). In Italy, the national government banned cruise ships of over 25,000 tonnes from the Venice lagoon in 2021 after that archipelago was threatened with being put on the 'danger list' of UNESCO's World Heritage Sites (Buckley, 2021). In the same year, the national government of French Polynesia limited daily cruise passengers visiting Bora Bora to 1,200, effectively banning large cruise ships. The message is clear: "Both in terms of capacity and size, [very large ships] are not suited for our destination" (Laird, 2021). In 2019, and with an eye towards reducing the heavy tourist flow, New Zealand introduced an International Visitor Conservation and Tourism Levy of NZ$35; its proceeds are invested in projects meant to create productive, sustainable, and inclusive tourism growth that supports and protects the environment and enriches New Zealanders' lives (Goodwin, 2023).

While noticeable and commendable, such measures have been the exception rather than the rule. And their results have also been mixed: in some cases, costs to tourists have increased, but tourist numbers have not declined. Powerful political-economic elites, even more salient and oligopolistic in small island settings, will lobby and argue, often persuasively, against any restrictions to tourism development – an industry that is 'too big to fail' – and inadvertently preserve that small island's vulnerability as a 'monocrop', tourism economy (Pryor, 1982). The wide democratic credentials of the industry – from which benefits can accrue to both large (hotels, airlines) and small operators (tour companies, tour guides, taxi drivers, Airbnb hosts) – can make it resistance-shy and reform-proof.

We can now state, with the advantage of hindsight, that a knock-out blow in the guise of a global pandemic was needed to floor the industry and render it unconscious, for at least two years. This was especially so in the smallest jurisdictions, where there was/is no domestic tourism worth speaking of (Connell and Taulealo, 2021). "Islands retreated into themselves; and seas made strong borders" (Connell and Campbell, 2021, p. 2). Small, mostly island, jurisdictions, with their open economies and fragile ecosystems, are especially vulnerable to external shocks. The onset of Covid-19 was especially damaging because it posed a clear danger to citizen livelihoods, unsettled what were already stretched public health systems, and obliged an international social distancing that throttled international tourism: by far the dominant industry for many islands, responsible for 30–60% of Gross Domestic Product (Connell and Campbell, 2021, p. 13).

But: with what long-term consequences? The June 2023 number of overseas visitor arrivals to New Zealand was (already) 84% of the (peak) pre-Covid-19 number of 213,500 in June 2019 (Trading Economics, 2023). Over three million tourists visited Malta in 2023, the first time this milestone has been reached, more than 8% higher than the previous record high of 2019 (Borg, 2024). By June 2023, the United Nations' World Tourism Organisation (UNWTO) was already estimating that tourist arrivals worldwide in 2023 may reach up to 95% of pre-pandemic levels: they were just 62% in 2022 (The Economist, 2023).

As many scholars have argued, now is a good time to assess the 'absorb, adapt or transform' effects of Covid-19 on tourism policies and practices, from both supply and demand sides of the industry (e.g. Roberts, 2023). It is therefore opportune to look out for any *new* approaches to inter-island tourism branding, marketing, transportation, multi-modality, competition, and so on, after the pandemic forced international (small) island tourism to shut down, and for so long … long enough for most tourism operators to have the time to soberly reflect long and hard on the future of their industry. The talk in social circles pre-Covid-19 was increasingly about 'overtourism', environmental degradation, marine litter, excess waste, and pressure on scarce resources on small islands and their social and natural ecosystems. Long-haul flights are also constituting a growing proportion of carbon emissions; obliging a re-think of how to possibly cut down or transform this industry towards greater sustainability. Come 2022, and the worst of the pandemic over in most countries (but not China), there have been calls for a new energy and commitment to do things better. Many held high hopes that 2022–2023 and beyond was "not merely a return to a 'normal' that existed before" (Lew et al., 2020); but would instead concoct a vision of how the world was changing, evolving, and transforming into something different and plausibly better … or, at least, being more cognisant of the need to build climate resilience at different scales. Meanwhile, many tourism-related operators are just keen to see tourist numbers grow as fast as possible, to pre-pandemic levels ... and beyond. This pragmatic disposition is posed as an existential issue, built around survivability concerns: sustainability, green/blue development, and decarbonisation just have to wait (yet again). Venturing into new areas, products and services comes with a serious caveat. Fantasha Lockington, CEO of the Fiji Hotel & Tourism Association (FHTA), captures this sentiment of caution and prudence succinctly:

> Replicating these [past relationships and frameworks] in new areas, or even venturing into new products and services, needs similar considerations of demand, available space and whether existing infrastructure is sufficient, with the added challenge of locating staff and the logistics if tapping transportation routes. Or building these from scratch.
>
> (FHTA, 2023)

Experimentation, and anything 'new', can follow ... but not immediately. After all, geo-political insecurity – including the uncertainty derived from the Russian aggression against Ukraine, and the complex conflict in the Middle East – along with staff shortages, and the potential impact of the cost-of-living crisis on tourism, are challenges that constitute real "downside risks" to an industry still in reset mode (UNWTO, 2023).

Four elements for the return of tourism

We are witnessing a bouncing back of tourism that has elements of recovery, revenge, rejigging and relaunch. *Recovery* from the supply side, for restauranteurs, travel guides, heritage sites and hoteliers to make up for two years or more of lost revenue, and especially so for those firms that would have held on to (most or all of) their staff, with or without government support, over the dark Covid-19 years (Bulchand-Gidumal, 2022). *Revenge* from the demand side, as individuals, families, and groups have finally started venting the many frustrations resulting from long periods of lockdown, 'cabin fever' and impaired movement, and have been going all out to return to and experience mobility, and in spite of higher transportation costs (Abdullah, 2021; Volger, 2022). *Rejigging*, as governments and industry service providers adjusted their offers to cater for a traveller who is desperate to travel, while presumably more conscious of safety and security than before 2020; and perhaps a tad more environmentally conscious, such that would impact on their choice of destination, accommodation, entertainment, transportation, or eating habits (e.g. Bangkok Post, 2022). And finally, *relaunch*, as industry players in particular invest in an industry that they see as continuing to expand and rope in still more travellers (Rodrigues et al., 2021), especially from the growing middle classes in South-East Asia: IndiGo, an Indian airline, announced in 2023 the purchase of 500 commercial airplanes from Airbus, the largest single airline order of all time (Hepher and Plucinska, 2023).

Core-periphery relations

The number of independent states grew more than three-fold – from 54 to 180 – with the wave of decolonisation that gripped the planet in the four decades immediately after the end of the Second World War (1945–1985). There were many who believed that sovereignty and the end of colonialism would unleash the development potential of these new, mostly small, countries. But hopes of a quick and smooth transition to modernisation by and large faded in the following decades: what came to be called 'dependency theorists' blamed capitalism and globalisation for locking the Global South in a tight embrace of systemic underdevelopment. The ensuing core-periphery relations were deemed both cause and consequence of an "unequal exchange" in trade, technology, talent, and investment. Deep, structural inequalities between First and Third World were being exacerbated, rather than attenuated, by the terms of trade; it was the periphery that was being bled of its natural, financial, and human resources to nourish, sustain, and develop the core; not the other way round (Baran, 1957; Frank, 1967; Rama and Hall, 2021). This approach found the support of radical thinkers and politicians from young independent states, notably in Latin America and the Caribbean (Beckford, 1972; Rodney, 1972). The argument has been strongly criticised for 'explaining away' the very possibility of a developmental 'take off', short of revolutionary change (Lewis, 1954); as well as for turning a blind eye to the governing malpractices of various developing countries; including the resort to graft, nepotism, corruption, militarism, and civil war. And yet, there remains something to be said for a lingering neo-colonialism and the real difficulty of 'moving up the value chain', especially for those countries that – like many small island states – continue to produce cash crops, such as bananas, copra, phosphate, sugarcane, nutmeg, cocoa, tobacco, and pineapples. Dependency theory has also reinvigorated the resort to a more radical, 'political economy' (rather than a neo-classical macroeconomic) approach and methodology towards understanding world affairs generally.

The political economy approach to tourism was pioneered by Britton (1982) who argued that tourism destinations in the Global South are largely managed and exploited by capitalist enterprises – airlines, hotel chains, food distributors, alcoholic drinks manufactures, credit card companies – with their head offices located in the metropole. Thus, most tourism revenue earned at home would quickly find its way back – as profits, tax breaks, import bills, senior management salaries – to the First World. Tourism had become the main exponent of the (now service-driven) 'plantation economy' in the twentieth century for many small island states and territories, and particularly after the artisanal producers of traditional cash crops – such as small-scale banana growers in the Caribbean – found it impossible to compete with more efficient industrial producers from Central and Latin America (Baldacchino and Bertram, 2009: Moberg, 2022). Tourism has been described as a contemporary version of imperialism (Sinclair-Maragh and Gursoy, 2015;

Higgins-Desbiolles, 2022). The (now longstanding) core-periphery narrative holds that "small islands are geographically and economically marginal entities, fated to spawn homogeneous tourism monocultures within contexts of persistent external dependency" (Weaver, 2017, p. 11).

A nuanced approach is called for. In a rare review of 'core-periphery' relations in archipelago tourism, Agius and Chaperon (2023) outline the basic tenets of this approach and its theoretical provenance, while also summarising its limitations and its critics. They argue that dependency theory has been chided for: (1) being too dogmatic in offering a one-sided interpretation of the tourism industry (Sharpley, 2022); (2) focusing too much on mass tourism and the movement of international visitors from the Global North to the Global South (McKercher, 2021); (3) adopting an overly general perspective of macro-structural dynamics at work, neglecting responses and flexibilities at household and individual level in the process (Baldacchino, 2011; Monterrubio et al., 2018); (4) failing to consider other, alternative, types of tourism such as eco-tourism, in addressing and redressing inequalities in developmental terms (Bianchi, 2015; Weaver, 2017); and (5) overlooking variations in local conditions, with their tendencies and strategies for local leadership, resistance, resilience and otherwise canny adaptation (Amoamo, 2021).

The idea of the archipelago

And so, this book explores the fascinating dynamics of inter-island/island-island tourism – also known as archipelago tourism – on the cusp of the post-pandemic epoch. It examines what novel approaches to tourism have been proposed, if at all, so as not to repeat past mistakes, and to nurture a more sustainable tourism business model. It looks at how the political-economic relationship between the main(is)land and outer island(s) may have changed during the pandemic and, if so, whether this shift has had a bearing on current tourism policy. The book also explores how these and other changes are reflected in the manner in which: (1) islands are branded; (2) island destinations are marketed; and (3) island transport logistics play out.

The most common, simple yet powerful definition of an archipelago is "a group of islands" (Stratford et al., 2011, p. 117). But this would be a land-based and land-biased characterisation that does not mete out proper justice to the complex term. As an alternative, consider the definition provided by the United Nations Convention on the Law of the Sea (UNCLOS). Here, an archipelago is defined as:

> A group of islands, including parts of islands, interconnecting waters and other natural features which are so closely inter-related that such islands, waters and other natural features form an intrinsic geographical, economic and political entity, or which historically have been regarded as such.

> (UN, 2024, Part IV, Article 46)

Even if necessarily legalistic, what stands out in this definition is the added emphasis on: (1) interconnectivity and interrelationships between the constituent members of the group (these being islands, but also islets, rocks, reefs); and (2) the equally critical interconnecting role of the sea and other 'waters'. It is the (fluctuating) *ensemble* of land and sea that comprise the intrinsic entity identified as an archipelago; and not just its island members.

The *very idea* of an archipelago has been tested and tweaked because of the pesky coronavirus, and its aftermath. Neighbouring islands with long traditions of connectivity that usurp national boundaries – like the island Caribbean - were 'de-archipelagised', obliged to suspend inter-island traffic and trade, and maintain those near-total lockdowns that are only possible and enforceable on small island jurisdictions, where entry and exit points are few, well known and well monitored (Connell and Campbell, 2021; DeShong et al., 2023). Meanwhile, and in sharp contrast, other islands found themselves impacted by the overflow of tourists from other jurisdictions, resulting in new dynamic archipelagos. For example, in French New Caledonia, the tourism industry outside the capital Noumea boomed like never before and extended to Wallis and Futuna: an unprecedented episode (Hoffer, 2021). Significant tourism industry players – airlines and cruise ships in particular – make it their business to make and unravel archipelagos, as they flexibly develop itineraries that attract clientele while beating the competition; dropping destinations as easily as building them (e.g. Douglas and Douglas, 2006). Small beach islands, like 200-acre ($0.8km^2$) Dravuni in Fiji, with its 150 residents, is entangled as a tourism island by virtue of being on cruise ship itineraries: there is no other way of getting there (Cruise Ship Karen, 2018).

Five attributes of archipelago tourism

This book reaffirms the five specific yet interrelated attributes of multi-island tourism, first proposed and tested in the earlier, *Archipelago tourism* volume: (1) visibility; (2) tweaked representation; (3) domination; (4) liminality or layering; and (5) differentiation (Baldacchino, 2015). It contextualises these in a post-Covid-19 age, with its conflicting thrusts towards, on one hand, decarbonisation, climate action, and sustainability; and, on the other hand, engineering a fast recovery of the tourism industry to pre-pandemic levels, and beyond.

Visibility

The *visibility* of islands has been generally enhanced during the coronavirus pandemic, and for various reasons. During the pandemic, small islands were amongst "certain spaces and geographies [that] were notably re-evaluated" (Burnett, 2023, p. 1). Being considered isolated, remote and 'far from', they – for some time – "occupied a heightened position of desirability and potential refuge" (ibid.). First, some islands became highly prized and sought-after destinations for anxious mainlanders wanting to 'escape' the virus. Those who had a second home on an island were the lucky ones, and an exodus to islands was witnessed in many countries as the massively disruptive nature of Covid-19 started sinking in, around Spring 2020. The situation was dramatically reversed within a few months, when the virus also appeared within island populations. What had been seen as a safe refuge suddenly became seen as a lethal trap. An exodus from the islands ensued.

Then, as the pandemic progressed, other islands found themselves 'on the radar' for having avoided the pandemic altogether, but only at the cost of a draconian and complete lockdown and a total suspension of air and sea services. Until August 2020, ten sovereign states, all in the Pacific, all islands or archipelagos, claimed not to have registered a single case of Covid-19: Palau, Federated States of Micronesia, Marshall Islands, Nauru, Kiribati, Solomon Islands, Samoa, Tonga, Tuvalu, and Vanuatu (Amos, 2020). Two years later, by the end of August 2022, all these Pacific states had reported infections; and the only virus-free places had been whittled down to just two, very small, non-sovereign jurisdictions, each with resident populations of less than 5,000: Tokelau (a territory of New Zealand) and St Helena (a UK Overseas Territory) (Hart, 2022). Ironically, the Pacific archipelago state of Tonga was Covid-19-free until the violent eruption of the undersea Hunga Tonga-Hunga Ha'apai volcano in early 2022: international aid workers who came to assist Tongans in the recovery from the ensuing destruction were found to have caught, and probably transmitted, the virus (Lyons, 2022).

Moreover, even before the end of the pandemic came in sight, and as governments of certain countries – such as Greece and Italy – were preparing to 'open up' to tourism, they may have prioritised small island residents for vaccination, in order for their small (and heavily tourism-dependent) communities to benefit early from the welcome financial injections that tourism brings (Dominioni, 2021; Craig, 2022). These moves were also early attempts to spread international tourism beyond the usual locations, which had been facing serious 'overtourism' issues right up to the onset of Covid-19, and eliciting a backlash from local residents (e.g. Sarantakou and Terkenli, 2021). But even here, miscalculations could prove costly: in the Bahamas and French Polynesia, the tourism industry lobby trumped public health concerns. These island jurisdictions reopened 'too soon' to tourism, with dire health consequences (McLeod, 2021; Heinzlef and Serre, 2021 respectively). By and large, the smaller the jurisdiction, the greater the hesitancy to re-open: other than French Polynesia, no other Pacific small island developing state (SIDS) reopened for tourism before 2023.

Tweaked representation

This last point takes us to the second attribute of archipelago tourism: *tweaked representation*. Islands found themselves, yet again, as "objects of representation" (Baldacchino, 2005), with texts being generated, appropriated, pivoted, and circulated by media actors (on and off islands), thus engendering, reinforcing, and *repositioning* various islands (and their residents) (Burnett, 2023; emphasis in original): singly; or collectively and differentially as archipelagos. The onset of the Covid-19 pandemic at first pushed islands to the forefront of desirability, with their geographical isolation purporting to serve as some moat that might prevent the virus from crossing over. But – apart from the few exceptions noted above – the world is so irrevocably interconnected, and all borders are porous. It was not so much a question of if, but when. When the virus *did* arrive – and accompanied by a blame game as to whether it was locals or foreigners who had brought the virus over – the same island places that had been attributed with positive features (safe havens, escapes, refuges) were dramatically transformed and straddled with negative ones (traps, cages, prisons). Gated communities, all-inclusive resorts and other enclosed spaces – sometimes islands nested within islands, as with Royal Caribbean's private resort on Labadee 'island', actually a peninsula, in Haiti (Nelson, 2020) – had a similar effect: boons as long as there were no infections reported 'inside'; banes if there were. There is perhaps no clearer and no better documented example of this dramatic, 'utopian to dystopian' switch than the *Diamond Princess* cruise ship fiasco – itself a floating island – with stories of passengers being locked in their cabins for weeks on end, contracting the disease or dying from it (Jimi and Hashimoto, 2020). Islands fared even more badly if they lacked, or were seen to lack, the medical personnel, epidemiological specialists or the sophisticated health infrastructure of the mainland. With (inter-) island transport services and supply chains weakened, cut down or suspended, access to such health services became even more challenging and precarious. As more people got sick, powerful centripetal forces came into play, as smaller and outer islands gave way to larger islands or to mainlands in so for as the availability of critical medical and paramedical facilities and epidemiological expertise was concerned. In early 2020, all talk was first about masks – which soon became quite readily available – but then quickly moved to ventilators, medical supplies and hospital bed availability: many small islands could not offer such services. The state-led regimes that managed and regulated our societies during the pandemic unwittingly (perhaps) reinforced and reasserted the power of the central state, and/or of the main island, versus the far-flung islands of an archipelago, and their respective local governments, if any. In such situations, some desperate islanders fell back on traditional medicine and local knowledge, while others sought explanations and solace in faith and religion. Each case would need to be examined on its own merits. But, one wonders: was this power imbalance ever restored once most of the world got back to a post-Covid-19 'normal'? In spite of paying lip service towards boosting island peripheries by privileging their vaccination schedules, how have the human and financial flows associated with tourism actually panned out post-Covid-19?

Power imbalance

Indeed, *power imbalance* is the third attribute of archipelago tourism; and the one most readily evocative of core-periphery relations. After all, a power inequality is *always* in play between mainland and island, and between islands. Hub and spoke transport logistics bear testimony to this. Differential representation in democratic institutions – such as Parliament and the Cabinet of Ministers – also speak to this imbalance. So do the decisions of airlines and cruise lines that decide to open up, restore, serve, or abandon a particular island destination but not another. This 'core-periphery', or 'domination-subordination' dynamic is manifest in the way in which, typically, one island in an archipelago grows – in terms of urbanisation, population, investment, retail footprint – at the expense of all the rest, haemorrhaging their youth, talent, entrepreneurs, and investment capital. The demographic trends tell the story.

So does the location of tourism infrastructure. By and large, this tends to follow the 'distance decay' principle: the interaction between two locales declines as the distance between them increases (McKercher and Lew, 2003) and – I would add – as the distance between them involves passage via water, and assuming no changes in inter-island transportation regularity, speed, affordability or plurality of offer (e.g. Pimentel et al., 2022, in the case of the Azores). Less than 3% of international tourists travelled 5,000 km or more before Covid-19 struck (McKercher and Mak, 2019). The percentage dropped during the Covid-19 lockdowns to practically zero; and it may not rise back as much or further post-pandemic, as families continue enjoying staycations and become more concerned with the rising cost of living (e.g. Anton Clavé, 2022).

Moreover, tourists are most likely to (continue to) gravitate towards, and spend money and time in, the island(s) served by the airports, seaports and cruise ship terminals where they are landed. Indeed, the zone of engagement is typically within one kilometre of these sites of dis/embarkation; and especially so for cruise ship passengers, with the concomitant crowding that this brings along (Coronato et al., 2021). After all, serious talk about 'archipelago tourism' must engage with the proverbial elephant in the room: the demand side of the equation. Most tourists have no intention to visit other islands. They may not even have an idea that there actually are other islands potentially worth visiting apart from the one they have disembarked on. And, let's face it: the owners of businesses on the main island would prefer to keep it that way.

Liminality

The fourth piece of the archipelago tourism puzzle is *liminality* or *layering*. Geographers remind us that, once we switch magnification and focus, what may appear to be just another island becomes the mainland to yet smaller, more peripheral islands. Here is a terse reminder that the definition of an archipelago is supple enough to generate multiple and nested identities and relationalities among its constituents, respectful and reflective of its geophysical and material fragmentation. "Smaller islands act as satellites of bigger ones" (Boumpa and Paralikas, 2021, p. 99).

Take Boughton Island, the subject of an autobiography and a historical study. It is located in Cardigan Bay, off central north-eastern Prince Edward Island (PEI), which is Canada's smallest (and only fully enisled) province. It is PEI's third-largest island, with an area of 2.4 km^2 (600 acres). In summer, the island is a popular destination for locals and tourists to visit for picnicking and hiking. Given the shifting sandbars in the area, it is sometimes possible to walk or wade across the 300-metre inter-island strait at low tide. Boughton is now uninhabited; but it used to have permanent residents until around seven decades ago. It boasted its own post office and school house. For its residents, Prince Edward Island was '*the* mainland': they would cross over for most of their purchases, for any medical issues, and to sell their produce (Holloway, 2002; King, 2005). This 'PEI as mainland' situation persists nowadays only for Lennox Island, the site and home of a 400-or-so Indigenous Mik'maki (First Nation) community, and which has been connected by a causeway to its 'mainland' since 1972 (Lennox Island First Nation, 2023). It is hard for any Canadian or Prince Edward Islander to imagine PEI as anything but a (small) island; and yet, this is not always the case. Size is relative and mainland-island relations exist among islands too. They also change with time, and modernity has typically put paid to many of the full-time resident communities of small islands worldwide. Perhaps the posterchild of this island abandonment, and the stuff of many a narrative, is St Kilda, in the Scottish Hebrides (Fleming, 2005); now the site of a nostalgic form of tourism.

What has Covid-19 done to liminality? The openness and closure of small islands, their ports and (especially) their airports, during the pandemic, has recentred and reconfigured some cores and peripheries, for better or for worse. Small volcanic (and therefore, high) islands in particular would be obliged to channel tourists and other visitors through airport infrastructure that, even on good days, may be only able to take aircraft of a certain size (because of short runways), or only on certain days and times (because of dangerous crosswinds). The airport in remote St Helena, when opened in 2017, faced serious wind shear issues (Buckley, 2017; Williams, You, and Joshua, 2020). Approaching St Barths' runway is a unique experience (and itself a tourist attraction best enjoyed from land!); Saba's 400 m strip may be the shortest commercial runway in the world (Chilton, 2022); Montserrat's runway is 600 m long; Barbuda's is 500 m; and the one in Heligoland, Germany, is 480 m. Although Madeira's runway is a more decent 1,600 m long, pilots require special training before they are allowed to fly there. Pitcairn, wanting to grow its own tourism industry, does not have an airport or seaport at all (Amoamo, 2011). Lockdowns and mobility restrictions have added layers and 'windows' of rupture and disengagement to an already difficult situation. Ensuring tourist flows may also only be possible with the concurrence of neighbouring countries' airport infrastructure, which can set up logistic dependencies.

Differentiation

This is the fifth and last dimension of archipelago tourism. Caught in these dynamics, (smaller) islands may find themselves as the incidental beneficiaries of tourism, with the bulk of receipts being creamed away by the main island(s). As a result, their marginality can give rise to creative attempts at product *differentiation*. They can become havens of exclusivity, authenticity, and more upscale tourism infrastructure, preferring value added and quality rather than bodies and quantity of visitors. They can be re-interpreted as more traditional societies, spared from the throes of modernity, living in a land that time (but also tourism) forgot. Their residents can be deliberately contrasted with those of the main island, or the mainland, a dangerous 'othering' tactic that can backfire by raising expectations about cultural exoticism and exceptionalism that can then be hopelessly dashed once the actual visitors arrive and see for themselves.

Post-Covid-19, some islands initially sought to use their natural assets as a marker of differentiation to restart tourism. Within the central Mediterranean, the Italian island of Pantelleria flaunts its national terrestrial park – which covers 80% of the island's land area – as its main attraction; something that the other neighbouring small islands do not have (Parco Nazionale, 2023). In other islands, as in Fiji, the tourism hiatus brought about by the pandemic offered an unexpected opportunity to return to *vanua* (tradition) and kinship with nature (Movono and Scheyvens, 2021). Which reminds us that the discourse about the locals' desperation to 'return to normal' may be tourism industry driven. This comes as no surprise when dealing with small islands where the locals may have no 'voice' in politics; where tourism is the economic mainstay; and where the local state may be captive to the tourism industry's lobby (Hampton and Christensen, 2007; Peterson, 2020).

The island narrative

The narrative here continues to peddle a political geography of space that unabashedly assigns value to materiality. Places have meaning; and in the throes of being socially constructed, they are allocated or assigned attributes that may make them more (or less) attractive to visitors (Urry, 2001). Thus, islands feature *qua* islands in this volume; their islandness is a *focus* of the investigation, and not just an accidental *locus* (Ronström, 2013). Island tourism – and not tourism on islands (Sharpley, 2012) – has, luckily, long been recognised as a legitimate field of research, education and policy thrusts (e.g. Baldacchino, 2006, 2012; Briguglio et al., 1996a, 1996b: Butler, 2012; Conlin and Baum, 1995; Dodds and Graci, 2012; Lim and Cooper, 2009; McLeod and Croes, 2018; McLeod, Dodds and Butler, 2021). Island tourism has a distinctive form, not simply for its powerful allure of (still) being able to "provide rest and relaxation to the mundane issues of life" (McLeod et al., 2021, p. 367; *also* Harrison, 2001). Islands are now also poised and presented as the icons, symbols, and posterchildren of the Anthropocene, becoming the objects of a new, contemporary fascination. They are living and empathetic "geographies of hope" (Turner, 2010) of how to build and nurture sustainable 'zero carbon energy' environments (e.g. Sperling, 2017); or of failing hopelessly at doing/being so (e.g. Munshi, Banerjee, and Chakraborty, 2022). They are natural laboratories of "entangled worlds", the most clearly available imbrication of nature and culture, of human and non-human living (Pugh and Chandler, 2021; Hall, 2010).

What beckons? 'Revenge holidays' may soon be eclipsed by the very real pinch of inflation and staff shortages; and the 'glorious summer' [in the northern hemisphere] of 2023 may turn out to be just a fluke, a 'one off' event; in contrast, the concerns over a warming planet are here to stay. Covid-19 has accentuated, but also adjusted, the five dimensions of archipelago tourism, revealing and revelling in a newly found fluidity, and extolling the merits of a newfangled centrality and peripherality. With climate change and rising temperatures, 'cold water islands' (Baldacchino, 2006) might come into their own again as preferred tourism destinations; while those 'warm water' islands basking in very high temperatures get progressively shunned and dropped off itineraries. And would the advance of urbanisation and modernity not make archipelagos even more lopsided?

Tourism is likely to remain a buoyant industry and a vehicle for economic prosperity. If so, researchers should persevere in observing and addressing the 'playing out' of tourism in special places, such as islands; sorry, archipelagos. Who benefits? Who loses? How and why? Many islands may be small; but they are *never* small enough to avoid the modalities of archipelago tourism and their nested core-periphery dynamics.

Organisation and contents of this book

Most islands would benefit from the freshness of being examined and critiqued using the archipelagic lens in these post-Covid-19 times. For this volume, authors have been contacted to provide cases studies that – with the exception of the Mascarene Islands – did not feature in the 2015 book. Additionally, and as articulated in the book's sub-title, there is a deliberate effort to expose and disambiguate some of the intriguing power dynamics that inform core-periphery relations on these islands, with respect to other islands as well as their respective mainland(s), near or distant.

This time, the material is organised into three geo-political categories: (1) sub-national island groupings that form part of countries that can either be exclusively islands (such as Miyako, part of Okinawa Prefecture in Japan; and Romblon province in The Philippines) or else form part of countries that consist also of mainlands (as with the islands of Croatia, Portugal and Spain, all in Europe; as well as the four-province cluster of the Atlantic Canada 'bubble', in North America); (2) complete, but single country, island groupings (as in the case of Grenada in the Caribbean Sea, Seychelles in the Indian Ocean and Vanuatu in the Pacific Ocean); and (3) regional, trans-national island groupings, which spill over various countries and multiple jurisdictions (such as the North-Eastern Caribbean, the Central Mediterranean and the Mascarene Islands in the Indian Ocean).

Note that the Atlantic Canada chapter is included in this volume even though it does not *stricto sensu* deal with an archipelago (i.e., a group of islands, girded by water). However the unique isolation regime that resulted from the Covid-19 pandemic did temporarily create a quasi-archipelagic regional grouping and identity in eastern Canada, part of which did indeed both consist of islands and was surrounded by water.

Part I of this book addresses core-periphery dynamics in sub-national archipelagos, where the protagonists comprise only part of the population of the country.

Chapter 1 deals with the islands of Croatia, a small European country with over a thousand islands, of which 50 are inhabited; none of its 21 adminis-trative divisions is an exclusive island unit. This is because all islands have low permanent populations, and are deemed to be relatively near the con-tinental coast, and so do not merit a separate administration. In this way, island life involves negotiation and coping with mainland dynamics, politi-cally and economically; meanwhile, island-island dynamics are rare, also because all islands compete as tourism destinations and for investment from the mainland. In terms of inward (mainly public) investment, the islands which fare somewhat better are those affiliated to the larger coastal cities, and their deeper pockets: Split, Rijeka, Osijek, and Zadar, in that order.

Chapter 2 takes readers to Miyako Island, part of Okinawa prefecture, the only island province of archipelagic Japan. Its triple insularity does not insu-late Miyakojima from mass tourism; while a strong communitarian culture amongst its islanders has increased its exceptional tourist appeal. Insights in this chapter are drawn insightfully from the vicissitudes of Machas, a com-munity grocery store that has been serving locals and tourists pre-, during, and post-pandemic, and has become unwittingly a beacon of authenticity.

Also facing mass tourism is the island quartet that forms the Balearic Islands, Spain, the focus of Chapter 3. The West Mediterranean islands of Majorca, Minorca, Ibiza, and Formentera have each smartly specialised in different types of tourism. The Covid-19 pandemic has: boosted domestic tourism; increased the number of peninsular Spanish and foreign residents who now live and work on the islands; and, as a response, energised social mobili-sation against mass tourism, also in connection with the right to affordable housing and labour rights.

The central focus of Chapter 4 is the autonomous Portuguese island region of Madeira, in the mid-Atlantic which – in spite of its name – actually com-prises a two-island unit: the smallest size for an archipelago. Smaller Porto Santo has a high reliance on the larger island of Madeira, for both summer tourists and strategic planning. No wonder Porto Santo finds itself in a posi-tion of 'double dependency' on its domineering neighbour, making it difficult for its tourism potential to be fully achieved. Indeed, the development of tourism in Porto Santo is more likely to be shaped by the needs of Madeira than its own.

For Chapter 5, the focus is on Romblon province, in Central Philippines. Here is an archipelago conceptualised and marketed as a space of seclusion and tranquillity to travellers, shaped by its geography, geology, politics, and transport logistics. Ironically, the very same features manifest a clear, nested 'core-periphery' relationship: between Manila, the Philippine capital, and the province; and, within the province, the rivalry for primacy between Tablas and Romblon, as well as the multiple insularity experienced by the rest of the archipelago.

We head to North America for Chapter 6. The four, largely-English speaking provinces of Atlantic Canada – Newfoundland and Labrador (where most of the population lives on the island of Newfoundland), New Brunswick, Nova Scotia (including the island of Cape Breton), and Prince Edward Island – experimented with being an 'Atlantic Bubble' for some months during the Covid-19 pandemic, if anything to salvage part of their tourism industries. The argument here is that, during the pandemic, archipelagic thinking allowed for more fluidity between and among the constituent pieces of the region. Nova Scotia and New Brunswick and their surrounding islands reconstitute the ideas of boundedness and connectedness that are the hallmarks of island living.

In all the above six examples, the cases comprise small proportions of the respective countries' populations and land areas. In contrast, Part II illustrates core-periphery dynamics that unfold over a complete sovereign state.

We start with the Caribbean, and the tri-island state of Grenada, as Chapter 7. The country takes its name from the main(is)land of Grenada, but the two, South Grenadine populated islands of Carriacou and Petite Martinique are integral components of the state. The latter two suffer from their geographical peripherality and the inordinate influence of the country's capital, St George's, resulting in their dealing with a limited (and mainly yacht) tourism. This marginalisation may also explain why their citizens do not depend as much on tourism for their livelihood and why these two islands and their citizens are represented as more authentic and traditional – that is, closer to their African roots – than mainland Grenada. Grenada may be the 11th smallest country in the world by land area; but, in spite of this scale, it still exhibits urban-rural and main island-outer island tensions.

Chapter 8 takes us to the Indian Ocean, and the archipelagic sovereign state of Seychelles. Here, archipelagicity, and the associated differential distances separating islands from Mahé, the main island, are fundamental concepts. The expression and manifestation of differentiation in the Seychelles tourism offer lies precisely in the progressive exclusivity provided by distance, and enabled by a plurality of island geographies. Hence emerges a meta-narrative where visitors are transported (pun intended) to more remote island worlds that can be coveted and enjoyed by progressively lower visitor numbers and higher natural content.

The archipelagic state of Vanuatu, formerly known as the New Hebrides before independence in 1980, in the Pacific Ocean, features in Chapter 9. As if to highlight the stubborn fixity of power relations, this chapter notes that it is not Covid-19 but an ever-present, ongoing normality that constantly changes, challenges, and frustrates the equitable development of tourism in this (and other) archipelagos. Vanuatu may consist of roughly 80 islands; but resorts and cruises emphasise the primacy of Port Vila, the capital, and its immediate neighbourhood, despite official commitments to the outer islands. Tourism makes minimal contributions beyond Vila and its suburbs.

Part III brings us to the end of our global review, this time with case studies that involve sweeps across borders, and usurp national boundaries. This section reminds us that archipelagos do not necessarily respect or correspond to neat, national, political borders.

The Central Mediterranean espouses a clutch of islands, with Sicily at its heart, but including the Maltese archipelago – its own sovereign state – and various components of the Italian state. Chapter 10 reviews how domestic tourism (and staycationism) has increased in this region post pandemic; and, in spite of lip service to sustainable tourism practices, the situation on the ground is very much one of 'business as usual'. There is also no change in the alleged neglect shown by central authorities: there is palpable discontent, especially by the residents and business community of the smallest islands, about efforts by local and regional governments in promoting their islands.

We return to the Caribbean for Chapter 11, this time looking at the significant role played by Dutch Sint Maarten as a North-Eastern Caribbean hub, being the main air and sea gateway to at least five other sub-national island jurisdictions. Here, tourism numbers have been seriously impacted not just by Covid-19 but by the exceptionally devastating effects of Hurricane Irma in 2017.

Finally, Chapter 12 revisits the Mascerene Islands of Mauritius, Réunion, and Rodrigues, a mixed bag of islands that had been reviewed in Baldacchino (2015). Almost a decade and a pandemic later, not much has changed. Mauritius is once again feeling the pressure of a tourism industry strongly dependent on coastal (beach and hotel) infrastructure. Meanwhile, the other members of the trio seek to make a virtue out of isolation. French Réunion continues to market its dramatic volcanic landscape for the few but high-end (mainly metropolitan French) tourists who can afford its relatively high cost of living. And Rodrigues, at the logistic mercy of its 'main (is)land', peddles its Creoleness to tourists as a vibrant cultural trait that remains more palpable there than in cosmopolitan Mauritius.

References

Abdullah, M. N. A. (2021). Revenge tourism: Trend or impact post-pandemic Covid-19? In A. Hudaiby, G. Kusumah, C. U. Abdullah, D. Turgarini, M. Ruhimat, O. Ridwanudin, and Y. Yuniawati (Eds), *Promoting creative tourism: Current issues in tourism research* (pp. 623–627). London: Routledge.

Agius, K. and Chaperon, S. (2023). The dependency-autonomy paradox: A core-periphery analysis of tourism development in Mediterranean archipelagos. *International Journal of Tourism Research, 25*(5), 506–516. https://doi.org/10.1002/jtr.2582.

Amoamo, M. (2011). Remoteness and myth making: Tourism development on Pitcairn Island. *Tourism Planning & Development, 8*(1), 1–19. https://doi.org/10.1080/21568316.2011.554035.

Amoamo, M. (2021). Brexit: threat or opportunity? Resilience and tourism in Britain's island territories. *Tourism Geographies, 23*(3), 501–526. https://doi.org/10.1080/14616688.2019.1665093.

Amos, O. (2020, 24 August). Ten countries kept out Covid. But did they win? *BBC.* www.bbc.com/news/world-asia-53831063.

Anton Clavé, S. (2022). Theme parks, staycation practices, and Covid-19: Opportunities and uncertainties. *Journal of Themed Experience and Attractions Studies, 2*(1), 21–25. https://stars.library.ucf.edu/jteas/vol2/iss1/6.

Baldacchino, G. (2005). Islands: Objects of representation. *Geografiska Annaler: Series B, Human Geography, 87*(4), 247–251. https://doi.org/10.1111/j.0435-3684.2005.00196.x.

Baldacchino, G. (Ed.). (2006). *Extreme tourism: Lessons from the world's cold water islands.* Oxford: Elsevier.

Baldacchino, G. (2011). Surfers of the ocean waves: Change management, intersectoral migration and the economic development of small island states. *Asia Pacific Viewpoint, 52*(3), 236–246. https://doi.org/10.1111/j.1467-8373.2011.01456.x.

Baldacchino, G. (2012). Island tourism. In A. Holden and D. Fennell (Eds), *The Routledge handbook of tourism and the environment* (pp. 200–208). London: Routledge.

Baldacchino, G. (Ed.) (2015). *Archipelago tourism: Policies and practices.* Farnham: Ashgate.

Baldacchino, G. and Bertram, G. (2009). The beak of the finch: Insights into the economic development of small economies. *The Round Table: Commonwealth Journal of International Affairs, 98*(401), 141–160. https://doi.org/10.1080/00358530902757867.

Baldacchino, G. and Ferreira, E. C. D. (2015). Competing notions of diversity in archipelago tourism: Transport logistics, official rhetoric and inter-island rivalry in the Azores. *Island Studies Journal, 8*(1), 84–104. https://doi.org/10.24043/isj.278.

Bangkok Post (2022, 23 September). Time for tourism rejig. Editorial. www.bangkokpost.com/opinion/opinion/2398233/time-for-tourism-rejig.

Baran, P. A. (1957). *The political economy of growth.* New York: Monthly Review Press.

Beckford, G. L. (1972). *Persistent poverty: Underdevelopment in plantation economies of the Third World.* New York: Oxford University Press.

Bertram, G. and Poirine, B. (2018). Island political economy. In G. Baldacchino (Ed.), *The Routledge international handbook of island studies: A world of islands* (pp. 202–246). London: Routledge.

Bianchi, R. V. (2015). Towards a new political economy of global tourism revisited. In R. Shapley and D. J. Telfer (Eds), *Tourism and development: Concepts and issues* (2nd edn) (pp. 287–331). Bristol: Channel View Publications.

Borg, N. (2024, 12 February). Malta received more than three million tourists in 2023, MTA says. *Times of Malta.* https://timesofmalta.com/article/malta-received-three-million-tourists-2023-mta-says.1083369.

Boumpa, A. and Paralikas, A. (2021). The Greek archipelago: A unique representative case-study of differential legal status and of double insularity. *Liverpool Law Review, 42,* 99–109. https://doi.org/10.1007/s10991-020-09265-w.

Briguglio, L., Archer, B., Jafari, J., and Wall, G. (Eds). (1996a). *Sustainable tourism in islands and small states: Issues and policies.* London: Pinter.

Briguglio, L., Butler, R.W., Harrison, R., and Leal Filho, W. (Eds). (1996b). *Sustainable tourism in islands and small states: Case studies.* London: Pinter.

Britton, S. G. (1982). The political economy of tourism in the Third World. *Annals of Tourism Research*, 9(3), 331–358. https://doi.org/10.1016/0160-7383(82)90018-4.

Buckley, J. (2021, 14 July). Venice bans cruise ships from the city center – again. *CNN.* www.cnn.com/travel/article/venice-cruise-ship-ban-government/index.html.

Buckley, J. (2017, 28 December). Why you should visit St Helena, home to the 'world's most useless airport'. *The Independent (UK).* www.independent.co.uk/travel/africa/st-helena-what-to-see-do-best-guideisland-distillery-tortoise-coffee-where-to-stay-consulate-hotel-a8132291.html.

Bulchand-Gidumal, J. (2022). Post-Covid-19 recovery of island tourism using a smart tourism destination framework. *Journal of Destination Marketing & Management*, 23, 100689. https://doi.org/10.1016/j.jdmm.2022.100689.

Burnett, K. A. (2023). Refuge or retreat: Resilience and the mediatization of Scotland's island space, *Scottish Geographical Journal.* https://doi.org/10.1080/14702541.2023.2242821.

Butler, R. W. (2012). Islandness: It's all in the mind. *Tourism Recreation Research*, 37(2), 173–176. https://doi.org/10.1080/02508281.2012.11081702.

Chilton, N. (2022, 8 July). What it's like to land on the world's shortest commercial runway. *CNN.* www.cnn.com/travel/article/saba-airport-shortest-commercial-runway/index.html.

Conlin, M. V. and Baum, T. (Eds). (1995). *Island tourism: Management principles and practice.* New York: Wiley.

Connell, J. and Campbell, Y. (2021). Aftermath: Towards the 'new normal'? In Y. Campbell and J. Connell (Eds), *COVID in the islands: A comparative perspective on the Caribbean and the Pacific* (pp. 517–528). London: Palgrave Macmillan.

Connell, J. and Taulealo, T. (2021). Island tourism and COVID-19 in Vanuatu and Samoa: An unfolding crisis. *Small States & Territories*, 4(1), 105–124. www.um.edu.mt/library/oar/handle/123456789/74986.

Coronato, A., Di Napoli, C., Paragliola, G., and Serino, L. (2021, June). Intelligent planning of onshore touristic itineraries for cruise passengers in a smart city. In *2021 17th International Conference on Intelligent Environments* (pp. 1–7). IEEE. https://doi.org/10.1109/IE51775.2021.9486648.

Couto, G., Castanho, R. A., Pimentel, P., Carvalho, C., Sousa, Á., and Santos, C. (2020). The impacts of Covid-19 crisis over the tourism expectations of the Azores archipelago residents. *Sustainability*, 12(18), 7612. https://doi.org/10.3390/su12187612.

Craig, V. (2022, August 21). Greece promotes smaller islands in post-Covid tourism push. *BBC.* www.bbc.com/news/business-62595814.

Cruise Ship Karen (2018, April 13). Dravuni island: Port guide. https://cruiseshipkaren.com/port_guide/dravuni-island/.

DeLoughrey, E. (2001). 'The litany of islands, the rosary of archipelagoes': Caribbean and Pacific archipelagraphy. *ARIEL: A Review of International English Literature*, 32(1), 21–51.

DeShong, H. A., Devonish, D., Grenade, W. C., and Roberts, S. (Eds). (2023). *Interdisciplinary perspectives on Covid-19 and the Caribbean, Vol. 1: The state, economy and health.* Cham: Springer International.

Dodds, R. and Graci, S. (2012). *Sustainable tourism in island destinations.* London: Routledge.

Dominioni, I. (2021, 10 April). Italy considers covid-free islands to save summer tourism. *Forbes.* www.forbes.com/sites/irenedominioni/2021/04/10/italy-on-the-loo kout-to-save-summer-tourismby-launching-covid-free-islands/.

Douglas, N. and Douglas, N. (2006). Paradise and other ports of call: Cruising in the Pacific Islands. In R. K. Dowling (Ed.), *Cruise ship tourism* (pp. 184–194). Wallingford: CABI. http://sherekashmir.informaticspublishing.com/625/1/9781845930486. pdf#page=206.

Ecott, T. (2019, 8 May). Sustainable tourism: Why the Faroe Islands closed for maintenance. *The Guardian (UK).* www.theguardian.com/travel/2019/may/08/faroe-isla nds-closed-maintenanace-voluntourists-conservation.

FHTA (2023, 3 March). FHTA tourism talanoa: Acknowledging the journey and the destination. www.linkedin.com/pulse/fhta-tourism-talanoa-acknowledging/.

Fleming, A. (2005). *St Kilda and the wider world: Tales of an iconic island.* Oxford: Windgather Press.

Frank, A. G. (1967). *Capitalism and underdevelopment in Latin America.* New York: NYU Press.

Goodwin, H. (2023, August 10). New Zealand shows the way to fight overtourism. *Travel Tomorrow.* https://traveltomorrow.com/new-zealand-shows-the-way-to-fight-overtourism/.

Government of The Philippines (2018, 26 April). *Proclamation No. 475.* Declaring a state of calamity in the Barangays of Malabag, Manoc-Manoc and Yapak (Island of Boracay) in the Municipality of Malay, Aklay, and temporary closure of the island as a tourist destination. www.officialgazette.gov.ph/2018/04/26/proclamation-no-475-s-2018/.

Hall, C. M. (2010). Island destinations: A natural laboratory for tourism: Introduction. *Asia Pacific Journal of Tourism Research*, 15(3), 245–249. https://doi.org/10. 1080/10941665.2010.503613.

Hampton, M. P. and Christensen, J. (2007). Competing industries in islands a new tourism approach. *Annals of Tourism Research*, 34(4), 998–1020. https://doi.org/10. 1016/j.annals.2007.05.011.

Harrison, D. (2001). Islands, image and tourism. *Tourism Recreation Research*, 26(3), 9–14. https://doi.org/10.1080/02508281.2001.11081194.

Hart, R. (2022, August 31). Covid has reached every corner of the world: But these three places claim to be virus-free. *Forbes.* www.forbes.com/sites/roberthart/2022/08/ 31/covid-has-reached-every-corner-of-the-world-but-these-three-places-claim-to-be-virus-free/?sh=241d8ef64a8a.

Heinzlef, C. and Serre, D. (2021). Did French Polynesia cope with COVID-19? Intrinsic vulnerabilities and decreased resilience. In Y. Campbell and J. Connell (Eds), *COVID in the islands: A comparative perspective on the Caribbean and the Pacific* (pp. 125–144). London: Palgrave Macmillan.

Hepher, T. and Plucinska, J. (2023, June 20). Airbus wins record 500-plane order from India's IndiGo. *Reuters.* www.reuters.com/business/aerospace-defense/airbus-win s-historic-500-plane-indigo-order-2023-06-19/.

Higgins-Desbiolles, F. (2021). The 'war over tourism': Challenges to sustainable tourism in the tourism academy after COVID-19. *Journal of Sustainable Tourism*, 29(4), 551–569. https://doi.org/10.1080/09669582.2020.1803334.

Higgins-Desbiolles, F. (2022). The ongoingness of imperialism: The problem of tourism dependency and the promise of radical equality. *Annals of Tourism Research*, 94, 103382. https://doi.org/10.1016/j.annals.2022.103382.

Hoffer, O. (2012). Covid-19 management in New Caledonia and Wallis and Futuna: A magnifying glass for local political and economic issues. In Y. Campbell and J. Connell (Eds), *COVID in the islands: A comparative perspective on the Caribbean and the Pacific* (pp. 145–161). London: Palgrave Macmillan.

Holloway, E. C. (2002). *Community decline on the offshore islands of Prince Edward Island: St. Peter's and Boughton Islands.* Charlottetown, PEI: University of Prince Edward Island.

Jimi, H. and Hashimoto, G. (2020). Challenges of Covid-19 outbreak on the cruise ship Diamond Princess docked at Yokohama, Japan: A real-world story. *Global Health & Medicine*, 2(2), 63–65. doi:10.35772/ghm.2020.01038.

King, D. C. (2005). *Memories of Boughton island.* Charlottetown, PEI: Museum and Heritage Foundation.

La Flamme, A. G. (1983). The archipelago state as a societal subtype. *Current Anthropology*, 24(3), 361–362.

Laird, S. (2021, November 1). As more ports ban mega cruise ships, what is the future of the largest vessels? *Condé Nast Traveller.* www.cntraveler.com/story/as-more-ports-ban-mega-cruise-ships-what-is-the-future-of-the-largest-vessels.

Lennox Island First Nation. (2023). *Lennox Island.* https://lennoxisland.com/.

Lew, A. A., Cheer, J. M., Haywood, M., Brouder, P., and Salazar, N. B. (2020). Visions of travel and tourism after the global Covid-19 transformation of 2020. *Tourism Geographies*, 22(3), 455–466, https://doi.org/10.1080/14616688.2020.1770326.

Lewis, W. A. (1954). Economic development with unlimited supplies of labour. *The Manchester School*, 22(2), 139–191. https://doi.org/10.1111/j.1467-9957.1954.tb00021.x.

Lim, C. C. and Cooper, C. (2009). Beyond sustainability: Optimising island tourism development. *International Journal of Tourism Research*, 11(1), 89–103. https://doi.org/10.1002/jtr.688.

Lyons, K. (2022 February 2). Tsunami-hit Tonga goes into lockdown after workers helping deliver aid catch Covid. *The Guardian (UK).* www.theguardian.com/world/2022/feb/02/tsunami-hit-tonga-goes-into-lockdown-after-workers-helping-deliver-aid-catch-covid.

McKercher, B. (2021). The periphery as a tourism market?. *Tourism Recreation Research*, 1–10. https://doi.org/10.1080/02508281.2021.2011592.

McKercher, B. and Lew, A. A. (2003). Distance decay and the impact of effective tourism exclusion zones on international travel flows. *Journal of Travel Research*, 42(2), 159–165. https://doi.org/10.1177/0047287503254812.

McKercher, B. and Mak, B. (2019). The impact of distance on international tourism demand. *Tourism Management Perspectives*, 31, 340–347. https://doi.org/10.1016/j.tmp.2019.07.004.

McLeod, M. (2021). The Bahamas: Tourism policy within a pandemic. In Y. Campbell and J. Connell (Eds), *COVID in the islands: A comparative perspective on the Caribbean and the Pacific* (pp. 219–230). London: Palgrave Macmillan.

McLeod, M. and Croes, R. R. (Eds). (2018). *Tourism management in warm-water island destinations.* Wallingford: CABI.

McLeod, M., Dodds, R., and Butler, R. W. (2021). Introduction to special issue on island tourism resilience. *Tourism Geographies*, 23(3), 361–370. https://doi.org/10.1080/14616688.2021.1898672.

Moberg, M. (2022). *Slipping away: Banana politics and fair trade in the Eastern Caribbean*. New York: Berghahn Books.

Monterrubio, C., Osorio, M., and Benítez, J. (2018). Comparing enclave tourism's socioeconomic impacts: A dependency theory approach to three state-planned resorts in Mexico. *Journal of Destination Marketing & Management*, 8, 412–422. https://doi.org/10.1016/j.jdmm.2017.08.004.

Movono, A. and Scheyvens, R. (2021). Tourism in a world of disorder: A return to the vanua and kinship with nature in Fiji. In Y. Campbell and J. Connell (Eds), *COVID in the islands: A comparative perspective on the Caribbean and the Pacific* (pp. 265–277). London: Palgrave Macmillan.

Munshi, S., Banerjee, S., and Chakraborty, I. (2022). Thilafushi a toxic bomb: Learning from an island's SWM (sustainable waste management) practice. *International Journal of Architecture and Infrastructure Planning*, 8(2), 1–12.

Nelson, M. (2020). Tropicality, purified spaces, and the colonial gaze: Exclusionary policies in cruise tourism and its impact on the Caribbean. In F. J. Riemer (Ed.), *Front and back stage of tourism performance: Imaginaries and bucket list venues* (pp. 153–168). London: Routledge.

Parco Nazionale. (2023). *Isola di Pantelleria Parco Nazionale*. [Pantelleria Island national park.] www.parconazionalepantelleria.it/Eindex.php.

Peterson, R. R. (2020). Over the Caribbean top: Community well-being and overtourism in small island tourism economies. *International Journal of Community Well-Being*, 1–38. https://link.springer.com/article/10.1007/s42413-020-00094-3.

Pimentel, P., Vulevic, A., Couto, G., Behradfar, A., Gómez, J., and Castanho, R. A. (2022). Maritime transportation dynamics in the Azores region: Analyzing the period 1998–2019. *Infrastructures*, 7(2), 21. https://doi.org/10.3390/infrastructures7020021.

Pryor, F. L. (1982). The plantation economy as an economic system. *Journal of Comparative Economics*, 6(3), 288–317. doi:10.1016/0147-5967(82)90039-7.

Pugh, J. (2013). Island movements: Thinking with the archipelago. *Island Studies Journal*, 8(1), 9–24. https://doi.org/10.24043/isj.273.

Pugh, J. and Chandler, D. (2021). *Anthropocene islands: Entangled worlds*. London: University of Westminster Press.

Rama, J. and Hall, J. (2021). Raúl Prebisch and the evolving uses of 'centre-periphery' in economic analysis. *Review of Evolutionary Political Economy*, 2(2), 315–332. https://doi.org/10.1007/s43253-021-00036-5.

Roberts, B. R. and Stephens, M. A. (Eds) (2017). *Archipelagic American studies*. Durham NC: Duke University Press.

Roberts, S. (2023). Absorb, adapt, or transform? An exploratory analysis of small tourism businesses' resilience strategies during the pandemic. In S. Roberts, H. A. F. DeShong, W. C. Grenade and D. Devonish (Eds), *Interdisciplinary perspectives on Covid-19 and the Caribbean, Vol. 1: The state, economy and health* (pp. 219–248). Cham: Palgrave Macmillan.

Rodney, W. (1972). *How Europe underdeveloped Africa*. London: Bogle- L'Ouverture Publications.

Rodrigues, M., Teoh, T., Ramos, C., de Winter, T., Knezevic, L., Marcucci, E., Lozzi, G., Gatta, V., Antonucci, B., Cutrufo, N., Marongiu, L., and Cré, I. (2021). *Relaunching transport and tourism in the European Union after Covid-19*. Brussels: Directorate-General for Internal Policies, European Parliament. www.europarl.europa.eu/RegData/etudes/STUD/2021/652235/IPOL_STU(2021)652235_EN.pdf

Ronström, O. (2013). Finding their place: Islands as locus and focus. *Cultural Geographies*, 20(2), 153–165. www.jstor.org/stable/44289601.

Sarantakou, E. and Terkenli, T. S. (2021). Non-institutionalized forms of tourism accommodation and overtourism impacts on the landscape: The case of Santorini, Greece. In C. Milano, M. Novelli, and J. M. Cheer (Eds), *Travel and tourism in the age of overtourism* (pp. 59–81). London: Routledge.

Sharpley, R. (2012). Island tourism or tourism on islands? *Tourism Recreation Research*, 37(2), 167–172. https://doi.org/10.1080/02508281.2012.11081701.

Sharpley, R. (2022). Tourism and development theory: Which way now? *Tourism Planning & Development*, 19(1), 1–12. https://doi.org/10.1080/21568316.2021.2021475.

Siegel, P. E., Hofman, C. L., Bérard, B., Murphy, R., Hung, J. U., Rojas, R. V., and White, C. (2013). Confronting Caribbean heritage in an archipelago of diversity: Politics, stakeholders, climate change, natural disasters, tourism and development. *Journal of Field Archaeology*, 38(4), 376–390. https://doi.org/10.1179/0093469013Z.00000000066.

Sinclair-Maragh, G. and Gursoy, D. (2015). Imperialism and tourism: The case of developing island countries. *Annals of Tourism Research*, 50(1), 143–158. https://doi.org/10.1016/j.annals.2014.12.001.

Sperling, K. (2017). How does a pioneer community energy project succeed in practice? The case of the Samsø Renewable Energy Island. *Renewable and Sustainable Energy Reviews*, 71, 884–897. https://doi.org/10.1016/j.rser.2016.12.116.

Stratford, E., Baldacchino, G., McMahon, E., Farbotko, C., and Harwood, A. (2011). Envisioning the archipelago. *Island Studies Journal*, 6(2), 113–130. https://doi.org/10.24043/isj.253.

The Economist (2023 June 15). *Holiday travel: Enjoy it while it lasts.* www.economist.com/business/2023/06/15/how-long-will-the-travel-boom-last.

Trading Economics (2023). *New Zealand tourist arrivals.* https://tradingeconomics.com/new-zealand/tourist-arrivals.

Trousdale, W. J. (1999). Governance in context: Boracay island, Philippines. *Annals of Tourism Research*, 26(4), 840–867. https://doi.org/10.1016/S0160-7383(99)00036-5.

Turner, C. (2010). *The geography of hope: A tour of the world we need.* Toronto, ON: Vintage Canada.

UN (2024). *United Nations Convention on the Law of the Sea.* www.un.org/depts/los/convention_agreements/texts/unclos/unclos_e.pdf.

UNWTO (2023, 9 May). Tourism on track for full recovery as new data shows strong start to 2023. *World Tourism Organisation.* www.unwto.org/news/tourism-on-track-for-full-recovery-as-new-data-shows-strong-start-to-2023.

Urry, J. (2001). The sociology of space and place. In J. R. Blau (Ed.), *The Blackwell companion to sociology* (pp. 3–15). London: Blackwell.

Visit Faroe Islands (2023). *Closed for maintenance, Open for voluntourism.* https://visitfaroeislands.com/en/closed.

Volger, R. (2022). Revenge and catch-up travel or degrowth? Debating tourism Post Covid-19. *Annals of Tourism Research*, 93, 103272. https://doi.org/10.1016/j.annals.2021.103272.

Weaver, D. B. (2017). Core–periphery relationships and the sustainability paradox of small island tourism. *Tourism Recreation Research*, 42(1), 11–21. https://doi.org/10.1080/02508281.2016.1228559.

Williams, C., You, J. J., and Joshua, K. (2020). Small-business resilience in a remote tourist destination: exploring close relationship capabilities on the island of St Helena. *Journal of Sustainable Tourism*, 28(7), 937–955. https://doi.org/10.1080/09669582.2020.1712409.

Part I

Internal, sub-national archipelagos

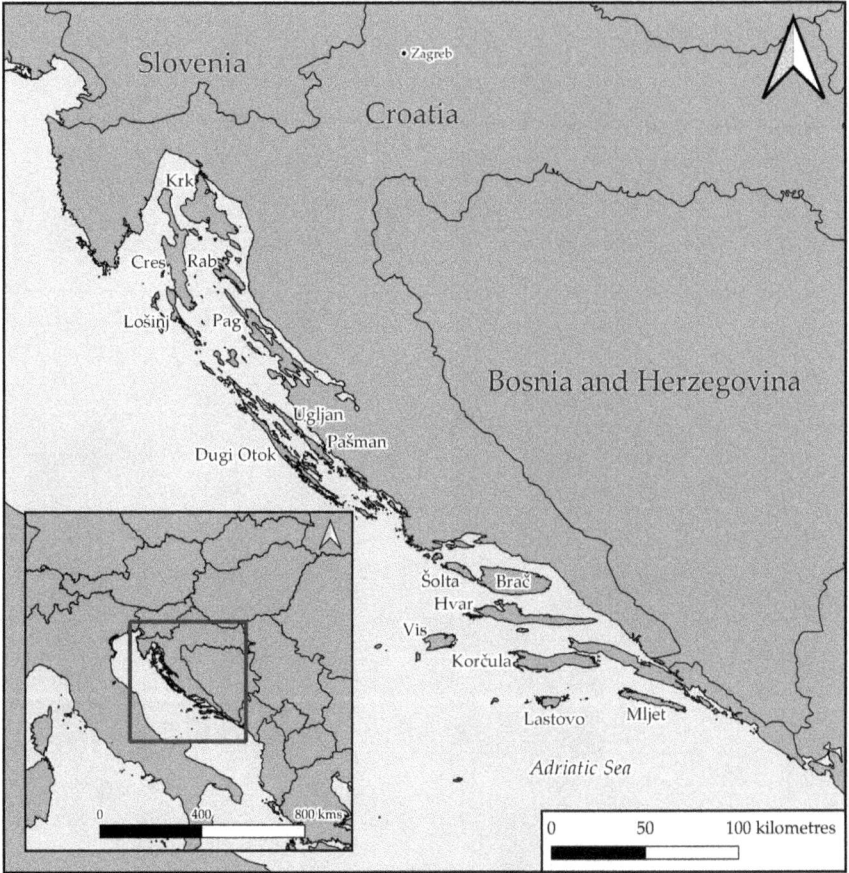

Figure 1.1 The islands of Croatia

1 Covid-19 and post-Covid times in the Croatian archipelago

Anica Čuka, Željka Kordej-De Villa and Nenad Starc

Introduction

The 1,244 Croatian islands are located along the eastern coast of the Adriatic Sea, a part of the Mediterranean that extends the furthest north towards the centre of Europe. With a surface area of 3,259 km^2 (Duplančić Leder, Ujević, and Čala, 2004), they account for 5.8% of the total surface area of Croatian territory. Fifty of these islands are inhabited. Bridges connect five, some can be reached by rowing, and the furthest inhabited island is 35 nautical miles away from the nearest coast. Among the inhabited islands, the largest covers 405.7 km^2, while the smallest covers 0.33 km^2. According to the latest census of 2021, Croatia had a population of 3.87 million, with 120,440 (3.8%) living on the islands (Census, 2021).

The islands' resources have been valued in various ways from the time of the Palaeolithic islanders to the present day. Over the centuries, some 40 smaller islands have developed an insular economy that began to connect more significantly with broader spatial-economic systems only about 50 years ago. Ten larger islands had been connected with these systems centuries earlier, allowing for shipbuilding, maritime activities, and trade development. The turning point in the development of both occurred in the second half of the twentieth century (Faričić and Čuka, 2020a). From the 1950s onwards, slowly, and then increasingly and seemingly inevitably, tourism became the dominant activity (Zlatar Gamberožić and Tonković, 2015). Traditional activities were neglected, and despite increasing tourism revenue, the islands' population decreased from one census to the next throughout the millennium. The islanders also became older (Klempić Bogadi and Podgorelec, 2020).

In July and August, during the peak season when there can be five, ten, or even 40 guests per islander, it is hard to believe that it all started at the imperial level, the highest possible. At the end of the nineteenth century, the court of Franz Joseph, Emperor and King of Austria-Hungary, cast its eyes on the north-west, 14-island cluster of Brijuni. This prompted the very first large tourist investor in the Adriatic to buy the then marshy, malarial, hardly habitable sub-archipelago in 1893. Austrian industrial magnate Paul Kupelwieser (1843–1919) drained the islands from its swamps – and thus eradicated

DOI: 10.4324/9781003451037-3

malaria from there – and then built villas, hotels, and other accoutrements that might be needed for imperial relaxation. Veliki Brijun, the main island of the sub-archipelago, grew into the most luxurious resort on the Croatian coast and islands which kept attracting members of not only the Austro-Hungarian court (Fatovic-Ferencic, 2006; National Park Brijuni, 2023).

Thus, tourism on the Adriatic islands began to develop on the least attractive island. Its comparative advantages, proximity, and pleasant climate were not enough. It was necessary to eliminate comparative disadvantages such as malaria to gain a competitive edge. Other, more attractive islands caught up. Before World War I, accommodation could be found on almost all larger islands. During the war, connections with the mainland were mainly cut, and the islands went through the 1920s without tourists.

A less demanding version of tourism – but more mass-oriented than the elitist imperial variety – began to develop in the 1930s and grew slowly until the next conflict. The Second World War did not bring destruction to the islands; so, as so many times in history, they served as shelters. The islanders and refugees from the mainland survived the war on the verge of famine without thinking about the tourists who stayed at home. In the late 1940s, in the post-war period, priority was given to factories rather than guests, and it remained so until the mid-1950s when the end of post-war reconstruction was already in sight. Initially suspicious of the socialist regime that opened the doors for them, foreign tourists began to disembark on the islands in increasing numbers only at the end of the 1950s. There were outdated hotels and inns on the larger islands, and the existing connections with the mainland were initially sufficient. The increasing demand soon exceeded the capacities of the few island hotels, so anyone who had a room or two in the house that could be prepared for guests started getting involved in tourism. Therefore, tourism developed faster than any other sector of the island economy. Slower than in tourist-attractive places on the mainland, though only briefly. In the late 1960s, when ferry and later catamaran connections with the mainland became more frequent, mainland tourism began to reach the islands. The larger islands quickly caught up and reached the tourist standards of main-land destinations in the early 1970s. The connections with the nearby main-land were good enough by then and seemed sufficient. Around 40 smaller islands without roads and cars remained connected by boats only. All they could offer at the time was sun, sea, beaches, untouched nature and their amateur adapted homes, a local grocery shop and a week-old newspapers.

The hosts of the first guests who disembarked on the islands, especially those who hosted foreigners, could not even anticipate that they were taking on the role of driving the future primary sector of the island economy. Data on tourist visits and overnight stays in the 1960s indicate rapid growth. Still, they seem modest compared to the data from the 1970s and especially the 1980s (Pirjavec and Kesar, 2002; Vukonić, 2005). In just three decades, tourism dominated the island economy, which lost its insular characteristics. The growth of tourist demand surpassed the potential of island resources as early as the 1960s. Further development almost exclusively relied on delivering necessary goods from the mainland. The proximity to the mainland became an increasingly significant factor: closer islands, especially those connected by bridges, progressed in every aspect more and faster than the more distant ones (Starc, 2020).

In the 1970s, tourists also conquered the small islands of the archipelago. Since then, there is no longer an inhabited island in the Adriatic where tourism and remittances from the diaspora are not the primary sources of income. Having reached the lighthouse islets that offered accommodation, tourism grew into an all-island activity. It has continued to develop, intensively on the large islands and more slowly on the small ones.

Unlike other socialist countries, Yugoslavia gradually opened its borders to citizens of Western countries; by the same token, Yugoslav subjects were allowed to travel abroad. The country also sought to participate in global trade, starting from the 1950s. Tourism was recognised as a kind of export and an increasingly abundant source of much-needed foreign currency, strengthening its macroeconomic importance and decisively determining tourism policy. The sector continued to grow, so the socialist authorities found it easier to overlook ideological differences in the name of economic development. Opened ajar in the 1950s, the door was already wide open to foreign tourists at the end of the 1960s. However, this led to a gap between domestic and foreign tourists. The initial idea that tourism was an activity whose primary purpose was the relaxation of the working class seemed anachronistic and almost exotic by the 1970s. The dichotomy of domestic and foreign tourists was still felt until the 1980s, when there were no traces of socialist ideology during the summer hustle and bustle.

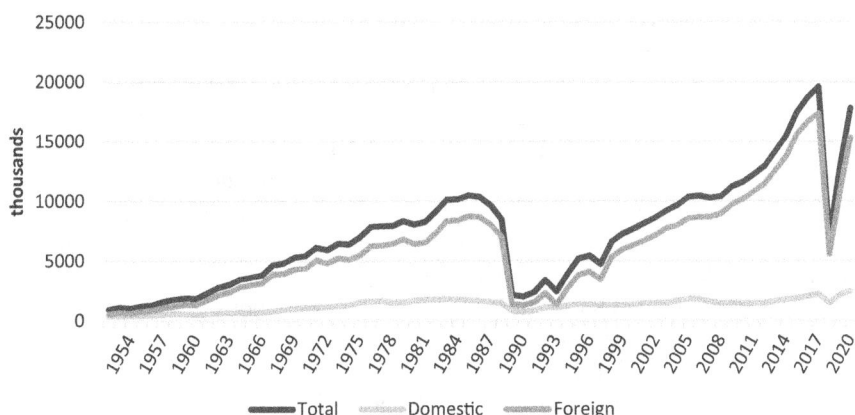

Figure 1.2 Tourist arrivals in Croatia in the period 1954–2022 (in thousands)
Source: Croatian Bureau of Statistics, Tourism – Review from 1954.

The gap between the growing demand for tourist services and the need to improve supply was also visible. Improvements occurred with delay and in a reactive manner, which was also the case with the tourism development policy prepared at the state and municipal levels. Demand management was lacking, so neither coastal nor island tourism developed profiles that would distinguish them from tourism in other Mediterranean countries.

In 1991, Croatia gained independence, democratised, and became a republic. The Homeland War began, lasted until 1995, and ended in victory. No sector of the economy experienced as many shocks as tourism during that time. The war started in the middle of the summer tourist season and caused an exodus. Coastal roads were congested with vehicles, and boats on the Adriatic routes were filled with foreigners leaving the islands with traumatic experiences. By the end of 1991, tourist agencies were left without business, and hotels and apartments were without guests. The hospitality industry was reduced to serving a few domestic guests. All activities dependent on tourist demand were more or less affected.

No sector recovered more successfully, however. Tourists began to return as early as the end of the war in 1995. By the end of the millennium, foreign tourists seemed to need to remember why they hadn't come four or five years earlier. In their third post-war recovery, coastal and island tourism again demonstrated resilience. It turned out, however, that recovery almost exclusively depends on the slow but persistent growth of tourist demand and the increasing need for more and more people to rest far enough from home and work (Sirgy, 2019; Dolničar, Lazartevski, and Yanamandram, 2013). Not too far, however, the proximity of countries most foreign tourists come from also appears as a factor of resilience (Arhivanalitika, 2021). This apparently unstoppable constant has determined tourism development on the Croatian coast. It has also determined the way of life, especially on smaller islands.

In the first two decades of the new millennium, coastal and island tourism reached and significantly exceeded pre-war figures, revitalising their macroeconomic significance. However, an analysis of the sector brings both good and bad news. First of all, since the independence of the Republic of Croatia in 1991, nine national tourism strategies and eight action plans for tourism development have been adopted. None of these development documents addresses the regional aspect of tourism, including the islands; sober reflections on island tourism have been left to the islanders. The latest official document – the Sustainable Tourism Development Strategy for 2022–2030, adopted by the Croatian Parliament (OG 2/2023) – praises the continuous growth in tourists, overnight stays, and investments. It also warns, however, that tourism's direct and indirect contribution to Croatia's GDP in 2019 increased to 25% (Croatian Bureau of Statistics – CBS, 2022, First Release – Tourism Satellite Account for the Republic of Croatia, 2019): all other 27 European Union member states have a smaller share of tourism in their GDP. It is also noted in the Strategy that the share of private accommodation accounts for about 60% of total tourism revenue (Ministry of Tourism and Sport of the Republic of Croatia, 2020); while in

Italy, Greece, and Spain, which are considered Croatia's main competitors, it does not exceed 25%. The increasing dependence of the economy on tourism indicates a weakness in the overall economic system. The more resilient tourism, the less resilient the overall economy.

The archipelago and its tourists

The Croatian Archipelago is located in the (relatively small) Adriatic Sea, where the criteria of near, distant, large, and small are different from those in the oceans. Most inhabited Croatian islands are considered small, with 32 having an area smaller than 25 km². None of the 26 smallest islands have more than 250 inhabitants each. Many have fewer than 100 residents. This diversity, scale, proximity to the mainland, and pleasant climate characterise Croatian islands and ultimately determine their tourism development.

The proximity to the mainland and differences in size have also determined the administrative division of the islands. Croatia is administratively divided into 21 counties and 556 municipalities and towns – in Croatia, the differences in administrative structure and prerogatives between municipalities and cities are negligible – which are further divided into a large number of local boards. Seven counties have a coastline, and only six include inhabited islands. Only five islands constitute separate and distinct single municipalities. Some islands have two to eight municipalities/towns; some belong to municipalities/towns run from mainlands or other islands. Some islands belong to part of a municipality as well as part of one or two mainland towns, and the island of Pag in the northern part of the archipelago is divided into two towns in two different counties. None of the small islands is a municipality. They belong to municipalities/towns on the mainland or some larger island. They are, thus, administratively determined at the lowest level; they establish local boards with too few prerogatives to manage the development by themselves.

This awkward administrative division reflects the weak inter-island connectivity. Islands in the northwest of the archipelago have always had very limited contact with islands in the southeast. There is little interest in changing this situation even today. So, despite improving communication technology and experiencing faster and more extensive information flows, contacts remain limited. On the other hand, the connections of all islands with the nearest city on the mainland are becoming closer.

The reason for the weak connectivity within the archipelago is not only the discriminatory administrative division but also the similarities between the islands. Uniform and scarce island soils and similar climates have determined archipelagic flora, agricultural production, and Adriatic fish stock and fishing. This has also led to a uniform production structure along the archipelago, so there was no basis for mutual trade exchange. Even today, not a single island can offer another island something it does not already have. This applies to almost all resources of the archipelago: tourism, the sea, pleasant climate, tranquillity, and everything else that attracts tourists can be found on all inhabited and even some uninhabited islands. However, tourism developed differently everywhere. The homogeneity of agricultural and fishing resources along the archipelago has given way to a heterogeneous tourist offer.

Despite similar comparative advantages, larger islands were more successful tourist destinations from the very beginning. They built tourism infrastructure, developed competitive advantages, and gradually increased their visibility on the Mediterranean tourism map. Smaller islands could not do that, so they have always been internationally visible only to the extent that larger islands included them in their marketing. Visibility in the domestic tourism market is better primarily due to the purchase of real estate, which was very cheap in the 1950s and 1960s when islanders were leaving their homes. Domestic buyers from the mainland bought properties on the coast and all islands, thus becoming regular guests who established tourism on small islands. The uneven administrative division, which does not reach the small islands, is also evident here. The current development does not indicate that municipalities and towns adequately care about tourism on their small islands. The relationship between the centre and the periphery proves unfortunate for small islands, both in the 'mainland-small island' relationship and, especially, in the 'larger-small island' relationship.

On the other hand, in marketing, stratification is easily noticeable. Specific hotels or hotel chains, villas, and apartments outside coastal and island cities are presented and offered. Such advertising includes villas on small islands but not the islands themselves. The marketing shifts from geographically determined destinations (Croatia, Adriatic) to geographically undefined places whose insularity should guarantee peace and, above all, exclusivity. The potential tourist attractiveness of the archipelago, especially certain sub-archipelagos, remains on the sidelines, unarticulated on posters, brochures, flyers, and the world wide web.

An exception is nautical tourism. Some foreign and almost all domestic sailors spend their winters in Croatian marinas, which had 251 marinas with 2,907 dry berths and 8,950 berths in the sea before the 2023 season (Ministry of Tourism and Sport, 2023). Once they set sail, sailors do not have one or two specific destinations in mind but rather a route they chart on the archipelago map. It is in nautical tourism that archipelago tourism truly comes to fruition. The nautical offer, as advertised, includes all islands and channels between them, which are particularly attractive for cruising. Furthermore, the sailing season is significantly longer than for conventional tourists. Sailing takes place in late spring, summer, and early autumn.

Geographically homogeneous, administratively fragmented, well-visited, and promoted in various (albeit unequal) ways, Croatian islands have recently experienced another tourism crisis after three wartime disruptions. This time, it was a pandemic.

Covid-19 and after

The first case of Covid-19 in Croatia, caused by the SARS-CoV-2 virus, was recorded on 25 February 2020. The last anti-epidemic measures were lifted three years and two months later, on 12 May 2023. Covid-19 left behind 18,273 deaths and some measurable economic and barely measurable social damage in the country (Government of the Republic of Croatia, 2020).

Islanders were as unprepared for Covid-19 as mainland residents but with more concerns. In addition to the grim health perspective, there was a fear that tourism, on which so much depends, would significantly decline in 2020. Moreover, the islands experienced a rekindling of the crises that reached them from the mainland. After refugees who once fled to the islands to escape the plague and later from regional and global wars, islanders also encountered refugees driven away by the virus ravaging the world. This time it was different because the refuge was claimed by so-called 'weekenders': people from the mainland who own second homes on the islands. They brought their families, and many brought relatives and friends as well. The islands were once again seen as safe havens, protected from the evils of the mainland. Those who did not hurry were left on the mainland: on 19 March 2020, crossing the state border was banned, and two days later, regular catamaran services were suspended, leaving only ferry services to link islands to the Croatian coast and (exceptionally) to other islands. Only island residents were allowed to use them. Starting from 23 March 2020, leaving one's place of residence in the Republic of Croatia was prohibited. Weekenders on the islands found themselves staying unexpectedly for weeks.

Foreign boaters found the most convenient route in the resulting standstill and confusion, which numerous anti-epidemic civil protection headquarters knew how to cause. They avoided anti-epidemic measures in their home countries, crossed the Adriatic, and docked in ports on small peripheral islands of the archipelago, whose local boards were not authorised to implement measures. They behaved like no measures were in place, causing discontent almost everywhere. The pre-season, which lasts from mid-April to the end of June on the coast and islands, started poorly, foreshadowing an even worse season.

However, it soon became apparent that anti-pandemic and tourism policies were not mutually exclusive and that adjustments could be made to partially open the doors to awaiting tourists. In a country whose economy relies heavily on tourism, a modality needed to be found to invite tourists to come unrestricted while maintaining measures such as controlling the size of groups, ban on dancing in couples, mandatory social distancing between tables, and so on. This modality was found as early as April. On 19 April 2020, the ban on leaving one's place of residence in the seven coastal and island counties was lifted, and the bans were supplemented with exceptions and eventually replaced with recommendations. All relaxations and easings applied equally to the mainland and islands. Islanders, accustomed to regionally indiscriminate development policies, were not particularly surprised. The interior residents were surprised, as it remained unclear how it came about that the epidemic was receding at sea but not inland. Besides, implementing measures on small islands was supposed to be carried out from the mainland or larger islands. However, even a cursory review of the municipal websites of the islands, where there are not many mentions of the small islands, shows that islanders on the periphery of the archipelago learned about anti-epidemic measures only if there was a TV signal.

Anti-pandemic measures were adopted for the entire area of Croatia without respecting regional specificities. Everything prescribed for the mainland had to be implemented on the islands. In the long series of measures adopted during the epidemic, only two stand out as island specific. The first one was introduced on 25 March 2020, on the island of Murter, a bridged island with two municipalities. A strict 14-day quarantine was declared and applied to both municipalities, even though Covid-19 was detected only in one. In such cases, measures were only introduced on the mainland in the municipality where the disease appeared. Here, it was estimated that the measure would be effective and easily enforceable only if it applied to the entire bridged island, so it was enough to close the bridge (Zadar, 2020). Anti-epidemic measures were introduced on the island of Brač a month later. After the outbreak of Covid-19 in one of the island's municipalities, the county civil protection headquarters imposed a quarantine for the entire island on 9 May 2020. Leaving one's residence on the island was prohibited for 14 days, and connections with the mainland were limited to one ferry crossing per day, only for island residents. On Brač as well, it was assessed that islands are good shelters because the disease cannot cross the sea easily. However, it went unnoticed that, once it does cross, the virus poses the risk of implosion within the island and perhaps more severe consequences than on the mainland. Covid-19 refugees on the island feared this. The news quickly spread, and the perception of a safe refuge turned overnight into a perception of a closed trap. The lineup of vehicles at the ferry terminal surpassed those manifest with the largest summer crowds. Islanders stayed on their island and perhaps watched the daily reports from public health institutions with a certain smile. During those days and in the following months, it became clear that the risk of infection on the mainland was much higher.

By 15 May 2020, public maritime transport restrictions were lifted, and all usual island connections with the mainland were reinstated. The doors opened wide, and the 2020 tourist season could begin. It underperformed, as expected, but not as much as feared. Accounts tallied in the fall revealed that half as many tourists visited the coast and islands in 2020 compared to 2019. However, as 2019 was a record-breaking year, the results did not seem so bad.

The concern gradually dissipated during the winter, and rightfully so. The 2021 season witnessed close to the number of arrivals and overnight stays from 2019, and the 2022 season even matched it (Table 1.1). Once again, tourists, it seems, have quickly forgotten why they did not come during the previous two years.

Table 1.1 Tourist arrivals and overnight stays on islands in the period 2019–2022

Island	Surface (km²)	Arrivals				Overnight stays				Tourist arrivals per islander in 2021*
		2019	2020	2021	2022	2019	2020	2021	2022	
Cres	405.7	126,811	63,467	106,817	127,729	906,960	480,156	786,744	931,436	39.3
Krk*	405.8	883,013	432,824	713,405	902,758	4,963,234	2,609,840	4,165,240	5,118,347	35.8
Brač	395.4	269,195	104,590	195,868	249,859	1,707,421	740,369	1,297,887	1,603,782	14.2
Hvar	297.4	331,184	105,185	216,477	288,618	1,568,307	728,206	1,243,008	1,491,141	20.3
Pag*	284.2	427,843	219,182	366,250	456,306	2,805,834	1,595,344	2,502,055	3,042,689	43.9
Korčula	271.5	174,775	66,420	117,236	159,009	927,398	484,587	766,747	922,097	8.0
Dugi otok	113.3	30,787	19,765	29,839	32,954	205,741	141,482	213,324	223,569	17.1
Mljet	98.0	34,048	15,583	26,442	30,042	150,442	81,580	126,010	140,503	24.9
Vis	89.7	51,541	26,729	40,075	44,694	282,651	167,905	243,058	263,655	12.1
Rab	86.1	278,608	144,563	223,474	267,716	1,985,952	1,164,142	1,677,147	1,947,576	27.0
Lošinj	74.7	294,898	128,618	230,402	283,243	1,969,220	968,775	1,668,920	1,961,412	30.6
Pašman	60.1	31,192	26,102	31,578	34,580	273,542	235,504	284,895	306,642	10.9
Šolta	58.2	18,549	9,889	13,976	15,336	150,133	89,233	122,715	127,567	7.1
Ugljan	51.1	26,726	24,889	33,315	38,270	221,074	210,750	275,452	294,970	5.8
Lastovo	40.8	9,365	5,756	8,984	9,207	60,104	38,993	60,629	60,348	12.0
Brijuni	33.9	141,072	64,217	108,976	144,593	1,003,648	477,475	825,926	1,074,292	31.3
Čiovo*	28.1	92,793	44,873	73,214	89,072	665,260	349,077	531,970	628,074	24.4
Olib	26.1	236	388	346	n.a.	2,412	3,818	3,179	n.a.	3.1

Island	Surface (km²)	Arrivals				Overnight stays				Tourist arrivals per islander in 2021*
		2019	2020	2021	2022	2019	2020	2021	2022	
Molat	22.2	478	z	310	n.a.	3,914	z	2,988	n.a.	1.5
Vir*	22.1	95,413	56,313	77,788	90,952	718,423	435,234	594,417	685,947	25.5
Murter*	17.6	123,154	68,077	113,806	131,993	889,608	529,288	836,036	966,276	23.5
Unije	16.9	1,130	758	987	986	8,587	6,332	8,101	7,646	15.0
Iž	16.5	3,922		1,927	n.a.	31,471		14,024	n.a.	3.7
Šipan	16.2	3,651	204	505	n.a.	21,060	2,254	3,441	n.a.	n.a.
Sestrunj	15.1	z	z	z	n.a.	z	z	z	n.a.	n.a.
Silba	14.3	3,372	2,355	3,070	n.a.	28,851	20,431	26,812	n.a.	8.8
Drvenik veliki	11.7	972	z	742	n.a.	8,545	z	6,502	n.a.	4.3
Ist	9.7	543	453	556	n.a.	4,876	4,302	5,619	n.a.	3.8
Premuda	8.7	423	328	414	n.a.	3,498	2,912	3,448	n.a.	6.8
Zlarin	8.1	7,841	2,772	4,611	n.a.	31,063	17,552	21,855	n.a.	15.6
Kaprije	7.1	1,137	642	888	n.a.	8,181	5,323	6,618	n.a.	4.6
Biševo	5.9	46	n.a.	n.a.	n.a.	566	n.a.	n.a.	n.a.	n.a.
Ilovik	5.5	385	z	453	538	2,821	Z	3,680	4,336	4.3
Lopud	4.4	13,378	986	4,742	n.a.	73,376	7,756	28,842	n.a.	17.6
Zverinac	4.2	z	z	z	n.a.	z	z	z	n.a.	n.a.
Sušac	4.0	z	z	z	n.a.	z	z	z	n.a.	n.a.
Susak	3.8	1,682	955	1,420	1,334	11,870	7,036	10,117	10,182	10.2

Island	Surface (km²)	Arrivals				Overnight stays				Tourist arrivals per islander in 2021*
		2019	2020	2021	2022	2019	2020	2021	2022	
Rava	3.6	z	z	192	n.a.	z	z	1,976	n.a.	2.7
Drvenik mali	3.4	n.a.	n.a.	n.a.	n.a.	n.a.	n.a.	n.a.	n.a.	0.0
Prvić	2.4	3,401	1,928	2,856	n.a.	26,111	15,670	22,462	n.a.	7.1
Koločep	2.4	z	z	823	n.a.	z	z	5,224	n.a.	n.a.
Vrgada	2.3	171	124	z	n.a.	1,599	1,175	z	n.a.	n.a.
Krapanj	0.4	24,032	11,379	14,572	n.a.	169,177	86,450	109,931	n.a.	88.9
Palagruža	0.4	z	z	z	n.a.	z	z	z	n.a.	n.a.
Ošljak	0.3	360	z	334	n.a.	2,739	z	2,852	n.a.	9.0
Sveti Andrija	0.1	n.a.	n.a.	n.a.	n.a.	n.a.	n.a.	n.a.	n.a.	n.a.
Glavat	0.02	n.a.	n.a.	n.a.	n.a.	n.a.	n.a.	n.a.	n.a.	n.a.
Peninsula of Pelješac	348	163,474	75,940	130,401	165,588	1,111,026	613,427	967,150	1,115,430	17.6

Source: CBS – Cities in Statistics, Arrivals and Overnight Stays in Coastal Towns and Municipalities, Census 2021.

Notes: * bridged islands; z – data is not published for confidentiality reasons; n.a. – not available.

The 2022 season convinced islanders involved in tourism that the standstill was behind them, and the Ministries of Tourism and Finance breathed a sigh of relief. As for the numerous headquarters, their reaction was as prompt as in 2020: at the beginning of April 2023, just before the pre-season, all measures were lifted except those in healthcare institutions. By mid-May, the end of the epidemic was declared.

Representing the archipelago: Krk, Hvar, Dugi Otok

Drawing general conclusions in a heterogeneous archipelago like Croatia's is challenging. This holds for almost all sectors, particularly tourism. Over the past two decades, a spatial-structural matrix of islands with varying sizes and development potentials has emerged along the Adriatic coast. Different forms of tourism have developed on these islands, ranging from affordable stationary vacations to charters and services for boaters to exclusive hosting for the wealthiest. Therefore, singling out islands to represent the archipelago is equally challenging. However, the following three islands, each in its way, accurately reflect how tourism recovered after the epidemic: Hvar, Krk, and Dugi Otok.

According to the usual geographic classification, Hvar belongs to the group of larger islands (Duplančić Leder, Ujević, and Čala, 2004). It covers an area of 297.4 km^2, and its 28 settlements are located in four municipalities. The island had a population of 10,678, according to the 2021 census (Population Census 2021). The beginnings of its tourism date back to 1868, but tourism emerged as the primary catalyst for change and the most important economic activity only in the second half of the twentieth century (Kušen, 2001; Zaninović, 2003) (Figure 1.3). In addition to its natural beauty, Hvar is rich in cultural and historical heritage.

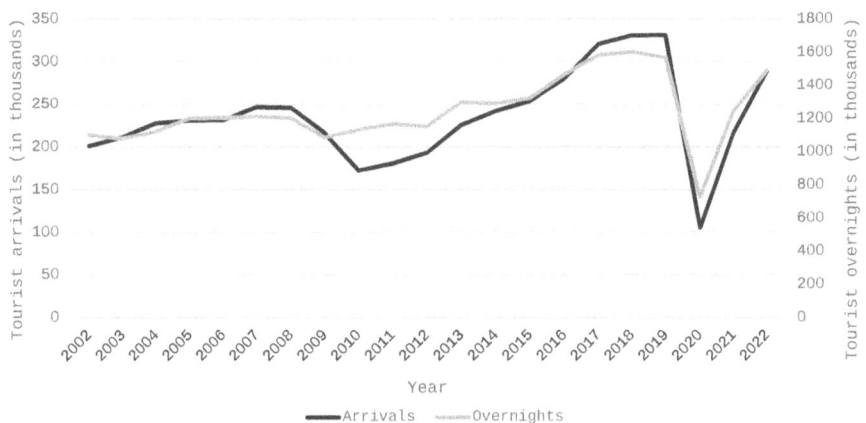

Figure 1.3 Tourist arrivals to the island of Hvar in the period 2002–2022
Source: CBS – Statistical Report Tourism (different years).

The bridged island of Krk is the northernmost and second-largest island in the Adriatic. It covers an area of 405.8 km^2 (Duplančić Leder, Ujević, and Čala, 2004). According to the 2021 census, it is home to 19,916 inhabitants, making it the most populous island. Krk has the highest tourism capacity and the largest tourism turnover of all the islands (accounting for around 23% of recorded overnight stays in 2022) (Figure 1.4). Thanks to the bridge that connected it to the regional centre of the city of Rijeka, Krk became a destination for secondary residences, characterised by intensive construction of private accommodation and spatial transformation (Opačić, 2008). The development of Krk demonstrates the importance of bridges. In 2022, five bridged islands – Krk, Pag, Vir, Murter, and Čiovo – accounted for 55% of tourist overnight stays in Croatia's islands.

Dugi Otok is located on the periphery of the archipelago and is reached by a longer and more challenging journey. With a total of 11 settlements, 1,691 inhabitants according to the 2021 census, and an area of 113.3 km^2 (Duplančić Leder, Ujević, and Čala, 2004), it also belongs to the group of larger islands. The momentum in its tourism development began with the opening of the first hotels in the late 1960s (Beverin and Armanini, 2009). Dugi Otok is also abundant in natural beauty and boasts one of two island nature parks in Croatia. It is also close to one of the three island national parks. The island has only recently been recognised as a more appealing tourist destination, so excessive construction has not yet occurred. It has smaller tourism capacities and lower tourist pressure than larger coastal islands. Among the observed trio, Dugi Otok is the smallest and least populated. So far, it has had the lowest tourism turnover (Figure 1.5). Its northwest and southeast coasts are characterised by deeply indented bays, making it a recognisable nautical centre. Thanks to boaters, it has been experiencing the highest annual increase in tourists for years. As early as 2008, boaters accounted for 71% of the total tourists on Dugi Otok. In the same year, boaters represented 27% of the total tourism turnover on Hvar and less than 2% on Krk.

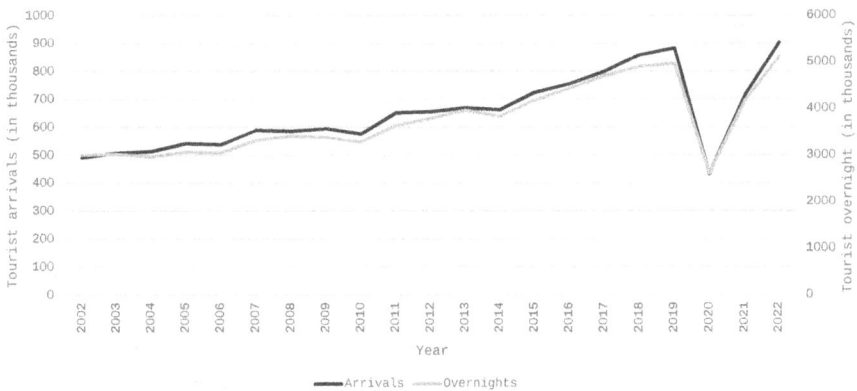

Figure 1.4 Tourist arrivals to the island of Krk in the period 2002–2022
Source: CBS – Statistical Report Tourism (different years).

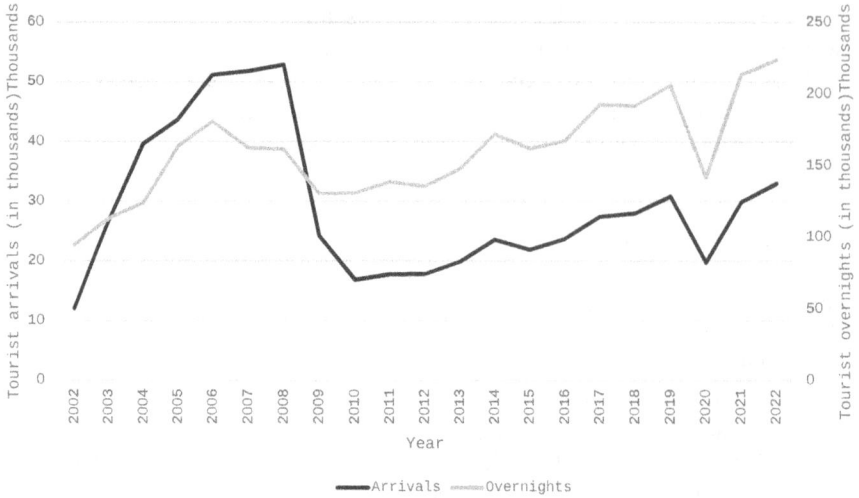

Figure 1.5 Tourist arrivals to the island of Dugi Otok in the period 2002–2022
Source: CBS – Statistical Report Tourism (different years).

The tourism industry in Hvar is based on beach tourism, rich cultural heritage, and a careful assessment of attractions to present to tourists. The town of Hvar has established itself as an essential meeting point for a wealthy yachting clientele. In 1998, the club *Carpe Diem* was opened, where young visitors seized their nights dancing and drinking and propelled the island as a popular party destination in the Mediterranean. The increasing number of restaurants with rich gastronomic offers has attracted boaters, including international celebrities. Hvar quickly became the most expensive Croatian island destination by attracting the affluent.

On the other hand, Krk has built its reputation on vacation tourism, mainly catering to families. Thanks to the bridge and its proximity to Slovenia and the continental part of Croatia, including its capital Zagreb, Krk is now the most developed Croatian island, with the highest tourism turnover. It accounts for about a quarter of the tourism turnover of all Croatian islands and has 21% of the total number of permanent beds on the islands (CBS). A significant portion of these capacities, about 49%, is represented by so-called non-commercial accommodation, that is, registered beds in second homes. For comparison, this accommodation capacity accounts for about 20% of Hvar and Dugi Otok.

Over the last two decades, these three islands have experienced more or less continuous growth in tourism turnover. The peak was reached in 2019. Hvar had the highest increase in the number of tourist arrivals during that period (+92%). Still, it had the smallest growth in the number of overnight stays (only 38%) (Figure 1.3). Consequently, Hvar had the lowest average length of stay for tourists, which, until 2019, had been declining and amounted to only 4.7 days. This is partly due to the development of party tourism. Young visitors who come for that purpose tend to stay much shorter than tourists on family vacations. It should be noted that the other two islands also experienced a decrease in the average length of stay, which can be attributed to changes in vacation patterns. In Croatian islands, as well as elsewhere, it has been observed that tourists want to see more destinations and engage in more activities, and they spend less time in a single location (Gössling, Scott, and Hall, 2018; Šulc and Fürst-Bjeliš, 2021).

During the pandemic, Hvar had the most significant decline in tourism turnover among the observed islands. One reason is its lesser accessibility compared to the bridged island of Krk. However, the guest structure regarding the countries of origin indicates that gathering and socialising restrictions that hindered party tourism and the suspension of air travel were crucial factors. Hvar is a destination where nearly half its foreign tourists come from distant countries. As much as 12% of Hvar visitors came from the United Kingdom and preferred to arrive by plane.

In 2020, Dugi Otok experienced fewer guests and overnight stays, but not for long. The pre-pandemic peak was already exceeded in 2021. Like Krk, private apartments predominate on Dugi Otok, attracting guests seeking isolation. Dugi Otok also has significantly fewer inhabitants than Krk and Hvar, which was an advantage during the pandemic. While most guests usually come from Germany, in 2020, there was a significant increase in the number of Slovenian tourists, who increasingly purchase properties there.

Krk's economy is much more diversified compared to those of Hvar and Dugi Otok. During the Covid-19 outbreak, the tourism industry suffered losses; but construction, shipbuilding, quarries, and transportation did not. The largest marina on the island also did not incur losses because its primary source of revenue is the annual berth lease paid by boat owners, regardless of how much and when they use it. Unlike the other two islands, the Krk economy still bears traces of the former insular economy, which had survived thanks to its diversified production structure and self-dependency.

The Covid-19 pandemic halved the records achieved in 2019 and proved to be a test of the resilience of island tourism. The sector passed the test, although at least two essential questions in their SWOT analysis remained unaddressed: how much were the opportunities seized upon, and how were the threats from the mainland managed or overcome? One of the advantages of Croatian islands concerning other European destinations was their relatively easy accessibility by car at a time when air travel was almost totally suspended. Additionally, private apartments, whose share in the total accommodation capacity is higher than in other Mediterranean countries, contributed to the advantage. In epidemiological conditions, hotels and hostels fared the worst. However, a study shows that islanders were not unified in easing measures, and residents of smaller islands were particularly critical of them (Kordej-De Villa and Slijepčević, 2022).

In the pandemic year of 2020, there was a decline in overall tourism turnover but an increase in the share of domestic tourists on all islands. This was due to a decrease in the number of foreign tourists not allowed to leave their countries and the entry restrictions into the countries that Croatians typically travel to. Prices of tourist services, especially accommodation, were also reduced. Tourists who visited the islands in 2020 stayed on average for more days than in pre-pandemic and post-pandemic years.

What's next?

Three wars, one recession, and one pandemic have not threatened the secular, slow-growing tourist demand for summer vacations on Croatian islands. The third generation of islanders, born and raised in tourism, tends to view it as a natural cyclical phenomenon. Having become rentiers to quite an extent with the centuries-old mentality of self-sustained islanders left behind, they were not particularly impressed by the incurable disease we have learned to live with. After all, the tourist infrastructure remained intact and ferries were readily waiting in their ports. Moreover, the Croatian tourism sector outperformed its main competitors in the Mediterranean during the summer seasons of 2020 and 2021 (Institute of Economics, Zagreb, 2022). Some of its structural characteristics, considered disadvantages for years, proved advantageous during the pandemic. Tourists visit the coast and islands during the three or, at most, four summer months, and the summer of 2020 and 2021 were epidemiologically more favourable than the rest of the year. Compared to other competing Mediterranean destinations, Croatia has a disproportionately high share of small scale, private accommodations (Ministry of Tourism and Sport, 2020), allowing for visits, vacations, and departures with minimal social contact.

Domestic tourists' role was essential for tourism's survival in 2020. Still, the revival of foreign attendance brought the sector back to track. Just two years after the pandemic, the islands, mainly thanks to foreign guests, approached record levels of tourist turnover in 2019. Krk and Dugi Otok achieved better tourism results in 2022 than in pre-pandemic years. Hvar joined them a year later.

However, the return of tourists was met with the same supply perspective as in previous years. Both larger and smaller islands continued where they left off, and the 2021 and 2022 seasons did not bring anything new (Šulc and Fürst-Bjeliš, 2021; Fürst-Bjeliš, Nel, and Pelc, 2022). After Covid-19, tourism development was approached similarly to the post-Homeland War period at the end of the last century: the most profitable sector of the economy had to be restarted as soon as possible. Urgency hindered innovations and improvements, and any serious debates on mass, excessive island tourism, and the islands' carrying capacity were postponed, perhaps indefinitely. Unfortunately, accommodation capacities are being built faster and more extensively, especially on the five bridged islands. The pressure on fragile local infrastructure is increasing, raising the question of how much tourism is too much for the tourist experience of those who consume it and for the quality of life of the hosting islanders during the peak summer months. Some studies indicate that many islanders do not see the positive effects of tourism on their islands' development anymore, although they are aware of its economic importance (Kordej-De Villa and Slijepčević, 2022). Therefore, special attention should be given to the participation of the local community in decision-making related to the sustainable development of islands, particularly in tourism (Zlatar Gamberožić and Tonković, 2015; Starc and Stubbs, 2014). Some proposed measures crucial for tourism recovery include strengthening domestic tourism and investing in the digitisation of the tourism sector (Okafor, Khalid, and Gopalan, 2022; Bulchand-Gidumal, 2022).

Despite the obvious, often ominous signs that things need to change, tourism on the larger islands continues to grow and pile up problems that overshadow what is happening on the smaller islands. Recently, proposals to connect the island of Krk with one more bridge regularly appear during the season of increasingly frequent traffic jams (Novi List, 2017). On the island of Pag, which was bridged in 1968 (Leksikografski zavod Miroslav Krleža, 2021), there are increasingly frequent water shortages at the height of the season. The water comes from the mainland because the island's sources are inadequate; requests for another pipe are becoming more frequent. This questionable approach, which seeks to solve problems by intensifying existing growth, is not applicable to small islands where no one would represent them anyway. There are few and increasingly elderly islanders and too few newcomers there. On small islands, it is not easy to find individuals who want and know how to represent the island on behalf of their local board and achieve anything in the municipalities/towns they belong to. Sometimes they threaten a referendum on which the islanders will decide on the transfer of the island to another municipality/town; sometimes, political party favouritism is invoked. The successes, however, are typically limited to modest increases in the small island's municipal/town budget item.

In all of this, small islands fare somewhat better if they belong to larger towns on the mainland, simply because the larger towns have deeper pockets (Faričić and Čuka, 2020b). As for managing tourism development, one is yet to witness appropriate plans for small islands prepared and implemented by a municipality/town and with the participation of the islanders concerned. The short-term stoppage of tourism due to Covid-19, which in any case amounted to a bad year, not even close to freezing the sector as seen during the Homeland War, could not shake these relations. Interest in changes (i.e. decentralisation) at the administrative level of counties, municipalities and towns has not increased. In contrast, expectations of higher revenues at the state level have risen.

On the other hand, the well-established activity of the state-owned and subsidised shipping company, which takes care of the island's connection with the mainland, stopped to some extent during the first few months of the epidemic and returned to the regular sailing schedule without any problems. Public investments, primarily in the port infrastructure, were not interrupted. These are multi-year construction projects financed from the state budget and increasingly from EU funds. For years they have been prepared in Zagreb, to a lesser extent in counties, rarely in municipalities/cities and never in local boards of small islands. The construction companies that get jobs in tenders could not even notice a standstill in hosting on the islands.

The question of 'what's next' is thus relatively easy to answer as the first years after Covid-19 demonstrate that almost nothing has changed and that it probably will not anytime soon. The question still needs to be asked because looking back over the shoulder indicates that something like this is not happening for the first time in the history of the Croatian islands. Some 120 years ago, the notorious phylloxera, a grapevine pest, ravaged European vineyards. When the scourge still seemed distant, Croatian islanders noticed that wine prices were rising and promptly specialised in viticulture and winemaking. However, phylloxera eventually arrived and destroyed the vineyards, and the islanders were unable to restructure their economy. Their only option was to leave the islands, often going very far away. The scorched grounds of former vineyards were etched in the collective memory among the islanders. Still, the lessons, it seems, were not learned. Half a century later, islanders were quick to specialise once again when tourists started disembarking on the islands in increasing numbers in search of accommodation. It remains to be seen whether history will repeat itself and, if things go awry, whether there will be enough strength for restructuring. It would be careless to assume that island tourism, especially in the Adriatic, will experience steady, long-term growth. Wars, global geopolitical realignments, ecological crises, and global warming can cause further setbacks and potentially permanently diminish the tourist attractiveness of the islands. The upcoming summers will show whether the secular growth of tourist demand will be obstructed by changes that emerge as insurmountable threats in the SWOT analysis of future tourism investors. At present, the Croatian tourism policy does not address these issues. Numerous consecutive state and regional strategies and action plans are based on demand trends that break every now and then but are still used for optimistic projections of steady growth. Strategic thinking thus primarily revolves around improving tourism supply and increasing national budget revenues. It would be much more useful if the next strategy devoted a page or two to whether the next crisis will be mild enough to be easily overcome yet severe enough to convince the island tourism sector to start caring about its long-term sustainability.

References

Arhivanalitika (2021). *Učinci turizma na hrvatsko gospodarstvo u pandemiji.* [The effects of tourism on the Croatian economy in the pandemic] Analysis prepared for Croatian Tourism Association. www.udrugaturizma.hr/media/1221/velimir-sonje-u cinci-turizma-na-hrvatskog-gospodarstvo.pdf.

Beverin, A. and Armanini, J. (2009). *Libar o Dugom otoku: Obilježja povijesne baštine.* [The book on Dugi otok: Features of historical heritage]. Zadar: Matica Hrvatska Zadar.

Brijuni National Park (2023). *Brijunski vremeplov: Novija povijest.* [Brijuni time machine: recent history]. www.np-brijuni.hr/hr/istrazi-brijune/brijunske-price/bri junski-vremeplov-novija-povijest.

Bulchand-Gidumal, J. (2022). Post-COVID-19 recovery of island tourism using a smart tourism destination framework, *Journal of Destination Marketing & Management*, 23, 100689. https://doi.org/10.1016/j.jdmm.2022.100689.

Dolničar, S., Lazartevski, K., and Yanamandram, V. (2013). Quality of life and tourism: A conceptual framework and novel segmentation base. *Journal of Business Research*, 66(6), 724–729.

Duplančić Leder, T., Ujević, T., and I Čala, M. (2004). Coastline lengths and areas of islands in the Croatian part of the Adriatic Sea determined from the topographic maps at the scale of 1:25,000. *Geoadria*, 9(1), 5–32. https://doi.org/10.15291/geoa dria.127.

eZadar (2020). *Zadar internet portal*. https://ezadar.net.hr/dogadaji/3796885/murter-ce-uskrs-provesti-u-karanteni/.

Faričić, J. and Čuka, A. (2020a). The Croatian islands: An introduction. In N. Starc (Ed.), *The notion of near islands: The Croatian archipelago* (pp. 55–88). Lanham MD: Rowman & Littlefield International.

Faričić, J. and Čuka, A. (2020b). The island and the city: The Zadar (sub)urban archipelago. In N. Starc (Ed.), *The notion of near islands: The Croatian archipelago* (pp. 135–152). Lanham MD: Rowman & Littlefield International.

Fatovic-Ferencic, S. (2006). Brijuni archipelago: Story of Kupelwieser, Koch and cultivation of 14 islands. *Croatian Medical Journal*, 47(3), 369–371.

Fürst-Bjeliš, B., Nel, E., and Pelc, S. (2022). Some conclusions about COVID-19's impact on marginality and marginalisation. In: B. Fürst-Bjeliš, E. Nel, and S. Pelc (Eds), *COVID-19 and marginalisation of people and places: Perspectives on geographical marginality* (pp. 233–238). New York: Springer. https://doi.org/10.1007/978-3-031-11139-6_16.

Gössling, S., Scott, D., and Hall, C. M. (2018). Global trends in length of stay: Implications for destination management and climate change. *Journal of Sustainable Tourism*, 26(12), 2087–2101. https://doi.org/10.1080/09669582.2018.1529771.

Government of the Republic of Croatia (2020). *Official government website for timely and accurate information on coronavirus*. https://www.koronavirus.hr/vazni-brojevi/56.

Institute of Economics, Zagreb (2022). *Sektorske analize: Turizam*. [Sector Analyses: Tourism]. www.eizg.hr/userdocsimages/publikacije/serijske-publikacije/sektorske-ana lize/SA_turizam_2022.pdf.

Klempić Bogadi, S. and Podgorelec, S. (2020). Demography of the archipelago: Migration as a way of life. In N. Starc (Ed.), *The notion of near islands: The Croatian archipelago* (pp. 89–110). Lanham, MD: Rowman & Littlefield International.

Kordej-De Villa, Ž. and Slijepčević, S. (2022). Impact of Covid-19 on Croatian island tourism: A study of residents' perceptions. *Economic Research: Ekonomska Istraživanja*, 1, 1–21, doi:10.1080/1331677X.2022.2142631.

Kušen, E. (2001). Hrvatski otoci u deset slika Prilog procjeni utjecaja turizma na razvoj hrvatskih otoka. [Croatian islands in ten pictures. Contribution to the evaluation of the influence of tourism on the development of Croatian islands]. *Sociologija i Prostor*, 39(1/4 (151/154)), 109–152.

Leksikografski zavod Miroslav Krleža (2021). *Hrvatska enciklopedija, mrežno izdanje.* [Croatian encyclopedia, online edition]. www.enciklopedija.hr/natuknica.aspx?id=42082.

Ministry of Tourism and Sport (2023). *Popis kategoriziranih turističkih objekata u Republici Hrvatskoj.* [List of categorized tourist facilities in the Republic of Croatia]. https://mint.gov.hr/pristup-informacijama/kategorizacija-11512/arhiva-11516/11516.

Ministry of Tourism and Sport (2020). *Turizam u brojkama 2019.* [Tourism in numbers]. www.htz.hr/sites/default/files/2020-07/HTZ%20TUB%20HR_%202019%20%281%29.pdf.

Novi List (2017). [Daily newspaper Novi list]. www.novilist.hr/novosti/hrvatska/p ogledajte-simulacije-novog-mosta-na-krku-bit-ce-dug-850-metara-zamisljen-je-ka o-viseci/?meta_refresh=true.

Okafor, L., Khalid, U., and Gopalan, S. (2022). Covid-19 economic policy response, resilience and tourism recovery. *Annals of Tourism Research Empirical Insights*, 3, 100073, https://doi.org/10.1016/j.annale.2022.100073.

Opačić, V. T. (2008). Ekonomsko-geografski utjecaji i posljedice vikendaštva u receptiv-nim vikendaškim područjima: Primjer otoka Krka. [Economic-geographical influences and consequences of the second home ownership in the receiving second home areas - the case study of the island of Krk]. *Ekonomska Misao i Praksa*, 17(2), 127–151.

Pirjavec, B. and Kesar, O. (2002). *Počela turizma*. [Principles of tourism]. Zagreb: Mikrorad i ekonomski fakultet.

Sirgy, M. J. (2019). Promoting quality-of-life and well-being research in hospitality and tourism. *Journal of Travel & Tourism Marketing*, 36(1), 1–13.

Starc, N. and Stubbs, P. (2014). No island is an island: Participatory development planning on the Croatian Islands. *International Journal of Sustainable Development and Planning*, 9(2), 158–176.

Starc, N. (Ed.) (2020). *The notion of near islands: The Croatian archipelago.* Lanham, MD: Rowman & Littlefield International.

Šulc, I. and Fürst-Bjeliš, B. (2021). Changes of tourism trajectories in (post)-covidian world: Croatian perspectives. *Research in Globalisation*, 3, https://doi.org/10.1016/j. resglo.2021.100052.

United Nations Conference on Trade and Development (2020). *Covid-19 and tourism: Assessing the economic consequences.* https://unctad.org/system/files/official-docum ent/ditcinf2020d3_en.pdf.

Vukonić, B. (2005). *Povijest hrvatskog turizma.* [The history of Croatian tourism]. Zagreb: Prometej.

Zaninović, V. (2003). The segmentation and promotion of selective forms of tourism on the island of Hvar: A fusion of authenticity and modernity. *Tourism and Hospitality Management*, 9(2), 271–288. https://doi.org/10.20867/thm.9.2.23.

Zlatar Gamberožić, J. and Tonković, Ž. (2015). Od masovnog prema održivom tur-izmu: Komparativna studija slučaja otoka Brača. [From mass tourism to sustain-able tourism: a comparative study of the island of Brač]. *Socijalna Ekologija*, 24(2–3), 85–102. https://doi.org/10.17234/SocEkol.24.2.1.

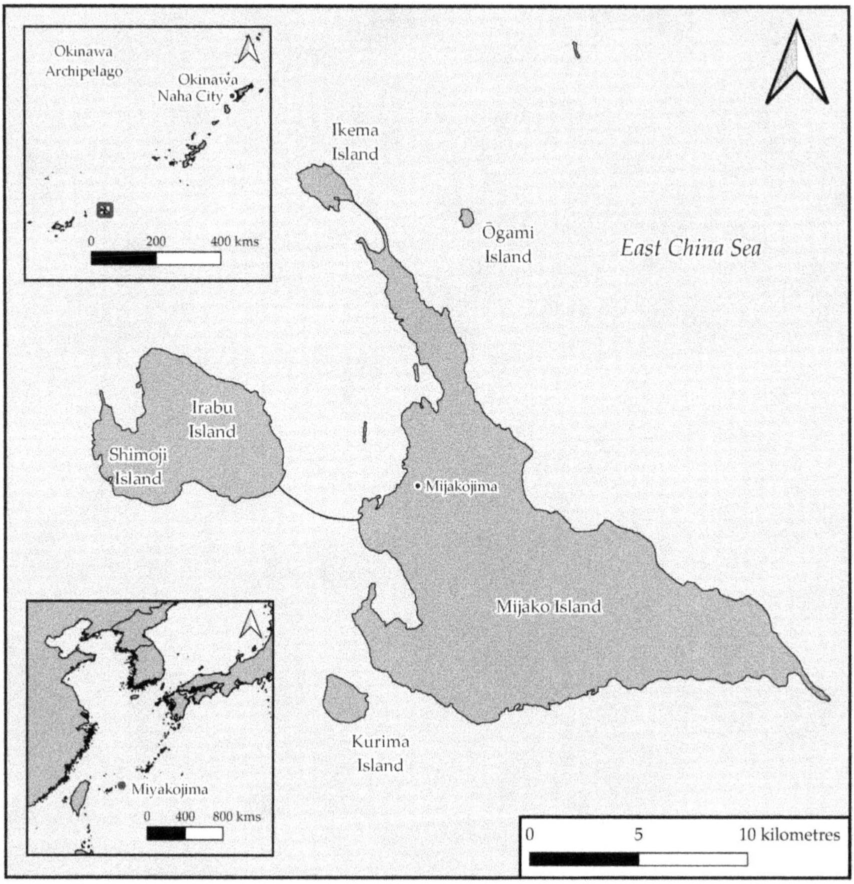
Okinawa
Archipelago
Okinawa
Naha City
0
200
400 kms
Ikema
Island
Ōgami
Island
East China Sea
Irabu
Island
Shimoji
Island
Mijakojima
Mijako Island
Kurima
Island
Miyakojima
0
400
800 kms
0
5
10 kilometres

Figure 2.1 Miyako Island, within the Okinawa archipelago, Japan

2 Beyond glossy tourist images

Miyako Island, Okinawa, Japan, through the stories of Machas, a small local grocery store

Ayano Ginoza and Atsushi Toriyama

Introduction

Archipelagos consist of multiple islands, which can be nested in multiple layers of core-periphery relationships. Smaller islands may depend on larger islands, which in turn depend on even larger islands. Perhaps the best examples of such nested islands are found in the world's four most populated archipelagic states, each of which has a population of at least 65 million: Indonesia, Philippines, Japan, and the United Kingdom. In such situations, transportation routes and logistics are likely to mirror the power assemblage of the country; main airports would be located close to the metropolitan centres and major urban areas – such as Jakarta, metro-Manila, Tokyo, and London respectively – which would be the natural airports of arrival for most international tourists. Once arrived, various tourists would also be considering domestic travel: the geographical extremes of such countries contain interesting sites and can nurture unique experiences: think Raja Ampat, Siargao, Okinawa, or the Shetland Islands respectively. Each of these are, in turn, their own archipelagos, with their own central and outlying islands.

In this chapter, we unpack island tourism by focusing on the role of a small local grocery store, Machas, in the Karimata district located at the northern-most tip of Miyako Island (宮古島, Miyakojima), with a land area of 159 km^2, located in the Okinawa Prefecture of Japan. Machas occupies the first floor of a 400 m^2 two-story building and sells daily necessities, including a variety food items such as eggs, vegetables, fruits, rice, bread, snacks, dairy, spices, drinks (including alcohol), as well as toilet paper, soap, and notebooks (Karimata Cooperative Union, 2009, p. 20). The second floor is a living space that traditionally accommodated the family of the president of the Cooperative. The store became known among tourists as one of the last remaining community-owned stores in the Okinawa archipelago, which survived the Covid-19 pandemic and thrived to attract even more tourists post-pandemic.

In this chapter, we first conceptualise Miyako Island in relation to Okinawa Island as well as its surrounding islands to map recent tourist development. Then, we trace the ways in which this community grocery store grew to attract and serve tourists over three periods: pre-, during, and post-pandemic.

DOI: 10.4324/9781003451037-4

In doing so, we argue that Machas expanded due to the idiosyncrasies of tourists and tourism while strengthening its service and care for the Karimata community. We contend that, while the development of large-scale shopping malls and the branding of island cultures to attract inbound foreign tourists has become common throughout the Okinawa archipelago, Machas seems to have bucked this trend, gradually increasing its tourist appeal precisely by *not* engaging in such deliberate island branding. Its mission and community identity helped the store to survive the pandemic and maintain its operations while consequently – and almost effortlessly – catching tourist interest for its rare value of locality, authenticity, and community-centredness in the Karimata district, distant from the advertised major tourist destinations on the island.

There is a dearth of academic research on Miyako Island and tourism. For this reason, our discussions draw from secondary data sources such as local newspaper articles, Miyako Island websites, and Karimata community archival sources, as well as an interview conducted with Karimata Community President, Yoshitaka Kuninaka.

Miyako island tourism

Okinawa is the southernmost prefecture of Japan. Miyako Island, along with the adjacent islands of Kurima, Ikema and Irabu islands, lie 300 km away from Okinawa Island and form an archipelago where the number of tourist visitors remarkably increased by 2.6 times in four years, from 430,000 in 2014 to 1,140,000 in 2018 (Nakamura, 2019, p. 18). The number of foreign visitors on cruise ships, which was only a small number in 2014, accounted for 40% of all incoming visitors in 2018. There are two major explanations for this sudden increase in tourists. The first was the opening of the 3.5-km-long Irabu Ohashi bridge in 2015, which connects Miyako and Irabu Islands. The view of the long bridge over the ocean has become a symbol of Miyako tourism. Three bridges – Irabu Ohashi, 1.6-km-long Kurima Ohashi (opened in 1995), and 1.4-km-long Ikema Ohashi (opened in 1992) – have become tourist attractions, from where visitors enjoy the beautiful 'Miyako Blue' ocean views. All together, these are called *The Bridges*, and they connect Miyako Island with its airport to the nearby, smaller island trio. Although Miyako used to be an inconspicuous tourist destination, its popularity as a resort island surged following the opening of the Irabu Ohashi because visitors could enjoy driving on the three bridges and around the islands. According to the website of Japan Tourist Bureau (JTB, 2023), the largest national travel agency, "Since there are no high mountains or rivers on Miyako, mud doesn't flow into the ocean even during rainfalls, so the high transparency of the blue ocean is always maintained." 'Blue Ocean' is a conventional term for promoting Okinawa as a resort destination. The ocean surrounding the Miyako Islands is particularly blue, given its distinct geographical and geological conditions. Therefore, using the term 'Miyako Blue' differentiates Miyako from the main island of Okinawa, which has an overwhelmingly larger number of tourists.

Another factor for the increase is a surge of foreign visitors because of port calls by cruise ships. Most tourists, mainly Chinese, spend only one day in Miyako. However, their shopping sprees for daily goods, electrical appliances, and pharmaceutical products at large-scale retail stores has a considerable economic effect on the local community, so much so that it exceeds the impact brought by more numerous, airplane-arriving tourists. With the increase in cruise ship port calls, the number of large-scale supermarkets on the island has also increased. A renowned, nationwide daily goods discount chain store has opened near Miyako airport, with sales-floor space equal to stores in metropolitan areas like Tokyo.

Such a change of consumption space on the island to meet the demand from increased foreign tourists has triggered a chain effect of opening stores to win over local consumers, with two ongoing processes: tourism-oriented development and standardisation of consumer space. As a result, small local stores that have long existed in the community could not and did not survive as they lost their competitive edge against larger franchised stores. This, in turn, has brought about changes in the distribution of commodities that supports the lives of local residents. The net effect has been that local, family run, 'mom and pop' shops were forced to close down, replaced by larger chain stores from the outside. This transition is a critical backdrop when considering the meaning and relevance of the communal store Machas in the Karimata community.

The population of Miyakojima City was about 57,000 at the turn of the twenty-first century, it then declined slightly for the following 15 years, but has begun a slight upward turn since 2015, now registering around 55,000 citizens. The revitalisation of the regional economy from increased visitor spending appears to have stemmed the population decrease.

On the other hand, this rapid tourist increase to the Miyako area has caused harmful effects to the community. Since 2015, when the number of visitors started to soar, construction projects for resort hotels followed one after another on the Miyako Islands. The number of hotels more than doubled in five short years: from 46 in 2017 to 104 in 2022. Hotel beds soared 2.4 times, from 2,432 rooms to 5,908, while the total number of tourists that hotels could accommodate at any time increased 2.7 times, from 5,868 from 15,870 (Okinawa Prefecture, 2023). A parallel construction rush of residential houses was also evident: construction of private rental accommodation surged from 314 in 2016 to 634 in 2017 and 1,843 in 2018 (Miyako Mainichi Shimbun, 2019a).

Although the sharp rise in construction costs has caused a decrease in private residential housing construction in Japan overall, housing construction in Myako increased by 5.7 times from 2015 to 2019, mostly to meet expected demand from hotel employees and construction workers (Miyako Mainichi Shimbun, 2019b). This construction rush and the purchase and sale of real estate for speculation purposes caused a sudden increase in land values in urban districts: up to ten times the values before the Irabu Ohashi opened. As a result, most islanders now cannot afford to purchase land on their own island (Nakamura, 2019, p. 18). It has become quite difficult for locals to secure new rental accommodation because of the rapidly rising rents and a shortage of rental stock despite the construction boom. According to Community President Yoshitaka Kuninaka, the number of people buying land for investment purposes is increasing, and the rental price of a house has gone from 60,000 yen a year to 70,000 yen a month. The Miyako bubble that started in 2015 also generates foreign investments for larger resort hotels, and the buying up of the shoreline makes it difficult for locals to access their familiar beaches (Nakamura, 2019, p. 18).

Since 2020, the Covid-19 pandemic has dealt a major blow to Miyako tourism. The number of tourists first declined sharply in April 2020. As a result, the occupancy rate of hotels decreased to less than 10%. A temporary closure of hotels followed (Miyako Mainichi Shimbun, 2020b). The mayor of Miyakojima City publicly announced a ban on non-islanders visiting the islands due to the fragility of the local medical care system (Miyako Mainichi Shimbun, 2020c). The Miyako Islands experienced the harmful effects of lengthy nationwide restraint in the tourist industry: the city had only 350,000 visitors in FY2020, down from 700,000 the previous year, while the number rose slightly to 430,000 in 2021.

In 2022, the number of tourists arriving by air recovered, reaching 730,000, the same scale as before Covid-19. However, the number of cruise ship and ferry passengers, which was 400,000 annually before the pandemic, had not yet recovered by 2023 (Miyako Mainichi Shimbun, 2023a). Only one cruise ship visited Miyako in 2020. With the recovery in visitors flying to the island, the number of rental cars increased. Consequently, the number of car-rental vehicles in 2022 was 1.5 times higher than the previous year and even 1.4 times higher than in FY2018, which recorded the highest number before the pandemic period (Miyako Mainichi Shimbun, 2023b). This is due to the rapid entry of new car-rental companies from outside the island. At the same time, many older taxi drivers retired out of fear of contracting the coronavirus, so the number of taxis on the islands had fallen (Miyako Mainichi Shimbun, 2021). Since taxis have played a pivotal role in public transportation on the Miyako Islands, which lack sufficient bus lines, there is a concern that the taxi shortage will negatively affect those with no means of transportation.

In the following sections, we will delineate three main periods for Machas: before, during and after the Covid-19 pandemic. We will use our interview with Karimata Community President Yoshitaka Kuninaka, conducted on 5 June 2023, to guide our discussion throughout the chapter.

Introduction to Machas: Before the pandemic

Machas is the nickname given to the local community grocery store in the Karimata district, located at the northern edge of Miyako Island. Machas was originally established on November 25, 1947, barely two years after the end of the Second World War, as *Karimata Kobai Kumiai*, or Karimata

Consumers' Association. The name Machas was taken after an Okinawan word *macha*, which means 'store', with an added 's' to reflect that the store is run by a team of Karimata community members (Miyagi, 2013). Its objective was to support people in Karimata when traffic infrastructure, electricity, and water supply systems were not yet developed, and the population had decreased after WWII (Ikema, 2009, p. 7). The Association opened the store using a corner of the district's Youth Association Building, which was symbolically built on the historic site of the former *bunmiyā*, the village guardhouse, which was responsible for the safety and security of the community. The store began with 271 district residents as investors (Kawamitsu, 2009, p. 1). Their sales profits were distributed to the Elder's Club, Youth Association, Crime Prevention Division, Athletic Association, and Firefighting Division, as well as financially supported events at the Karimata elementary and middle school and the Karimata Community Association (KCA). Hitoshi Ikema, a former KCA president, lists its three characteristics: first, it continues to play the role of village guardhouse, where you can acquire the names of all 212 households; it knows all the ceremonial occasions, travel plans, health matters, even some real estate information about its members; and serves as their 'lost and found' office. Second, it serves as a bank to exchange large bills for smaller bills and vice versa since bills are commonly gifted at major cultural events such as birthdays and funerals. Third, it is a place for social interaction, and there are always a few members of the community chatting, exchanging information, or sharing concerns, there. Some patrons visit the store more than once a day for interaction (Ikema, 2009, p. 3).

In April 1992, the store was renovated as part of the island's main road-widening construction work (Miyako Island Education Department, 2014). Since its establishment, Machas has provided daily necessities for community members, who have now increased to 212 households and approximately 460 people. The store is open all year from 7 am to 9 pm. As an example of mutual care and relationship based on trust, community members can obtain items at the store by leaving a record in the ledger and paying later (Miyagi, 2013). Among the population, 137 people are over 70 years old. One third of the Karimata population is aged 70 and above, 1.5 times the rate of Okinawa Prefecture; while one third of those aged 39 and under have left the island in the past 10 years (Ministry of the Environment, 2019). Growing together for over 75 years, those elders and Machas have established a close relationship and a mutual support system. The elderly in the community also serves as important caretakers of children while their parents are at work. In this sense, the community maintains traditionally established roles for each member to create a *yuimāru*, meaning a mutual support system (in Okinawan). Responding to the needs of the community has been the priority and the foundation of the store. The long and enduring relationship which the store has built with its community may be the reason for its attraction to tourists today; it may remind them of the 'good old days' of their own neighbourhood community stores, which probably no longer exist.

The KCA became the first community store in Okinawa Prefecture to sustain annual net sales of ¥100 million (around US$720,000). As stated earlier, the store witnessed its highest sales to date coinciding with the opening of the Ikema Bridge in 1992: over ¥200 million (Karimata Cooperative Union – KCU, 2019, p. 116). The store was officially registered as a cooperative, making it the first community store in the prefecture to become one. While its president and auditors have all been male, the employees have always ranged from one to six women since its establishment (KCU, 2019, pp. 31–33). The store facilitated the employment of women who otherwise would have continued serving as housewives, engaged in unpaid labour. According to Kuninaka, the employees are mostly married women whose children have become independent and who have more flexibility at home.

The main items the store carries include daily necessities, such as fresh vegetables, fish, meat, rice, bread, eggs, milk, condiments, snacks, drinks, beer, toilet paper, soaps, and detergents, as well as prepared meals known as *bento*. The grocery items are purchased at large supermarkets in the Hirara district of Miyakojima City. Machas employees keenly pay attention to sales flyers from the supermarkets to buy items in bulk and sell at their store for decent prices. Thus, most items are slightly more expensive than at larger supermarkets; but, for many locals, especially the elderly – termed *kaimono jakusha*, or shopping refugees – who may find shopping challenging because of ill health and/or lack of shopping facilities nearby, the grocery store is convenient because these residents are spared the time and expense of a taxi ride to a large supermarket (Nakamura, 2020). Moreover, many elders enjoy the time and space to converse with the familiar employees or neighbours who gather at the store to pick up items. As such, the community store has grown to occupy a central role in creating a safe and communal place for maintaining the health of the community.

Since its opening during the US military occupation of Okinawa (1946–1972), the store has served its community well, working closely with community members. The core philosophy of the store is to cater to its own community, manage the small store effectively while also strengthening the community's social fabric and maintaining its health and safety.

Our research identified three major characteristics of the store. First, it is community based and community first. The sustained mission since its establishment in 1947 has never changed, and the groceries it stocks continue to meet the needs and requests of its community members. Second, non-locals and tourists are attracted by the community-centredness that distinguishes Machas from the large, corporate chain stores that are replacing small stores. Thirdly, Machas' flexibility and willingness to sell, upon request by the community, fresh, locally grown produce as well as fresh fish caught daily by Karimata residents adds to the popularity of the store. Fresh fish catches for the store are announced immediately throughout the island using the island's FM radio network. All three aspects derive from their motto to serve their community customers and producers to the best of their ability, which in turn attracts tourists to the store: an unintended outcome. For Machas, members of the Karimata community are registered as local customers, who are distinguished from tourists, who are identified as any other customers from outside the Karimata community. Tourists, thus, include not only tourists from outside Miyako Island but are defined as those from other districts of the island and smaller neighbouring islands. While tourists have bolstered the store's success, the store remains driven and motivated by local needs.

We interviewed Yoshitaka Kuninaka, president of the Karimata Community Association. During our interview, he referred to an episode during a super typhoon in 2020 (Ryukyu Shimpo, 2020). This was a tropical storm that knocked down electric poles, tore the roof off the airport control tower, caused power outages for the entire island. This event highlights the character of Machas' role in the community, as Kuninaka says, "driven by a sense of mission":

> There was a huge typhoon, around three years ago. During the strong winds, the store was closed and the metal shutters were pulled down to protect the store windows. But during the two hours when the island was in the eye of the typhoon (a period of mostly calm), the store president called the Karimata Neighbourhood president and asked him to announce through the community microphone the reopening of the store while the island was in the eye of the tropical cyclone until the wind got strong again. We could do this because it is a small, concentrated community. So, at around 9 pm, the locals came out one by one to purchase bags of ice, beer, and other vital items. The nimbleness to respond and act instantaneously is its advantage.

This episode exemplifies the ways in which Machas strives to fulfil its function to respond to the needs of the community in the midst of a disaster by quickly reopening its store. Only the community members who lived through many typhoons knew the exact window when they can safely invite custumers to the store. The sales generated during that window weighed less than the importance of providing access to the store, comfort and a sense of security. Machas is also flexible in undertaking services upon request from their regular customers; for instance, driving to a larger supermarket to purchase items they do not carry when requested. These actions show care for their own community, which mainly consists of 137 elders over 70 years of age, who are treated like family by store employees.

The opening of the Ikema Bridge in 1992 permitted tourists to rent a car and enjoy a scenic drive along prefectural road 230, which runs through the Karimata district and in front of Machas. Kuninaka suspects that the location of the grocery store has contributed to its commercial success. In fact, many tourists have asked the store to expand its limited, three-space parking lot to expect increasing numbers of tourists.

During the pandemic

The Covid-19 pandemic affected the island significantly. In 2020, a prefecture-wide state of emergency was declared to contain the spread of the virus: local schools were closed for weeks; restaurants and stores were asked to restrict their hours and close by 8 pm and to limit the number of customers; and stores were asked to enforce safety measures such as wearing masks, sanitising hands, and taking temperatures at the entrance. Many local and communal events were cancelled, including: the scheduled Olympic torch relay; local schools' annual sports events; and cultural, academic, and local traditional cultural activities (Mayor of Miyakojima, 2020). The tourist industry was also heavily impacted. At a time of high infection rates, Japan Transocean Air (JTA) suspended direct flights between Tokyo Haneda and Miyako Island, which were usually 90% booked for their two daily roundtrip flights; and both All Nippon Airlines (ANA) and Ryukyu Air Commuter (RAC) reduced their flights in and out of Miyako (Miyako Mainichi Shimbun, 2020a).

The pandemic enhanced, rather than diminished, Machas' role in Karimata. When most stores shortened their opening hours or closed for days during the pandemic, Machas never did, in order to carry on with its role as the caretaker and food provider for the community. Kuninaka remembers an event from 15 June 2021:

> About two years ago, I went to ask Machas about an elder, Mr. K, as he did not show up on time for a scheduled pick up to attend an association event. He regularly visited Machas four times a day. It turned out he hadn't come to the store that day, and we visited his house to find him collapsed and suffering from heatstroke. As this was the second time that he was found unwell at home, his neighbours brought him some *yushi dofu* (tofu curd) and vegetables to help him recover. When he did, and returned to the store to buy beer, the store employee warned him about drinking too much. So he settled for one [can of beer] instead of the two he originally intended to buy.

This episode shows a well-established relationship between the store and Mr. K as the store could have benefited from Mr. K's purchase of another beer, and Mr. K could have simply insisted on his right to purchase what he wants. The employee was comfortable enough to speak about an uncomfortable matter out of concern, and the community member was open to accepting the employee's advice about his health. Thus, Machas serves less as a profit-driven grocery store than as a conscientious caretaker of its community. This interaction can only occur at a local, long-established, community-focused store, but less likely at an outside vendor, or large franchise stores.

Machas also exhibited its unique character of community service during Okinawa Prefecture's state of emergency, when most stores were advised to limit their hours and restrict customers. During that time, many stores were closed for the safety of their employees and customers and qualified for government financial support. Machas wasn't one of them. Its commitment to the community, however, brought more customers to Machas from neighbouring communities whose local chain stores were closed. Machas never closed because its customers depended on the store. It carried staple items such as toilet paper, rice, soap and condiments; but also offered fresh seafood and vegetables brought in by community fishers and farmers who suffered from the closure of major retail stores where they used to sell their produce before the pandemic. Providers knew that Machas would carry their produce if their usual retail stores were closed. Thus, the store also provided a venue to support their livelihoods, offering a valuable sense of security, reliability, and relief. During the pandemic, the mutual support between the community and the store was strengthened by its contributions to workers whose income had suffered during that time.

The mutual support between the producers and sellers at Machas led to increased customers from outside Karimata, who came looking for fresh sashimi and produce that were then hard to find anywhere else on the island. Kuninaka was one of the few fishers who still practised drive fishing using a net and belonged to the only remaining drive-fish team on the island, Tomori Group, led by Tetsuo Tomori, who was 83 years old when featured in the local paper in 2021 (Okinawa Times, 2021). Around ten fishers work as a team to trap fish encircled in a net. This traditional fishing method has been passed on to Karimata middle school students for over thirty years and was awarded the forty-eighth Okinawa Times Education Award in 2009 (Okinawa Times, 2009). One day, Kuninaka made a public announcement on Miyako FM radio announcing his drive fishing catch, which received a good response. He suspected it might work to not only spread information for that day but would also increase the expectation that Machas would have fresh fish again. Having an effective and already established information relay route could facilitate both community and non-community members' purchase of fresh fish and produce. They established a system where fishers would call Community President Kuninaka via the radio in their fishing boat from the ocean to inform him when their fresh catch would arrive at Machas. Then, Kuninaka would call the Miyako Island FM radio station to share the information. By the time the seafood was unloaded at the store, many customers from across the island had arrived, thrilled to purchase the rare treat of freshly caught seafood that was unavailable at larger chain stores because their regular purchase routes were closed during the pandemic. The store's community-centredness made its presence on the island well known during the pandemic. Increased numbers of islanders undertook the half-hour drive to visit Karimata to find fresh produce and seafood. Islanders knew if they went to Machas, there would be something worth their drive, when many usual activities were being discouraged and self-restrained during the pandemic. In doing so, the store acquired repeat customers outside the Karimata community and became known for its stability and reliability and the freshness and variety of the produce they provided, even when other large stores could not. President Kuninaka claims that "Machas turned the pandemic that affected the entire island to their advantage." In doing so, Machas gave its local customers and tourists a sense of security for daily necessities, including the fresh and inexpensive treat of sashimi.

Post-pandemic

Machas' community-centredness has been further fostered by working in collaboration with the Karimata Cooperative Union (hereafter, the Cooperative), which was officially registered in November 2022 as the first of its kind in Okinawa. Their actions immediately followed the enactment of the Worker Cooperative Law that became effective the previous month to encourage workers to invest in their own businesses, supporting their community, working together, and being involved in their company's management (Ministry of Health, Labour and Welfare, 2022). The Karimata Cooperative consisted of 12 members by June 2023, including fishers, female homemakers, and Karimata community officials who utilise their area of expertise. The Cooperative took over from the former Karimata Community Centre's work, such as making and selling lunches for kindergarten children and elders, providing transport services for elders who commute to hospitals, and transport to high school students who commute to school in different island districts (Okinawa Times, 2023). Kuninaka told us that "working on community development suits Karimata's character and will be a benefit for the Community Association. I hope that it will make the community even livelier and more prosperous." In our interview, he called the Cooperative "an organization of its own enterprise". He shared the following episode with us:

> Working at the recently updated kitchen of the Karimata Community Centre located directly behind Machas, two women homemakers practise their cooking skills. The idea was developed by listening to the needs of mothers with children at kindergarten who had to prepare *bento* box lunches for their kids; but the mothers were simply too busy to prepare them daily. While private companies cater lunches for kindergartens in larger communities, supplying the small Karimata kindergarten with six pupils is not profitable for such private firms. Kuninaka explains that, although it is not profitable, "we have Machas," perhaps implying that assisting community members in need comes before profit.

The box lunches made at the Cooperative are filled with a generous assortment of dishes using local ingredients. According to Kuninaka, the Cooperative in this manner creates jobs for women whose developed skills can provide them with critical revenue. Their box lunches are sold at Machas. Due to their convenient location next door, the Cooperative's cooks can quickly respond to demand for more lunches. Until recently, the store purchased lunches three times a day from Hirara City. Now, the Cooperative's own cooks, located nearby, can adjust the supply of popular items. Previously, the Machas Secretary would commute to Hirara to purchase bento boxes; but, thanks to the Cooperative, customers can enjoy their favourite dishes picked up locally while still warm, and seeing their own locally grown produce utilised in each dish, highlighting Karimata's community-centredness.

Interestingly, the *bento* box lunches originally produced for mothers with kindergarten children, are now more in demand by tourists, who started to return after the pandemic and board boats from Karimata's large harbour. Machas is excellently located just a three-minute drive from the harbour, and without any competition close by. Karimata Port offers spacious docking for dive ships. Especially on weekends, tourists who visit nearby beaches or go snorkelling or diving, purchase large quantities of local food. According to Kuninaka, such tourists enjoy discovering undeveloped, non-touristy beaches, which exist because Karimata continues to reject applications for tourist development. He adds:

> Although a beach is private or community property, I acknowledge that it comes down to a community's decision; while, in other communities, such as Irabu Island, beaches are privately owned and developed. We have received several offers and invitations for costal tourism but have not accepted any so far because we share the same values and consciousness, to a certain degree, to discuss what we want to see happening in our community and our land.

Thus, many inbound tourists, dozens at a time, stock up at the store before going to unspoiled beaches that are not in tour guidebooks or before taking an ocean boat ride. Proximity to a relatively large leisure boat harbour and being the nearest store for tourists are major factors in making Machas a mini-economic hub. The Cooperative's cooking division proved itself successful during a recent major holiday, with ¥270,000 (around US$2,500) in sales, ten times more than the previous year. Kuninaka believes the Cooperative creates jobs for mothers whose children have become independent and who are willing to utilise their cooking skills as a source of income.

Kuninaka explained that the large increase in box lunch sales is also related to returning off-island domestic tourists who are still not comfortable travelling overseas, and sales are also boosted by tourists to the indoor Miyako Island Underwater Park, located just a five-minute drive away from the store (Miyakojima Kaichu Koen, 2023). The park attracts tourists, regardless of the weather, but receives more tourists when rain prevents outdoor activities. Park visitors reached record numbers at the end of March 2023, at the close of the fiscal year. Having maintained stable sales during the pandemic, Kuninaka anticipates a large increase in customers at Machas in 2024 as more tourists pass by on the way to aquatic leisure activities during the long summer break and the pandemic restrictions are almost fully lifted for the first time since 2020.

Conclusion

Here is a different way of diagnosing archipelago tourism: by focusing on one of the nodes of human relationships that imbricates both locals and visitors, hosts and guests. The quantity and quality of these relations, and how they change with time, is a manifestation of the dynamic nature of nested, core-periphery relations in this little corner of Japan.

Machas' relationship with the community, established over the past 75 years, and their favourable location with little competition nearby, have created a unique suburban-type of tourism, or tourism for locals. The character of the small, rural fishing community of Karimata attracts non-community members since it is a short enough drive to get fresh fish but far enough to enjoy the feeling of a short excursion from their homes, creating regular visitors. The business has been sustained because it successfully meets the needs of the community with its solid 'community first' objective, which has been recipro-cated by patronage and mutual local support. Community Association President Kuninaka believes such relationships explain how the small community store survived and continues to thrive during and after Covid-19. By fostering support for local fishers and farmers, the store unintentionally made itself known island-wide as one of the only stores to remain open full time while others restricted their hours at a time when fresh food was scarce. The store has a strong foothold in the local community but also meets the needs of other island residents and visiting tourists. Machas thus expands its tourism possibi-lities in the broadest sense. Kuninaka foresees challenges in the decreasing population in the community and expects no dramatic increase in population in the near future. However, he anticipates an expansion of the store's business with a growing demand by inbound tourists for products such as beverages (including alcohol) and food products in the summer months. With some flex-ibility and communal support, Machas should be able to continue to provide much needed services to its patrons for many more years. In spite of suffering from multiple layers of insularity, Machas has somehow thrived from this marginalisation, and has developed its appeal precisely by sticking to its loca-lised authenticity and not succumbing to trending globalisation.

This chapter also illustrates how new transport infrastructures, such as bridges and other fixed links, can bring about shifts in the quantity and quality of archipelago tourism and encourage island-island visitations by tourists, at least until the 'novelty effect' wears off.

References

Ikema, Y. (2009). A congratulatory address. *A money tree: Sixty years of history* (pp. 6–7). Karimata Cooperative Union 60th Anniversary Organizing Committee.

JTB (2023). Miyako jima de hazuse nai Osusume Kanko Spot & Tanoshimikata 20 sen [Must see spots on Miyako Island. Twenty recommended ways to enjoy Miyako and tourist spots]. *Japan Tourist Board.* https://article.jtb.co.jp/kokunai/okinawa/rem ote-island/miyako-island/miyakojima-sightseeing/.

Karimata Cooperative Union (2019). *The history of the community store. In: A money tree: Sixty years of history* (pp. 20–23). Karimata Cooperative Union 60th Anniversary Organizing Committee.

Kawamitsu, Y. (2009). A ceremonial address. *A money tree: Sixty years of history* (pp. 1–2). Karimata Cooperative Union 60th Anniversary Organizing Committee.

Mayor of Miyakojima (2020, August 3). *Miyakoshi*, Miyako shicho messagi. [Message from the Mayor] Press Conference. www.city.miyakojima.lg.jp/gyosei/mayor/oshira se/press_conference.html.

Ministry of Health, Labour and Welfare (2022). *Rodoshakyodokumiaiho toha.* [What is the Workers Cooperative Act?] www.roukyouhou.mhlw.go.jp/about.

Ministry of the Environment (2019). *Okinawaken miyakoshi*, ekoisland Miyako jima ni okeru chiikijunkann kyouseiken kouchiku jigyo. [Project to establish regional recycling and symbiosis zone on eco-island of Miyako.] www.env.go.jp/content/900498533.pdf.

Miyagi, N. (2013, October 24). Karimata Mâchâs. *Lequio.* Volume 1490. https://im g01.ti-da.net/usr/k/y/o/kyoudoubaiten/20131024KarimataScan.jpg.

Miyako Island Education Department (2014). *Miyako Island Application Ayanatsu.* https://miyakojimabunkazai.jp/bunkazaiinfo95/.

Miyakojima KaichuKoen Website (2023). *Miyakojima Kaichu Koen [Miyako Island Underwater Park].* https://miyakojima-kaichukoen.com.

Miyako Mainichi Shimbun. (2019a, 29 May). 18 nen hotel jyuugyouin ya sagyoin muke. [Breaking ground for the construction of housing up 2.9 times compared to the previous year of 2018: Miyako district for hotel employees and construction workers] www.miyakomainichi.com/news/post-120345/.

Miyako Mainichi Shimbun. (2019b, 3 November). 4 nen de 5.7 bai. [Breaking ground for construction of housing: Up 5.7 times in four years]. *Miyako Mainichi Shimbun Web.* www.miyakomainichi.com/news/post-125325/.

Miyako Mainichi Shimbun. (2020a, 16 April). *Haneda-Miyakosen unkyuhe*, JTA. [Suspension of Haneda-Miyako Line/JTA.] www.miyakomainichi.com/news/post-130383/.

Miyako Mainichi Shimbun. (2020b, 26 April). Rinji Kyugyo Aitsugu Tounai hotel. [Temporary closure of hotels follows one after another.] *Miyako Mainichi Shimbun Web.* www.miyakomainichi.com/news/post-130506/.

Miyako Mainichi Shimbun. (2020c, 28 April). GW wa Miyako ni Konaide Kankou-kyaku ya Shusshinsha ni Uttae [Don't come to Miyako during Golden Week: Miyakojima city mayor. Appeal to tourists and visitors coming from outside Miyako.] *Miyako Mainichi Shimubun Web.* www.miyakomainichi.com/news/post-130527/.

Miyako Mainichi Shimbun. (2021, 3 February). Kadou 3 wari, 100 nin ga Taishoku Miyako no Taxi Gyokai. [Operating rate 30%: 100 drivers retired in Miyako's taxi industry]. *Miyako Mainichi Shimbun Web.* www.miyakomainichi.com/news/post-136905/.

Miyako Mainichi Shimbun. (2023a, 21 April). Nenkan Nyuuiki Kanko Kyakusuu 22 nendo wa 737,729 [737,729 tourists in 2022 to Miyako]. *Miyako Mainichi Shimbun Web.* www.miyakomainichi.com/news/news-175180/.

Miyako Mainichi Shimbun. (2023b, 2 May). Rental car daisuu ga Kako saitani. [The number of rental cars highest in Miyako history]. *Miyako Mainichi Shimbun Web.* www.miyakomainichi.com/news/news-175393/#:~:.

Nakamura, R. (2019, 14 September). Ohashi Kaitsu de hotel rush Okinawa – Miyako jima kanko no baburu. [Miyako Island Kanko Bubble: Hotel construction rush caused by the opening of the Irabu Ohashi]. *Shukan Toyo Keizai.*

Nakamura, R. (2020). Kyodobaiten niokeru chiikifukushi no yakuwari: Juminshutai no chiiki kea shisutemu. [The role of community welfare in community stores:

Regional care system centered on the residents] In *Kyodobaiten no Aratanakatachi wo motomete: Okinawa ni okeru yakuwari, kadai, tenbou.* [*Searching for a new role for community stores: Their roles, issues and prospects in Okinawa*] (pp. 351–396). Henshu Koubou Touyou Kikaku.

Okinawa Prefecture (2023, 28 March). Shukuhaku shisetsu ni kansuru tokei data. [Statistics for accommodation]. www.pref.okinawa.jp/site/bunka-sports/kankoseisaku/15853.html.

Okinawa Times (2009, 14 March). 2 shi, 1dantai ni taimusu kyoikusho, 25 nichi ni zoteishiki shukugakai. [Two individuals and one organization receive Times Education Award, award ceremony on the 25th].

Okinawa Times (2021, 29 August). Oikomiryo katkizuku Miyako Karimata 350 kilo wo mizuage. [Drive fishing is active in Karimata, Miyako: 350 kg unloaded]. www.okinawatimes.co.jp/articles/-/821679.

Okinawa Times. (2023, 2 February). Kenhatsu no kyodoroudou, bentoseizo, sogei, tokusanbutsuhanbai nado jigyoka, miyakojima, karimatakyodokumiai. [The first 'worker cooperative' in the prefecture: Commercialization of lunch production, transportation and sales of local products at Miyako Island Karimata Cooperative Association.]

Ryukyu Shimpo (2020, 10 August). Taifu5go honto tsuka: sorano 83 bin nado kettko [Typhoon 5 hit the main island: 83 flights cancelled]. https://ryukyushimpo.jp/news/entry-1171493.html.

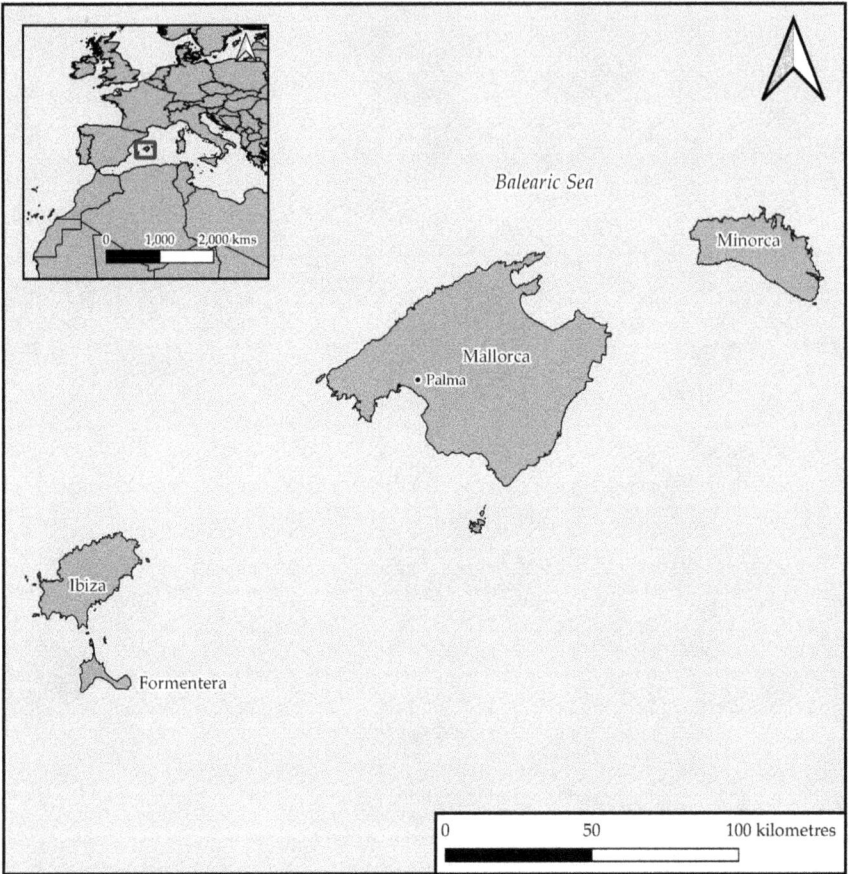

Figure 3.1 The Balearic Islands, Spain

3 Nightlife, well-being, nature, and a lighthouse

Differentiation and convergence of the post-pandemic images of the Balearic Islands through tourism

Aina Gomis and Mercè Picornell

Introduction

On their 2011 album *Lamparetes*, the most widely known music group from the island of Mallorca, Antònia Font, included one of their biggest hits, titled 'Islas Baleares'. The song, which, like all the other songs by the group, is written in Catalan, offers an island-by-island description of the different iconic features, ranging from specific locations to traditional festivities. To link them all together, there is a refrain in Spanish with a 'foreign' accent (i.e., stressing the American English sound of the 'r' and ending the phrase with an "oh yeah! oh no!", also in English). The refrain affirms that no two of the Balearic Islands are alike.

For the local islanders, the very concept of 'Illes Balears' could only ever take on meaning as a single cultural grouping when uttered in the Spanish language, as the particular identity of the residents of each island eludes any commonality with those of the others. On the islands, very few would define themselves as 'Balearic'; if they were to do so, it would primarily be either from political positions associated with Spanish ultra-nationalism or from a perceived notion of homogeneity that is projected onto the islands from the mainland city centres (Barcelona, Madrid). As a whole, the song embodies both a celebration of the local characteristics and a parody, with the portrayal of a tourism unit that is paradoxically constructed both as a shared brand and through the pursuit of the distinctly specific traits of each island setting.

For this reason, it is at once complex and necessary to speak of the Balearic Islands as a tourist archipelago, as the external projection of an image of the islands and their unity in the form of an archipelago may contribute to reinforcing local imaginaries, while at the same time rendering them more uniform (Baldacchino, 2015, p. 3). This chapter seeks to analyse the tension between singularity and uniformity in the tourist image of the Balearic Islands, understanding as a key moment the attempts to redefine the tourism scenario in the wake of the economic crisis caused by Covid-19 in the summer of 2020. Several studies have already examined this issue from the perspective of the economy (e.g. Gómez, 2020; Adamiak, 2021). We are more interested

DOI: 10.4324/9781003451037-5

in exploring the discourses and images surrounding the promotion of island tourism, to assess how the presentation of one or several tourist identities shapes the international and local imaginaries of this group of islands. Therefore, our analysis will be based on audiovisual tourism promotion material, which we will combine with other sources, primarily with a view to understanding the reactions of the local people. Ultimately, we will assess the evolution of the heterogeneity of the image of the islands individually and as a group, and we will evaluate how those changes have affected the islands' self-images. To do so, we will base our observations on the premise that part of the projection for future visitors indirectly targets a group of islanders who are worried about the social and environmental consequences of naturally limited regions that are becoming increasingly overcrowded.

When we speak of the Balearic Islands, we are referring to an archipelago made up of four populated, and quite unequally sized, islands: Mallorca (3,620 km^2 and a population of 940,470), Menorca (695 km^2 and a population of 99,380), Eivissa (Ibiza, in English) (572 km^2 and a population of 158,620), and Formentera (83 km^2 and a population of 11,250). Then there is the island of Cabrera, currently an uninhabited natural park; and several islets, including Sa Dragonera – originally earmarked for luxury tourism, but declared a national park in 1995 – and Es Vedrà, which are worthy of mention as sites of ecological resistance against urban tourism development. Others, such as the islet of Tagomago, off the coast of Eivissa, one of the few privately owned islands in the Mediterranean, have small tourism establishments. The archipelago is divided into two sub-archipelagos, one for the Pityusic Islands (Eivissa, Formentera and their neighbouring islets) and the other for the Gymnesians (Mallorca, Menorca, Cabrera, and other nearby islets). The island of Mallorca houses the capital city of the Balearics, the city of Palma, which is home to 415,940 inhabitants of the archipelago's total population and accommodates the highest percentage of island infrastructures. However, it cannot be considered the gateway to the rest of the islands: Menorca and Eivissa – but *not* Formentera – each have their international airports, accessible from both the Spanish mainland and from many places around Europe.

The four main islands moreover retain several distinctive features, some of which are rooted in history, as is the case of Menorca, for example, which was under French and British occupation during the eighteenth century. Other unique characteristics are associated with overseas trade and the movement of people, as can be seen in the historical connection between Eivissa and Valencia (Ribes 2014, p. 18). Another example is the connection between the south of France and the town of Sóller, in the north of Mallorca, which for centuries was isolated by the mountains; a connection that has even affected the local vocabulary and phonetics, which have become slightly Frenchified. In fact, the differences among the Balearic Islands take the form of recognisable dialectal variants of Catalan, specific folkloric traits (in terms of traditional dances and songs, for example) and particular distinguishing features (i.e. the almond and olive trees in Mallorca, the cows and cheeses from

Menorca, the white houses in Eivissa or the fig trees and lighthouses of Formentera). Indeed, it is at the very least challenging to think of a broad set of
cultural traits that could serve as a shared identifier for the archipelago. Even
the existence of two types of definite articles (el/la, es/sa, stemming from the
Latin *ille* and *ipse*, respectively), which in the informal register sets the island
dialect apart from that of most of the Catalan-speaking mainland, is not
shared in certain towns, such as Pollença, in the north of Mallorca.

Despite the proximity among the islands, they have not always formed part
of the same administrative unit. Today they comprise a single Spanish autonomous community. This means that their government lies in the hands of a
single institution, the Government of the Balearic Islands, although each island
has its own area of local government that is known as an island council (Consell Insular). As we shall see, this is important for the creation of brands with
varying degrees of uniformity or uniqueness in terms of tourism promotion.
The Government of the Balearic Islands has even implemented measures aimed
at promoting a sense of 'Balearic' oneness, including the creation of a Balearic
slogan, 'Quatre illes, un país, cap frontera' ['Four islands, one country, no borders'] (1999), and the proclamation of a Balearic Islands Day, celebrated
throughout the archipelago. In the institutional images, apart from the motto
proclaiming unity, different visual strategies can be seen to bridge both the
maritime and social separation, whether by means of abstraction (four crossed
lines) or through the image of cooperation (four hands interlaced together).

Uneven tourism development

Although the four islands share a location that renders them attractive to
tourism, with extensive coastlines and warm yet not extreme climates, their
respective tourism development has not been uniform. In fact, at least up
until the post-pandemic period, there was a tendency to create different tourism brands, also perhaps an attempt at "contrived complementarity" (Baldacchino and Ferreira, 2015). Generally speaking, those brands have
associated Menorca with a nature-focused family tourism; Eivissa and Formentera are suggestive of a somewhat alternative, counter-cultural lifestyle
and are also closely tied to images of a vibrant nightlife; while Mallorca
comes with a typical 'sun and beach' scene that can vary more than the
others, given the larger size of the island and its more diverse topography.
Sociologist Joan Amer (2010, p. 72) already noted this in his thoughts on the
hotel business in the Balearic Islands:

> We must bear in mind both the lack of tourism studies and social
> research on the four islands as a group, as well as the fact that each island
> is marketed separately, because that is how they are known. The promo-
> tional material found in a travel agency in Great Britain or Germany
> speaks of Majorca, Minorca or Ibiza, rather than of the Balearic Islands.
> (Author's translation)

The differences in the island tourism models correlate with the disparate economic development of the islands and with the timing of their tourism development processes, which in all three cases relates to what is known as the Spanish tourism 'boom'; in other words, the Franco regime's promotion of Spanish coastlines, in particular, as a tourist attraction and a means of economic development. This took place as of the 1960s and brought about a major transformation in the island regions. The change has also been manifested demographically, with the arrival of a large workforce that primarily came over from the south of mainland Spain, leading to a cultural and linguistic exchange. Additionally, this boom would expose the local people to the activities and customs of the Europe of the 1960s, to which the locals had been denied access by Franco's autarchic and authoritarian dictatorship.

Before explaining the trends in the tourist promotion of the islands, we will provide a brief overview of the different ways in which they have evolved. There are many studies that have documented the importance of Mallorca as a destination for both elites and bohemians since the early twentieth century (Moyà, 2017). Milestones include the opening of the Grand Hotel in 1903, a luxurious *Modernista* building in the centre of Palma; the creation of the first car rental companies at the turn of the century, embodied by Garage Balear (1907); the emergence of the first travel agencies, such as Club Mallorca (1912); and the publication of tourism guidebooks such as Frederic Chamberlain's guide for Anglophone travellers, with a print run of 5,000 copies (Chamberlain, 1927). Moreover, the creation of the tourism business promotion body 'Foment del Turisme' dates from 1905, and remains active (Vives, 2005). The tourism industry was becoming an increasingly more prominent industry, though it would be interrupted by the Spanish Civil War (1936–1939). Soon after the war, attempts were made to revive it, as can be seen, for example, in the 'Luna de miel en Mallorca' ['Honeymoon in Mallorca'] campaign, which would take off with the tourist boom of the 1960s, bringing mass tourism to the island (Barceló i Pons, 2020).

The evolution of tourism in the Balearic Islands, and specifically in Mallorca, can be explained based on a series of economic recessions stemming from geoeconomic phenomena. The result of this dynamic is the reiterated reinvention of tourism development areas, which is linked to new periods of economic growth (Murray Mas, Yrigoy Cadena, and Blázquez-Salom, 2017). According to Pons and Rullan (1998), the first boom began in 1960 and lasted until the 1973 oil crisis, resulting in the creation of a mass tourism model. The second boom, which ended with the Gulf War in 1990, was characterised by extensive urban development and the generation of new forms of tourism, in the form of residential apartments. The third boom, ending in 2001, was marked by residential tourism. Rullan (2019) and other authors have identified a fourth boom, with the emergence of the tourist rental market, as well as the tourist colonisation of isolated areas previously uninhabited by visitors, such as the heart of the city of Palma and the purchase of real estate in rural areas by wealthy foreigners as second homes or for use as holiday rentals. As is apparent in the promotional images, the recession triggered by the Covid-19 pandemic did not reverse this process, whereby everything on the island, from the sea to cultural uses, could be transformed into a tourist attraction. This would be further compounded by more assiduous stops on the island by tourism cruise ships, a major source of environmental pollution, yet also a source of revenue in the city centre – particularly in the streets most accessible from the Port of Palma – where local or traditional businesses have been replaced by souvenir shops, ice cream parlours and brand-name franchises that cater to visitors with high purchasing power.

The promotion of Eivissa as a tourist brand occurred later than that of Mallorca: Eivissa's airport opened to commercial traffic in 1958, but it was not until 1966 that it opened to international traffic. The Franco regime and the local business sector exploited the arrival of artists to the island as of the 1930s, that is, before the Spanish Civil War, in order to promote an attractive and doctored image of Eivissa (and, by association, Spain) as a place of freedom and tolerance (Tomillero and Cardona, 2020). The island thus started to serve as a symbol for an open and cosmopolitan image of Spain that was projected internationally by the fascist dictatorship and viewed by the government and part of the local population as both a threat and an appealing source of income (Bayart, 2015; Cerdà and Rodríguez, 1999). The legacy of the hippies, who in the late 1960s were known by the authorities as "nonconformist tourists", continued to be bandied about in the promotion of the Pityusic Islands, in connection with the craft markets and the 'Adlib fashion'. In the 1980s, Eivissa also became a destination associated with nightlife, with the creation of major clubs such as Pacha and Amnesia, as well as a sort of "experience" for young adults that generated its own rituals, including the sunset at the Café del Mar and a visit to the lighthouse, Far del Cap de Babaria, leading to a description of the island as "une image presque parfait d'un hédonisme où se fondent experience du Plaisir, immersion dans la transe, liberation de la repression et dépaysement" (Michaud, 2012, p. 5) ["An almost perfect image of hedonism where the experience of pleasure, immersion in trance, liberation from repression and expatriation combine"].

Formentera, on the other hand, has not developed a tourism linked to nightlife, although, according to Cardona (2017), it operates in a somewhat supplementary mode to Eivissa, as though it were Eivissa's daytime territory, as evidenced by the constant yacht traffic between the two islands that are connected by a regular ferry service only during the day. Nevertheless, it must be noted that the resident populations of Eivissa and, above all, Formentera, have the highest percentages of people not born in the Balearic Islands: 31% for Eivissa and 33% for Formentera (IBESTAT, 2022). It is also odd that given the diverse range of foreign nationalities that visit Eivissa (British, German, and other European destinations), since the late 1980s, Formentera, has become a primarily Italian tourist destination. Some people explain it as a strategy of the Italian tour operators, who were looking for an alternative to the Balkan destinations, which at that time were unstable due to the wars in the former Yugoslavia. The company Going promoted Formentera as 'L'isola assoluta, l'ultimo paradiso' ['The quintessential island, the last paradise'], thereby reiterating the cliché of the paradise island on Europe's southernmost Mediterranean Spanish island. As we will see below, over time, many of these tourists have become part-time residents, either working remotely or running businesses on the island, to such an extent that today almost one-sixth of Formentera's population is of Italian origin. The tendency of Northern Europeans to purchase second homes has become common on all the Balearic Islands in connection with summer holiday use or as residences from where they could work remotely. The onset of Covid-19 has reinforced these trends.

In Menorca, several circumstances led to an even slower rate of touristification, although today this process has reached an economic dependency on the tourism sector similar to that of the other islands. In the late nineteenth and early twentieth centuries, the cosmopolitanism of Menorca, which had been under British and French rule, appeared to disappoint visitors who arrived there in search of a traditional local flavour and authenticity that they would instead find in Mallorca and Eivissa. In addition to this, travellers complained about the island's harsh climate, such as the wind (which is very frequent on the island) and the barrenness of the countryside. This negative image had an impact on the frequency of maritime communications. In contrast, the three main attractions of the archipelago's northernmost island were the port of Maó, the archaeological sites, and the organ of the Church of Santa Maria in Maó (Méndez, 2017). The coastal area, nowadays the attraction *par excellence*, did not seem to be particularly important at the time. Nevertheless, it is also paradoxical that today one of the most highly valued features of Menorca, especially in contrast to the homogeneity of Mallorca and Eivissa, is precisely its 'authenticity'. The fact that it has retained part of its original essence is due in part to its delay in joining the international tourism circuit, and in part to the public resistance and opposition to certain development projects from the 1970s onwards. If the flow of visitors to the island had been insignificant prior to the Spanish Civil War, Menorca's Republican (and therefore vanquished) position during the war would result in a lack of funding for the island during the post-war period, with direct repercussions on its ability to build infrastructure: the airports of Mallorca and Eivissa were officially opened to international traffic in 1960 and 1966 respectively, whereas that of Menorca would not open until 1969. Moreover, the existence of time-honoured industrial and agricultural activity on the island generated a certain apprehension regarding a change in the sector, leading to the maintenance of a certain economic diversification for several decades. The landowners enjoyed social prestige thanks to their vast country estates, which were the primary source of their income. Therefore, they were not apt to sell them off. Moreover, the major investments needed from outside the island in order to develop the tourism industry would not arrive until later, once those other tourist areas were full and there was a need to exploit pristine locations (Farré-Escofet, Marimon i Sunyol and Surís i Jordà, 1977, p. 228; Fullana and Seguí, 2012). Partially due to the oligopolistic nature of supply, with a few foreign tour operators, prices remained higher than those on the other islands (Farré-Escofet, Marimon i Sunyol, and Surís i Jordà, 1977, p. 258) and the most prevalent type of accommodation was residential, with limited hotel beds. The inevitable arrival of mass tourism in the 1980s led to the *Balearisation* of Menorca – a term that refers to an urban development model based on intensive and uncontrolled construction in coastal areas – with the indiscriminate building of large numbers of relatively inexpensive apartments (Méndez, 2017). However, the relative conservation status of the island was a well-known condition and this would spur and culminate in the declaration of Menorca as a Biosphere Reserve in 1993. According to Beltrán (2015), this valuable natural asset has led to the protection of the natural environment:

> not only for ethical and moral reasons, but also for economic reasons, as it can serve as a powerful competitive advantage by enabling the offer of a different, high quality product with its own specific identity.
>
> (Beltrán, 2015, pp. 16–17, author's translation)

Indeed, the Menorcan tourism model benefits from its status as a Biosphere Reserve, a differentiating feature (one example is the large number of agrotourism resorts that have opened recently) and a factor that makes it possible to offer an authentic, quality experience, with family and nature-based leisure and recreation activities, far removed from the nightlife, alcohol and partying that abound in other parts of the Balearic Islands. There is only one exception to this: the Sant Joan festivities, which attract huge numbers of young people – chiefly Mallorcans – who flock to a festival that is no longer local.

Difference and similarity in tourism promotion after the Covid-19 pandemic

The process of change in the promotion of the Balearic Islands' tourism image after the pandemic cannot be considered without taking into account that it occurred concurrently with an administrative change, specifically the Balearic Islands Government's transfer and devolution of tourism promotion powers to the different island councils: Eivissa in 2015, Formentera, Menorca, and Mallorca approved in 2018 and implemented the following years. This differentiation would spur the creation of distinct brands for each island, a requirement that has been expressly stipulated in different tourism management plans (e.g. Fundació Foment del Turisme de Menorca, 2023; Ajuntament d'Eivissa, 2022; Fundació Mallorca Turisme, 2021). Oddly enough, as we discuss below, despite the appearance of specificity in the promotional material that emerged with the change – the creation of individual slogans and the pursuit of emblematic landscapes and events – we have identified a convergence in the promotional imagery of the islands, which we believe correlates to at least two causes: (1) the desire to diversify the local supply; and (2) the paradigm shift stemming from the fundamental need to quickly re-establish tourism in the post-pandemic era.

According to the Pla Integral de Mesures per a Reactivar la Temporada Turística a les Illes Balears (Conselleria de Model Econòmic, Turisme i Treball, 2020), the number of tourists visiting the archipelago in 2020 dropped by 81%, primarily due to Covid-19 restrictions on foreign visitors. The consequences of this decrease were not, however, entirely negative. During the toughest months of Covid-19 in the Balearic Islands, like elsewhere in the world, there was an unusual and unintended beneficial effect embodied by the return of nature to areas formerly occupied, trampled upon and weighed down by humans. In the Balearic Islands, there were reports of dolphin sightings along the coasts and near harbours, and birds such as kites and kestrels could be seen in the skies above the most urban areas of Palma, the capital city. The absence of tourists also allowed the local residents of the islands to recover areas that had been previously overcrowded by visitors. The most easily accessible beaches were virtually deserted, with crystal-clear water and space for peace and quiet. All things considered, amid the unrest and concern, the islanders suddenly had a small oasis of peace, which was also caused by the forced unemployment of those who worked in the hotel and services sector during the summer months. The recovery of natural areas for tourism and recreation was followed by a groundswell of ideas regarding the right to reclaim these areas for the residents. Examples of this can be seen in two specific actions. In August 2020, two journalists organised a tourist route through Magaluf, the most demonised British and German tourist party area, to enable locals to explore the area from a critical and historical perspective. The name of the route was 'Via guiri: Un viatge al fetge turístic de Mallorca' ['Tourist lane: A journey to Mallorca's tourist liver'], making an obvious reference to the heavy consumption of alcohol by the tourists who frequent the area (Hernández, 2020). The second initiative, which was more festive, involved the transfer of a traditional festivity of the town of Sineu, a newer tradition known as the Much, which had nonetheless become extremely popular among the local young people, from the interior of the island, where this type of activity was not yet authorised, to the area of Magaluf, where the festivity was permitted. During the celebration, the tourists' most stereotypical behaviours were mimicked in parody, even engaging in *balconing* into a plastic children's swimming pool. This is a dangerous, and sometimes fatal, practice by young tourists who attempt to jump from their hotel balconies into swimming pools below or over to other hotel rooms.

The effect of the pandemic exacerbated the social perceptions of tourism in the Balearic Islands. On the one hand, the economic and employment crisis generated by the sudden disappearance of the main source of income on the four islands led to concern among both employers and workers in the sector. It even manifested in the campaign promoted by hotel owners, yet widely supported by small business proprietors, which, under the slogan 'SOS Turisme', aimed to force the institutions to open up the sector sooner and relax the health measures. The movement even led to the composition of a sort of hymn, a song entitled 'El turismo eres tú' ['You are tourism'], written and interpreted by Richard Memories, which speaks to the beauty of the islands and the need to recover tourism.

> Piensa en que todo va a ir bien
> que va a favor el viento
> y que las estrellas nos iluminan
> hasta caminan:
> el turismo eres tu
> Piensa en volver a empezar
> no hay ningun rival,
> no hay quien nos pare.
> Tenemos playas, hasta montañas
> pero, el turismo eres tu.

> Think that everything is going to be okay
> that the wind blows in your favour,
> and that the stars illuminate us,
> they even walk:
> tourism is you
> Think about starting over,
> there is no rival,
> no one to stop us.
> We have beaches, and even mountains
> but, tourism is you.

The song's chorus, which serves as its title, includes the resident subject as the tourist subject, thus implicitly pushing for the complete touristification of the island space. If the residents themselves are 'tourism', then everything is tourism: there is no possible alternative, and there is no remaining space isolated from the visitor's use and gaze. We will see below how this very notion is shaping the post-Covid-19 tourist imaginary in the Balearic Islands.

The other reaction emerged precisely out of the awareness of the islands' vulnerability in the face of the tourism monoculture. This critical view came as an addition to the shift in the environmental movement in the Balearic Islands in recent years, which was a pioneer at the national level in Spain. Its scope of action had morphed and broadened from environmental conservationism to more across-the-board positions, addressing the social consequences of the tourism monoculture and overtourism, for example, in terms of education (training and school drop-out rates resulting from easy access to precarious and poorly paid jobs in hotels) and access to basic rights such as housing, which is one of the main problems of island residents, thanks to both tourist rentals and the acquisition of homes by wealthy foreigners. Before the pandemic, documentaries such as *Tot inclòs* (2018) and *Overbooking* (2019) had brought to the screens the need to put an end to overtourism and the imperative to spur a change of direction in the islands' economy. The applause at the hotel entrance, welcoming the first tourists who were allowed to visit Mallorca in June 2020 despite the prevalent social distancing measures throughout Spain at the time, was prescient of the fact that the discourse in favour of limiting tourism or changing the economic paradigm was just rhetoric, and would not bring about any real change. Nevertheless, it would at least make a difference in the tourism projection of the islands.

If we examine the tourism promotion campaigns created by both the Balearic Islands Government and the Island Councils as an indicator of such projections, as well as the image of the islands in other types of advertisements launched by the private sector, there are at least two, non-opposing aspects that could be sensed prior to the pandemic yet which have gained traction in recent years. The first relates to the diversification of the tourism supply of each island. Such expansion probably stems from the desire to attract visitors beyond the classic 'sun, sea and sand' island lure (e. g. Péron, 2004), which nevertheless continues to be their main attraction. This can be seen in the strategic Action Plans of each island, which were drafted in the wake of the pandemic. The Action Plan of Eivissa is intended to promote a "multi-thematic, emblematic and highly innovative supply", in addition to remote working and "micro-holidays", presumably drawing on the small size of the island. The multi-themed nature of the supply can easily clash with its representational focus, which necessarily showcases the most emblematic and significant features of each island destination. Mallorca's 2020 Action Plan focuses on the need to diversify what the island has to offer. Menorca's latest Action Plan (2023) shares with the other islands the need to break down the industry's stubborn seasonality and diversify its supply, while at the same time emphasising the aim to continue to be a benchmark for sustainable tourism and to promote luxury tourism, a combination embodied by the new 'Ecoluxe' label:

> Luxury tourism focused on ecotourism has become Ecoluxe, a tourist product that can afford Menorca a solid advantage, since it is rooted in quality, authenticity, sustainability and the guarantee of finding natural resources in their most pristine state.
> (Pla d'Actuació 2023, Fundació Foment del Turisme de Menorca, 2023)

The audiovisual advertisements for Menorca, Mallorca, Eivissa and Formentera combine nature with the rural settings, gastronomy with traditions, and sport with well-being. With the intention of showcasing everything each island has to offer, this diversity of settings leads all the islands to resemble one another, to a certain degree. There are only a few distinguishing features (the focus on the nightlife in Ibiza, the caves and monuments in Menorca and the golf and event tourism in Mallorca) in compositions that draw on collage to present a comprehensive, wide-ranging offer for everyone, yet within a necessarily limited space: that of the islands. One successful slogan in promoting Eivissa was 'Totes les illes en una' ['All the islands in one']. In the case of Menorca, the slogan for 2023 was 'La isla de los pequeños placers' ['The island of little pleasures'] where the plural serves to invite the viewer to seek out the diversity in a small place (Fundació Foment del Turisme de Menorca, 2023). Along the same lines, the audiovisual promotion of Mallorca for 2023 is based on a phrenetic video, 'Travel to Mallorca in 30 seconds', which promises to display the vast diversity offered by an island self-purported to be the 'island of calm' (Mallorca Tourism, 2022). The promotional video of Formentera is based on the dichotomies, such as tradition/modernity, calm/party, among others, that can be found on a total island.

To sum up, the 'new' tourist island model that these spots depict is that of a complete and diverse region, like a diorama that presents everything that a tourist could wish to find is on offer: whether it be sport or family life, tradition or nightlife, culture or beach, the mountains or the sea, nature or festivities, tradition and modernity. The island, already a world unto itself, is a complete tourist attraction; and presumably comes with a universal appeal. No wonder then, that the title of the generic spot advertisement for the Balearic Islands in 2022 was 'I want it all' (Agència d'Estratègia Turística de les Illes Balears, 2023). The final stage in this process of the island region's total touristification is when the visitor becomes a resident. There are different strategic plans for tourism in the Balearic Islands that mention remote working as an attraction for foreign visitors. It is likely that the creators of those plans had in mind the residents of wealthier and less sun-filled European countries who could convert their second homes into more permanent places of residence. The identification of the Balearic Islands as a 'home' or 'family place' for Europeans is frequent in promotional speeches under explicit slogans such as 'Balearic Islands: Welcome to your new home' (Agència d'Estratègia Turística de les Illes Balears, 2018). Here, the ownership of the island becomes 'shared by the tourist' in this island 'house' where each island is its own room:

> Arriving at the Balearic Islands is like feeling at home. In every room, the experience you want to live, and in all of them your home.
> (Agència d'Estratègia Turística de les Illes Balears, 2018)

In 2019, under the slogan 'Illes Balears, always yours', different European tourists were presented, expressing their special bond with Mallorca, which was incorporated into their life geography year after year. This concept was also expressed in the Menorcan spot where, once again, it is the recurrent visitors who describe the island for prospective future visitors.

We must add that, in the Comprehensive Plan of measures to revive the tourist season in the Balearic Islands (Conselleria de Model Econòmic, Turisme i Treball, Agència d'Estratègia Turística Illes Balears, 2020), there was an express desire to promote inter-island mobility for the first time, both for tourists and, above all, for residents. That objective materialised in the creation of a tourist discount that aimed to encourage residents to travel among the islands. This is a key factor that explains the increase in domestic tourists – which probably consisted mainly of Mallorcans – visiting Minorca, and to a lesser extent, Eivissa. These subsidies offered €100 per person to the islanders who wished to travel to the other islands. They were advertised under the inclusive slogan 'som illencs' ['we are islanders'] and, therefore, from the perception of a shared island identity, rather than a perspective focused on the specificity of each island region.

The other aspect mentioned above refers to the reinforcement of the environmentalist or ecologist discourse, as well as the social discourse, within the framework of tourism promotion, a trend that was already identified by Bardolet (2001) and Royle (2009) from the perspective of public planning. In the tourism plans mentioned above, there is often a need to cater to a new type of tourism that is more in tune with environmental needs, local health safety and the well-being of workers and residents. This is manifest, for example, in the Integral Plan of Measures to revive the tourist season in the Balearic Islands (2021), where the decrease caused by Covid-19 was viewed as an opportunity to improve the tourism model "making environmental, economic and social sustainability an objective, an opportunity that we were already considering before the Covid crisis", alongside the fulfilment of the UN's Sustainable Development Goals. In the tourism advertisements for Menorca and Formentera – the two destinations that have remained the most preserved for family and nature tourism or for visitors that at least seek out the special individual features of the islands – the resident is portrayed and given the role of the guardian of the island's traditions, which they cherish to the point of wanting to share them with the tourist. On an island like Formentera, where it is almost impossible for young people or workers to rent a home in the summer season, the advertisement is (at the very least) somewhat misleading. In other advertisements, such as the one accompanying the slogan 'El turisme del futur' ['The tourism of the future'], which appeal to the need to build an image of the islands as places where the land is preserved and where workers' rights are respected, the well-being of local residents, it seems that the target audience is locals rather than actual or potential visitors. This is clear in the advertisements that intend to portray the improvement in the labour conditions of hotel workers, which has been one of the most significant issues on the public and political agenda, particularly as a result of the recent unionisation of hotel room cleaning staff. Images of environmental awareness and respect have even been carried over into advertisements for the Estrella Damm beer brand, which in 2009 began to launch a series of promotional videos under the slogan 'Mediterràniament' ('Mediterraneanly') (BBC Story Works, 2023). Each year, those spots present small, apparently local stories, although in some cases they can feel contrived to the local people: here, the locals spend the summer amid idyllic surroundings, going from beach to beach and from dinner to dinner. The setting around them is both rural and coastal; nobody gets up early to go to work in a town where many of the inhabitants work in the hotel sector, especially during the summer season. In fact, the number of workers in the tourism sector in the Balearic Islands increased by 17% year-on-year during the first quarter of 2023, with few variations on the different islands. After several advertisements with an amorous tone or aimed at a young audience, the brand's latest adverts, filmed in Menorca, conversely take on an epic ('Act III: Estrella Damm commitment', 2020) or comical tone ('Love at first sight', 2021) and focus on environmental concerns as a central theme to find other ways of living in peace (while, of course, drinking the brand's beer).

Local reactions to overtourism

As we have seen, the pandemic pushed to extremes the positions of two previously existing discourses on the effects of tourism in the Balearic Islands. Those positions are the opposite sides of the same coin: the dependency on a single model of economic development. As Murray Mas, Yrigoy Cadena, and Blázquez-Salom (2017) have observed, it was not the first time that an economic crisis had brought about a restructuring of the tourism sector, though it had never done so with such intensity. As has occurred in other recessions, the tourism sector has emerged stronger in the Balearic Islands. On the one hand, this is due to their geostrategic location, being just south enough to retain its status as a climatically and environmentally attractive island for visitors from the north, and just European enough to position itself as a 'safe' destination. That 'safety' had been a trump card when armed conflicts threatened the market position of the main competitor destinations in North Africa and the Balkans during the 1990s. In the wake of a global pandemic, the concept of safety has been redefined in terms of health safety and the construction and projection of an image of well-being (healthy food, sport and nature).

On the other side of the coin, the effective opportunity to recover the island's spaces for the locals, even if only for just one summer, was an experience that also left its mark. At times it has done so in the form of a superficial complaint: the islanders' resentment regarding the impossibility of being able to go and relax at the "usual" beach, the hordes of German cyclists who fill the roads, and the massive crowds of visitors who flood the streets of Palma on the days that the smoke-spewing cruise ship arrives in the harbour. Nevertheless, there has been a consolidation of a broadening of critical thought around the tourism monoculture from the perspective of ecology, to a more far-reaching debate on the very quality of life of the island's inhabitants, as well as a genuine concern regarding the future and sustainability, not only of tourism, but of the island itself, which moreover, ties in with the plight of global ecology. At the local level, it links environmentalism to the community's resistance to gentrification (examples of this include the group 'La ciutat per qui l'habita' ['The city for its residents']), as well as the movement for the regulation of the cruise ship arrivals, which connects with other mainland areas that face similar problems, such as Lisbon and Venice. Even before the pandemic, a media and political campaign by the sectors interested in the promotion of tourism development defined these critical positions under the term 'tourismphobia'. The presence in the islands of graffiti that speaks out against tourism, with messages such as 'Tourists go home. Refugees welcome'; citizen activism in the streets, with the inauguration of a 'tourist lane' to allow residents to walk peacefully around the city; and 'Operation Confetti', a street performance protest with flares and confetti on Palma's pier, which led to the legal prosecution of the young people who organised it, served the conservative sectors to construe and present tourismophobia as a sort of proto-terrorism that threatened tourism, which for them embodied the goose with the golden eggs (Blanco-Romero et al., 2019).

All in all, the debate on the regulation of tourism in its many facets, including urban planning, though also in connection with the right to housing and labour rights, is now a central issue of the political agenda throughout the Balearic Islands. In the cultural sectors, this matter has generated unique imaginaries stemming from the precise geography of the islands and their location between Europe and Africa. For the moment, the aim is not to present alternatives to the currently existing overtourism, but rather to portray the virtually dystopic consequences of not slowing such growth. As we have already explored elsewhere (Picornell, 2020) in Mallorca, critical artists and social organisations are reformulating the postcard imagery of the island to show what is just beyond the frame of the photo: the exploitation of both labour and the environment. The same awareness process is also taking place in Menorca. Thus far, the island has been somewhat protected as a Biosphere Reserve by virtue of which it has marketed itself as a green tourism destination, although its local people are beginning to feel the effects of both overcrowding and the purchase of second homes by wealthy Europeans (mainly, French and Italians). In Eivissa, the local newspapers reported on the creation of a new platform known as 'Eivissa es rebel·la' [Eivissa rebels] and formed by local historical environmentalist movements and global groups including Extinction Rebellion, Ibiza Conciencia and Rebelión Científica (Redacción, 2023). Hence, once again, it seems that each territory is defending itself separately, guaranteeing the specificity of each island, despite the shared nature of what the islands are facing.

Conclusion

The pandemic has exacerbated a shift in the tourism promotion of the Balearic Islands that had already been in the making for some years. The total touristification of the island space seeks to satisfy the desires of a visitor who apparently does not want to feel like a tourist, but instead wants to enjoy an exclusive experience while, at the same time making the very most of their visit. The locals, on the other hand, have also become accustomed to seeing their island depicted from an outside point of view. Examples of this are the

Estrella Damm advertisements which, despite being filmed on location and targeting a 'Mediterranean' public, nevertheless feature characters from the mainland who have come to vacation on the islands. Justified by the economic recession caused by Covid-19, promotion has only intensified and created new strategies, such as the diversification of the supply and the dissemination of an image of sustainability, safety, and well-being. In contrast, the residents, who were able to enjoy their islands without crowds for a few months and who later witnessed the problems of access to housing and massification, have in some cases responded forcefully, with campaigns against tourism that have often been subjected to almost criminalising scrutiny.

Although the islands of the archipelago have different tourism histories and trajectories, certain common processes have come into play in their public promotion. Whatever the case, the contemporary tourism projection of the Balearic Islands is displaying a tendency to diversify the supply of each island, which paradoxically leads to an increasingly more similar promotion of the four islands abroad. We have also identified convergences among the local reactions to tourism: concerns for its continuation; the maintenance of growth (and, therefore, support for economic and political sectors that favour the creation of more urban infrastructures or the maintenance of, or even increases, in the number of hotel beds); or in the form of critical thought and action on the harmful effects of overtourism, both environmentally and socially. Finally, we must note that the existing studies of these processes in the Balearic Islands often take a local context as their starting point, which is not particularly conducive to comparative work. This makes it difficult to identify island specificities or situations that are shared throughout the archipelago, and which this book addresses and should thus help to resolve.

References

Adamiak, C. (2021). Cambios en la oferta de Airbnb durante la pandemia Covid-19. *Okonomics: Revista de economía, empresa y sociedad*, 15, 1–11. https://doi.org/10.7238/o.n15.2107.

Agència d'Estratègia Turística de les Illes Balears (2018). *Balearic Islands. Welcome to your new home*. Conselleria de Turisme. Govern de les Illes Balears. www.illesbalears.travel/en/baleares/.

Agència d'Estratègia Turística de les Illes Balears (2023). *Illes Balears, "I want it all"*. Illes Balears official promotion. Conselleria de Turisme, Govern de les Illes Balears. www.illesbalears.travel/article/ca/illesbalears/illes-balears-i-want-it-all.

Ajuntament d'Eivissa (2021). *Pla estratègic de turisme ciutat d'Eivissa*. https://turisme.eivissa.es/wp-content/uploads/2021/02/PLA-ESTRATEGIC-CIUTAT-EIVISSA-2021-2023.pdf.

Amer, J. A. (2010). Turisme de masses i societat. Les Balears com a paradigma. *L'Espill*, 35, 72–78.

Baldacchino, G. (2015). More than island tourism: Branding, marketing and logistics in archipelago tourist destinations. In G. Baldacchino (Ed.), *Archipelago tourism: Policies and practices* (pp. 1–18). Farnham: Ashgate.

Baldacchino, G. and Ferreira, E.C. D. (2015). Contrived complementarity: Transport logistics, official rhetoric and inter-island rivalry in the Azorean archipelago. In G. Baldacchino (Ed.), *Archipelago tourism: Policies and practices* (pp. 85–102). Farnham: Ashgate.

Banal-Juaneda, E., Farré, M., Llinàs, X., Oliver, J. Ll., Mulet, T., Mesquida, J., Domènech, S., González, J., Rico, F., Serra, D., Ramírez, I., Hernández, N., De Echave, P., and Blanquer, N. (2018). *Tot inclòs: Danys i conseqüències del turisme a les nostres illes.* Palma: Col·lectiu Tot Inclòs.

Barceló Pons, B. (2000). Història del turisme a Mallorca. *Treballs de la Societat Catalana de Geografia,* 50, 31–55.

Bardolet, E. and Sheldon, P. J. (2008). Tourism in archipelagos: Hawai'i and the Balearics. *Annals of Tourism Research,* 35(4), 900–923.

Bardolet, E. (2001). The path towards sustainability in the Balearic islands. In D. Ioannides, Y. Apostolopoulos, and S. Sonmez (Eds), *Mediterranean islands and sustainable tourism development: Practices, management and policies* (pp. 193–213). London: Continuum.

Bayart, P. (2015). Hippies, peluts i turistes disconformes a Formentera. *Eivissa,* 58, 35–45.

BBC StoryWorks (2023, 1 June). *Made by the Mediterranean. Brewing ambition. BBC StoryWorks.* [Video]. YouTube www.youtube.com/watch?v=lUIZYbhDmio.

Beltrán, C. (2015). *El model turístic menorquí: Mite o realitat (1960–2015).* Mahón: Edicions Documenta Balear.

Berrozpe Martínez, A. (2016). Ibiza como marca de destino turístico. *Opción: Revista de Ciencias Humanas y Sociales,* 11, 111–120.

Blanco-Romero, A., Blázquez-Salom, M., Morell, M., and Fletcher, R. (2019). Not tourism-phobia but urban-philia: Understanding stakeholders' perceptions of urban touristification. *Boletín de la Asociación de Geógrafos Españoles,* 83, 2834, 1–30. http://dx.doi.org/10.21138/bage.2834.

Cardona, J. R. (2020). Arte y artesanía en el imaginario y la oferta turística: El caso de Ibiza. *El periplo sustentable: Revista de turismo, desarrollo y competitividad,* 38, 150–173.

Cardona, J. R. (2017). Peculiaridades del turismo de Formentera. *International Journal of Scientific Management and Tourism,* 3(1), 555–578.

Cardona, J. R. and Serra Cantallops, A. (2014). Historia del turismo en Ibiza: Aplicación del Ciclo de Vida del Destino Turístico en un destino maduro del Mediterráneo. *Pasos: Revista de Turismo y Patrimonio Cultural,* 12(4), 899–913.

Cerdà Subirachs, J. and Rodríguez Branchat, R. (1999). *La repressió franquista del moviment hippy a Formentera.* Palma: Edicions Res Publica.

Chamberlain, F. (1927). *The Balearics and their peoples.* London: John Lane.

Conselleria de Model, Econòmic, Turisme i Treball, Agència d'Estratègia TurísticaIlles Balears (2020). *Pla Integral de mesures per a reactivar la temporada turística a les illes Balears, 2021.* Govern de les Illes Balears. www.caib.es/pidip2front/adjunto?codi=2636100&locale=ca.

Dioscórides, Á. (2019). *Overbooking.* [Documentary video]. Palma: Mallorcadocs.

Estrella Damm (2020, July 11). *Acto III. Compromiso. Estrella Damm 2020.* [Video]. YouTube. https://youtu.be/3SsbRoyJLH0.

Estrella Damm CAT (2021, June 16). *Amor a primera vista. Amb Mireia Oriol i Mario Casas. Estrella Damm 2021* [Video]. YouTube. www.youtube.com/watch?v+BQa z2dRsfQM.

Farré-Escofet, E., Marimon i Sunyol, R., and Surís i Jordà, J. M. (1977). *La via menorquina del creixement.* Palma: Banca Catalana, Servei d'Estudis.

Fullana, A. and Seguí, M. (2012). El turisme a Menorca. Retard en el seu desenvolupament dins de l'àmbit balear. *Revista de Menorca*, 91, 125–162.

Fundació Foment del Turisme de Menorca (2023). *The island of small pleasures*. www.menorca.es/es/Campanya_Menorca_por_los_pequeos_placeres/15604.

Fundació Mallorca Turisme (2021). *Pla d'actuació 2022*. Consell de Mallorca. https://fundaciomallorcaturisme.net/wp-content/uploads/2021/11/Pla-dactuaci%C3%B3-2022-PDF.pdf.

Hernández, Ll. (2020, August 8). Mallorquines en Punta Ballena. *Diario de Mallorca*. www.diariodemallorca.es/sociedad/2020/08/07/mallorquines-punta-ballena-9013697.html.

IBESTAT (2022). *Población por isla y municipio de residencia, sexo y zona de nacimiento*. Ibestat tables: https://ibestat.caib.es/ibestat/estadistiques/4504e4e2-5e94-484e-9e5d-44f6be85f490/8052bd68-c6a0-4f0e-b25f-3149b1a99c57/es/pad_res03_20.px.

Mallorca Tourism (2022). *Travel to Mallorca in 30 seconds*. [Video]. www.youtube.com/watch?v=AWwvNicCNVs.

Méndez, A. (2017). *A la recerca del paradís: Història del turisme de Menorca*. Mahon: Institut Menorquí d'Estudis, Consell Insular de Menorca.

Michaud, Y. (2012). *Ibiza mon amour: Enquête sur l'industrialisation du plaisir*. Palma: Nil.

Moyà, E. (2017). *Journeys in the sun: Travel literature and desire in the Balearic islands (1903–1939)*. New York: Peter Lang.

Murray Mas, I., Yrigoy Cadena, I., and Blázquez-Salom, M. (2017). The role of crises in the production, destruction and restructuring of tourist spaces: The Balearic Islands. *Investigaciones Turísticas*, 13, 1–29. https://doi.org/10.14198/INTURI2017.13.01.

Péron, F. (2004). The contemporary lure of the island. *Tijdschrift voor Economische en Sociale Geografie*, 95(3), 326–339.

Picornell, M. (2020). The back side of the postcard: Subversion of the island tourist gaze in contemporary Mallorcan imaginary. *Island Studies Journal*, 15(2), 291–314. https://doi.org/10.24043/isj.109.

Pons, A. and Rullan, O. (2014). The expansion of urbanisation in the Balearic Islands (1956–2006). *Journal of Marine and Island Cultures*, 3(2), 78–88.

Redacción (2023, February 23). "Eivissa es Rebel·la" inicia campaña contra los jets privados dándole visibilidad en el carnaval. *Nou Diari Eivissa i Formentera*. www.noudiari.es/noticias-ibiza-formentera-sidebar/eivissa-es-rebel%C2%B7la-inicia-campana-contra-los-jets-privados-dandole-visibilidad-en-el-carnaval/.

Ribes, E. (2014). La petjada valenciana en la toponímia d'Eivissa i Formentera. *Jornades d'Antroponímia i Toponímia*, 27, 318–328. https://slg.uib.cat/gabinets/go/publicacions/#12.

Royle, S. A. (2009). Tourism changes on a Mediterranean island: Experiences from Mallorca. *Island Studies Journal*, 4(2), 255–240. https://doi.org/10.24043/isj.236.

Rullan, O. (2019). Islas globales y paisajes culturales postmodernos en las Islas Baleares. In *Paisaxes nacionais no mundo global* (pp. 37–62). Palma: Grupo de Análise Territorial (ANTE).

Tomillero, E. and Cardona, J. R. (2020). Promoción institucional de la marca turística Ibiza. *Grand Tour: Revista de Investigaciones Turísticas*, 21, 164–184.

Vicens Gómez, J. M. (2020). Impacto de la pandèmia por SARS-COV-2 en la economía de Baleares. *Medicina Balear*, 35, 82–87.

Vives Reus, A. (2005). *Història del Foment del Turisme de Mallorca (1905–2005)*. Palma: Foment del Turisme de Mallorca.

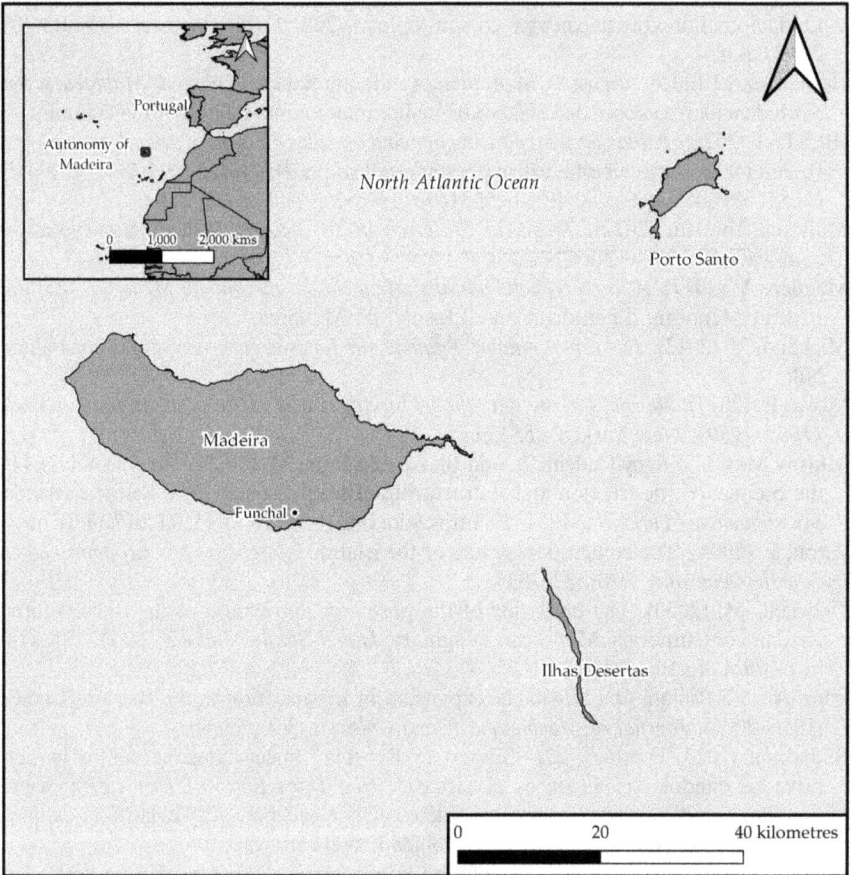

Figure 4.1 Madeira and Porto Santo, Autonomy of Madeira, Portugal

4   Tourism in Madeira and Porto Santo

Resigned subordination or partnership?

António Manuel Martins de Almeida and Brian Garrod

Introduction

One of the recurring themes in the study of islands is the suggestion that they tend to find themselves on the wrong side of a dominant-subservient relationship with other, more powerful geographical entities. A range of theoretical perspectives have been employed to explain this tendency, of which the core-periphery model has attracted much attention (Boissevain, 1979; McLeod et al., 2017; Butler, 2017; Weaver, 2017; Amoamo, 2018; Agius and Chaperon, 2023). This model, which has typically been employed at an international rather than domestic level, posits that "an inherently exploitative relationship is seen to exist between wealthy tourist-generating 'more developed countries', which comprise the economic core, 'and relatively poor tourist-receiving "less developed countries", which form the economic periphery' (Weaver, 1998, p. 292). Weaver (1998) goes on to describe some tourism practices as representing the modern equivalent of the plantation model, based on the exploitation of natural resources and landscapes (particularly coastal areas and beaches), by rapacious multinational corporations and foreign investors from the more powerful, developed countries. Many islands have been relegated to being mere observers of the inexorable advance of the so-called "pleasure periphery" (Turner and Ash, 1975), as tourism enters, occupies and ultimately devours places.

Tourism development is, under such conditions, highly likely to be shaped, in size and form, to the advantage of countries at the core. This inevitably disadvantages those located at the periphery. These are said to be in a 'development-dependent' state, insofar as their economic progress is conditioned, sometimes almost entirely, by the needs of the countries of the economic core. Dependency theory asserts that the economic development of the periphery under such circumstances tends to be "cosmetic, involving no concomitant change in the existing systemic pattern of wealth extraction from that periphery" (Weaver, 1998, p. 293), which is inherently costly, subservient, and deeply inimical to the interests of the local population. In this vein, Harrigan (1974) considers tourism development in the Virgin Islands to represent nothing less than a new form of colonial subservience. Oppermann

DOI: 10.4324/9781003451037-6

(1993) recognises the supportive role of local elites in the dependent territory in providing political backing to foreign capital. Weaver (1998, p. 293) observes this in the case of Trinidad and Tobago, suggesting that the "domestic involvement in the national tourism sector was somehow assumed to be either implicitly benign, or negligible".

According to Weaver (1998), the core-periphery model also offers a perspective on relations at the national level, specifically among larger and smaller islands within an archipelago. Larger islands may dominate smaller ones, as part of a local-level core-periphery relationship. Where the larger island is also in a subservient core-periphery relationship, the position of the smaller island can be thought of as on a 'double periphery' and potentially in a state of 'double-subservience'. The smaller islands may be disadvantaged not only by being on a local periphery but also on an international one. These islands are destined to become among the most disadvantaged in the developing world. Weaver's examples include Trinidad and Tobago, and Antigua and Barbuda, with the latter in each pair being peripheral to the former.

This chapter examines the relationship between the islands of Madeira and Porto Santo, focusing particularly on the economic potential of the tourism sector on Porto Santo and the challenges associated with its achievement. The discussion is set within the framework of Weaver's double periphery model, which begs the question: should the relationship between Madeira and Porto Santo be shaped as a resigned subordination of the needs of one to the other, or as a flourishing partnership between the two (Boissevain, 1979; Sultana, 2018)?

The core-periphery relationship

The received wisdom on the issue of core-periphery dynamics in the small island context suggests that such territories are bound to become externally dependent and "economically marginal entities" trapped in a tourism monoculture (Weaver, 2017, p. 11). The positioning of islands in a neoliberal world system has been described and explained as being one of dependency, underdevelopment, and political subservience. The islands are located at the economic and political periphery of a system that serves the interests of the wealthy countries at the core (Baldacchino, 2012; Chaperon and Bramwell, 2023; Pleasant and Spalding, 2021).

The effect of this core-periphery relationship is for the core countries to exploit the politically weaker and more distant tourism dependent-territories, often by pulling the strings of their former colonial links (Bianchi, 2002; Osagie and Buzinde, 2011). As Lea (1988, p. 10) observes, tourism "has evolved in a way that closely matches historical patterns of colonialism and economic dependency". Sharpley (2002), quoting the literature on dependency theory, remarks that under such conditions it is no surprise that small islands tend to mimic the political, legal and economic macro-structures of the former colonial masters. The upshot is that they assume the yoke of dependence. Autonomous development is not possible without the permission of the countries located at the core of the system, be that permission formal or tacit. Based on the case of Cyprus, Sharpley (2003, p. 248) contends that economic growth is a fundamental prerequisite to address the "pollution of poverty" in Cyprus, and that tourism contributes significantly to a higher standing of living, reduced poverty, the empowerment of women, meeting people's basic needs and providing local autonomy of choice and political freedom. Yet, Cyprus is challenged by the negative consequences deriving from its overdependence on tourism, including a highly seasonal market that is heavily dependent on UK tourists, a negative image associated with the clubbing/nightlife scene in Ayia Napa, and the contribution of tourism to environmental damage and the depletion of natural resources. These conditions imply that Cyprus is unable to attract sufficient numbers of upmarket tourists to withstand the fierce competition from other cultural-rich environments in the area or to develop other sectors of the economy. Their only option, therefore, is to adhere to its "firmly established image as a mass, summer-sun destination" (Sharpley, 2003, 261).

Peripherality implies that many islands are trapped in a vicious cycle of poverty, geographical irrelevance, a slower pace of life, and a lack of employment opportunities and emigration. Nevertheless, the condition also presents a number of advantages to peripheral places as tourist resorts that serve holidaymakers from the core. Such places embody the ideals of discovery, escape, paradisiacal landscapes, pleasure, exhilaration and exotic allure. Peripheral islands also often have climatic, cultural, historical, and ecosystem features that are distinctive and unavailable in the core. Weaver (2017, p. 13) asserts that islands may therefore benefit from a "virtuous periphery syndrome", with their marginalisation ironically offering them an opportunity to identify and lock on to a secure niche in the worldwide dynamics of tourism development (*also* Agius and Chaperon, 2023).

Baldacchino (2014, p. 10) further contends that, instead of being simply victims of 'pseudo-development strategies' that follow the pattern of boom and bust that is typical of niche market development – often followed by failure, decline and demise – islands have the potential to excel through the "judicious management of extra territorial opportunities" based on the exploitation of their geostrategic advantages. As a result of "both necessity and endowment", and spurred by the development of competitive tourism sectors proficient in providing peak experiences to a relatively large number of visitors, islands have been able to exhibit an impressive degree of "resilience and innovation" and high levels of economic and human development within the context of "balanced autonomy and cultural distinctiveness" (Weaver, 2017, p. 13).

Tourism development in Madeira and Porto Santo

Located in the Atlantic Ocean, the archipelago of Madeira comprises the larger island of Madeira, the smaller nearby island of Porto Santo, and many other smaller uninhabited islands and islets. Madeira is one of Portugal's two autonomous regions (the other being the Azores archipelago). In Portuguese, it is known as the Região Autónoma da Madeira (RAM). Like many small island territories, Madeira's economy is driven primarily by the public sector (25.5% of Gross Value Added, or GVA) and tourism (16.6% of GVA). The territory is thus heavily dependent on tourism to support its social and economic development.

The island of Madeira has a land area of 758.4 km^2 and is home to the capital city, Funchal. It has a population of around 255,000 people. The island has an extensive system of roads, mountain tunnels, harbours, and marinas, a port for large cruise liners and the region's airport: Funchal Airport (IATA Code: FNC), built in 1964, was re-named in 2016 after Madeira-born footballer Cristiano Ronaldo. The original 1,600 m runway was later extended to 2,780 m by means of a platform over the sea to accommodate larger and more modern aircraft. Over 30 airlines have scheduled flights to and from the airport, although some of these are only seasonal. The most popular route is to Lisbon, Portugal's capital city, which records around 1 million passengers per annum (DREM, 2023a). In total, the region welcomed just over 2 million tourists in 2022 (DREM, 2023b, 2023c). Of these, 75% were international visitors (DREM, 2023e). There is no ferry service with the mainland.

Porto Santo is a municipality of the region of Madeira. It is located 43 km to the northeast of the main island of Madeira. Porto Santo is sparsely populated, with 5,150 permanent inhabitants (2021 Census), on a land area of 42.6 km² (DREM, 2023f). Its airport (IATA Code: PXO) actually predates the one near Funchal, and has a 3,000 m runway. Flights are available between Porto Santo and Funchal and between Porto Santo and Lisbon on at least a daily basis. The *Visit Madeira* website suggests that Porto Santo is just a 90-minute flight away from Lisbon (Visit Madeira, 2023). This is technically correct, even though direct air connectivity is restricted to a small number of flights a week. Apart from TAP (the national carrier), various smaller airlines, such as easyJet, offer flights less frequently. There are also charter flights linking Porto Santo directly to other cities in Europe. These operate, however, almost exclusively in the summer months. Other air travellers must make a stopover, usually in Funchal or Lisbon. Porto Santo is thus considerably dependent upon its larger neighbour for access to its overseas tourist markets.

The tourism sector on both islands is highly reliant on the natural environment (Ismeri Europa, 2011). Their tourism offers are very different, however, and a major reason for this is their contrasting physical geographies. There is a virtual absence of natural sandy beaches on the island of Madeira, the coastline being particularly rocky on the north coast (Ismeri Europa, 2011). The only large bathing beach is an artificially created one near the resort of Calheta. The Natural Park of Madeira, located mainly in the rural hinterland, contains a central mountain range running from east to west, punctuated by rugged volcanic peaks (Leal et al., 2020). According to Visit Madeira (2023), the central mountain range includes the highest peaks of the islands (Pico Ruivo at 1,862 m and Pico do Arieiro at 1,818 m) and the highest plateau (Paul da Serra, with an altitude of 1,400 m), along with other prominent peaks. The central and northern parts of the island also host the 'Laurisilva' laurel forest, a UNESCO World Natural Heritage site (WHC, 2023). The island of Madeira has a warm year-round climate due to its proximity to the Gulf Stream and the Canary Current (Barton, 2001; Avelar et al., 2020). Rainfall depends largely on altitude, with the central range catching the moisture of the oceanic currents. The Laurisilva forest absorbs a large amount of water from this rainfall, which then flows into groundwater reserves and, eventually. into an extensive system of human-made irrigation channels known as 'levadas' (Leal et al., 2020).

The physical geography of Porto Santo is very different. The island is flatter, and the average altitude is just 112 m above sea level. The island largely comprises two types of landscape: the mountainous northeast and the coastal plain in the southwest. The 9-km-long 'Beach of Porto Santo', a white sandy beach, which is the best-known landscape feature of the island, is located to the south. The northeast part of the island hosts an area with low-altitude peaks such as Pico do Castelo (437 m), Pico da Juliana (447 m), Pico da Gandaia (499 m) and Pico do Facho (517 m). Spectacular views of the island can be had from its highest points. The relatively low altitude of Port Santo leads to a semi-arid climate with low levels of rainfall. Porto Santo therefore depends on water desalination technology to meet the freshwater needs of its inhabitants. Visitors can here explore one of UNESCO's World Network of Biosphere Reserves, totalling 27,310 hectares of protected area (UNESCO, 2023), as well as several Natura 2000 Network sites.

Largely because of the different tourism offers, the two islands contrast greatly in terms of their tourism dynamics. In principle, therefore, they could be developed and marketed as completely different destinations, each with its own appeal and offer. The island of Madeira has a well-established brand image as a relatively upmarket, year-round tourist destination (Almeida and Garrod, 2022). It is known for its warm climate, spectacular natural landscape and year-round calendar of cultural festivals. While fluctuating in recent decades, tourist numbers remain high (DREM, 2023b). Porto Santo, meanwhile, is much less well-known as a stand-alone tourism destination. Its tourist offer is based on the classic triad of 'sun, sand, and sea' (3S) upon which much mass tourism in Europe has traditionally been built (Ismeri Europa, 2011). It draws a small number of international tourists (around 46,000: authors' own estimate based on length of stay by municipality), most being domestic tourists from the island of Madeira (Duić and Carvalho, 2004). Overall, tourism numbers remain small. The pattern of demand is also highly seasonal, with a strong peak in the summer months and a very small number of arrivals in the winter.

The islands do, however, have similar vulnerabilities in terms of their small market size, which makes economies of scale hard to achieve, combined with being a substantial distance (typically thousands of kilometres) from their main international source markets. Both are heavily dependent upon tourism due to a lack of economic diversification. While the tourism offer on both islands is based heavily on scarce natural assets and fragile ecosystems, neither biodiversity loss nor climate change is a major political issue, particularly on Porto Santo (Assembleia Legislativa, 2020; Visit Madeira, 2023). The lack of attention given to the natural environment in either islands' politics may be explained by the territory not having yet suffered the loss of key ecological assets and historical monuments. A warning sign, however, comes from the growth in the stock of accommodation provided through the sharing-economy, known locally as 'local lodgement': this is leading to a massive increase in the number of beds (DREM, 2023c). The aesthetic appeal of the island of Madeira is still strong on the north coast and rural areas; but it is recognised that both islands must avoid the excessive construction of housing and hotel facilities along the coast (Ismeri Europa, 2011; Rodrigues, 2017).

The shared vulnerability of the two islands can also be seen in their experiences of the Covid-19 pandemic, which followed a similar pattern. As Table 4.1 shows, the islands of Madeira registered a decrease of 64% between 2020 and 2019 in terms of arrivals. The corresponding figure for Porto Santo was 59%. The strong recovery of the number of tourists observed in 2021 was rather similar, with year-on-year growth of around 85% and 91%. For 2022, the number of arrivals in Madeira as a whole increased by a factor of 2, while the number of guests in Porto Santo increased by a factor of 2.3, compared to ten years before.

Figure 4.2, meanwhile, shows that tourist arrivals to the region of Madeira as a whole and to the municipality of Porto Santo experienced a similar pattern. This is the case also during the Covid-19 pandemic, with Porto Santo experiencing a slightly smaller reduction in arrivals during the crisis (58.6% in Porto Santo compared to 64.4% for Madeira as a whole) as well as a similar recovery following the lifting of travel restrictions.

Porto Santo was partly protected from Covid by its domestic tourism markets during the Covid-19 pandemic. In both territories, the share of Portuguese national increased in this period. The share of Portuguese national (including both residents in the mainland and in the archipelago) rose from 12.5% to 17.8% in Madeira and from 50.0% to 59.5% in Porto Santo. Figure 4.3 shows, however, that Porto Santo experienced an increase in its share of arrivals to Madeira over the same period.

Tourism development in Porto Santo and its relationship to Madeira

Since the 1980s, Madeira and Porto Santo have both seen consistent growth in terms of the number of guests and overnight stays. This served to reinforce the region's heavy dependence upon tourism. Most tourists to the region were from a mere handful of origin countries that were well established in the global economic core, including the UK, Germany and France, as well as mainland Portugal. The sector recorded only minor annual fluctuations until 2010 when, on 20 February that year, a major natural disaster struck the region. Heavy rainfall resulted in extensive mudslides and flash floods that affected much of

Table 4.1 Index of tourism arrivals to Madeira (RAM) and Porto Santo (PS) (base = 2012)

	2012	2013	2014	2015	2016	2017	2018	2019	2020	2021	2022
RAM	100.0	109.7	117.5	126.8	146.6	160.3	160.2	158.8	56.6	104.5	203.6
PS	100.0	118.3	138.6	152.5	196.7	217.6	209.9	209.2	86.7	165.5	229.2
RAM	-	9.7%	7.1%	7.9%	15.6%	9.4%	-0.1%	-0.9%	-64.4%	84.6%	94.8%
PS	-	18.3%	17.1%	10.0%	29.0%	10.6%	-3.6%	-0.3%	-58.6%	91.0%	38.5%

Source: Own calculations based on data provided by Statistical Office (DREM, 2023e).

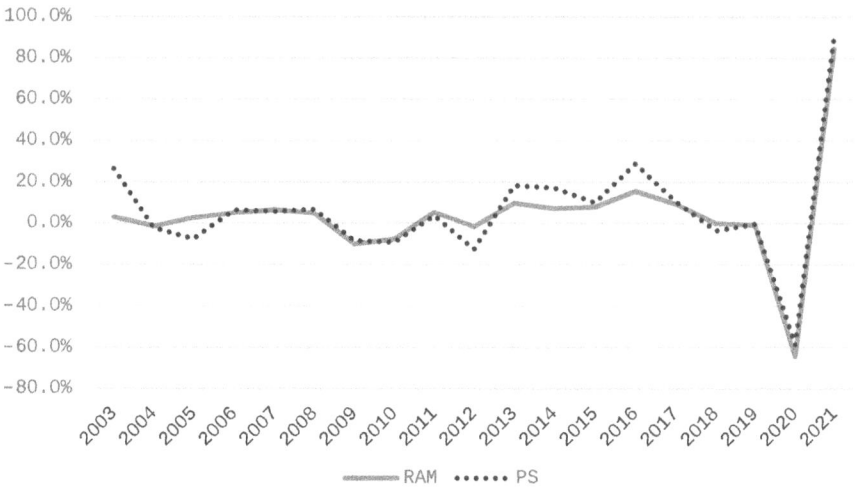

Figure 4.2 Annual growth rate of tourist arrivals in Madeira (RAM) and Porto Santo (PS)

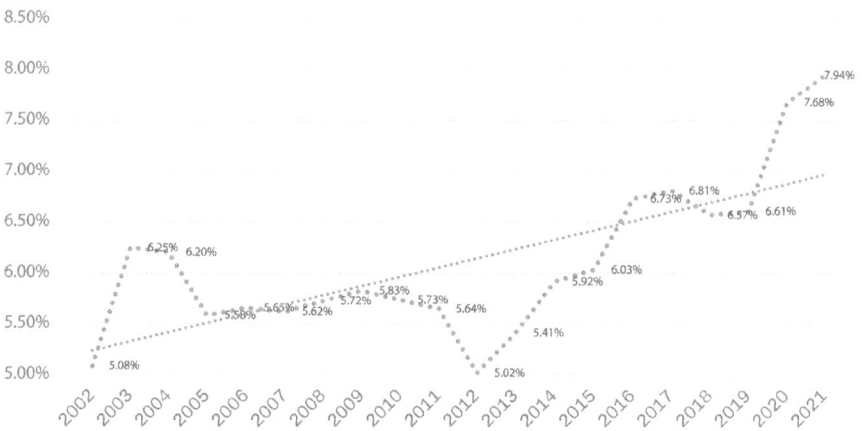

Figure 4.3 Share of Porto Santo in total arrivals to Madeira, 2002–2021

the built-up southern coast of the main island. Much infrastructural damage resulted, including the closing of the airport and many bridges being washed away (Fragoso et al., 2012). Porto Santo was hardly affected by this event. Such episodes serve to remind key players of the potential consequences of a high dependency on tourism. In particular, such islands have "high vulnerability to the erratic and uncertain movements of tourism" (Ismeri Europa, 2011, p. 128), including catastrophic events and the ongoing effects of climate change (Fróis, 2019). Climate change is, for example, associated with an increasing incidence of episodes of disruptive weather at the airport.

Recognising the economic importance of tourism, the regional government prioritised recovery. From 2012 to 2019, the region returned to steady growth, with the number of overnight guests rising from 1 million to 1.6 million for the region as a whole, and from 50,470 to 103,760 specifically for Porto Santo.

Tourism in Porto Santo is less well-established than on the island of Madeira. A report commissioned by the European Union (Ismeri Europa, 2011, p. 135) describes Porto Santo as "a newcomer to tourism ... with small hotel capacity and offers a different product – sea, sun and sand (yellow sand, contrary to Madeira)". Porto Santo accounts for just 5.4% of the land area in the Madeira region as a whole, 4.7% of the total number of establishments, 7.3% of the total number of rooms and 8.4% of the total accommodation capacity. Funchal, the main and capital city of the main island and the region, accounts for two thirds of the supply of accommodation; while Porto Santo ranks as the fourth best-endowed municipality (after Funchal, Calheta, and Santa Cruz).

Table 4.2 provides data on overnight stays in tourism accommodation, by municipality and by country of origin. Porto Santo is clearly more dependent on the Portuguese market (almost 60% of all overnights recorded in 2019), and which includes large numbers of visitors from Madeira Island (Opção Turismo, 2020). The second largest origin market for Porto Santo is UK nationals followed by Danish nationals, which can be explained by specific agreements made with tour operators to open new routes fed by chartered flights.

Data for the region of Madeira indicate that there are around 190 hotel establishments with a total capacity of 37,000 beds. This corresponds to 10% of the total national supply of beds. The region has a higher-than-average capacity in 4- and 5-star hotels, which is a key feature of the tourism industry of Madeira. Around 63% of the region's accommodation capacity is in Funchal. In most recent years, alternative and cheaper types of accommodation (local lodgement) have been developing rapidly, as is also happening in mainland Portugal. Data for 2020 indicates a total of 401 establishments in the tourism sector in the Madeira region, of which 207 (52%) are categorised as "local lodgement" according to data released by the Statistical Office. The impact of this market niche in the total number of guests is around 20%. The tourism sector started developing rapidly since the opening of the main airport, with Funchal and surroundings areas as the key areas in terms of tourism development. Tourism dynamics perform well in the rural hinterland; but Funchal still attracts around two thirds of the demand.

Figure 4.4 shows that the issue of seasonality is much more pronounced in Porto Santo. Around 79% of the guests arrive between May and October; the corresponding figure for Madeira is 59%. Meanwhile, around 39% of the visitors arrive to Madeira in the summer months (June to September); the corresponding figure for Porto Santo is 59%.

In recent years, the economy of Porto Santo has benefited from an improvement in the economic related indicators of the sector (DREM, 2023a). However, the impact of seasonality in the generation of income is evident: around 80% of the revenue per available room (RevPAR) is generated in the peak season (DREM, 2023a).

Table 4.2 Overnights by country of origin and municipality (2019)

Country	Madeira (RAM) (%)	Calheta (%)	Câmara de Lobos (%)	Funchal (%)	Machico (%)	Ponta do Sol (%)	Porto Moniz (%)	Ribeira Brava (%)	Santa Cruz (%)	Santana (%)	São Vicente (%)	Porto Santo (%)
Total	100.0	100.0	100.0	100.0	100.0	100.0	100.0	100.0	100.0	100.0	100.0	100.0
Portugal	17.8	11.6	12.7	16.8	13.5	10.4	16.5	11.4	12.1	11.3	15.4	59.5
Germany	20.2	33.9	18.0	16.8	15.3	38.2	22.5	33.9	33.9	34.5	19.4	7.1
Austria	1.2	1.4	1.0	1.1	1.2	2.3	1.7	0.9	1.9	1.5	0.9	0.4
Belgium	1.7	2.3	1.7	1.6	2.4	2.6	3.0	2.7	2.1	4.2	2.0	0.2
Denmark	2.0	0.7	0.7	1.9	0.6	2.3	0.8	0.7	0.7	0.3	0.5	11.0
Spain	2.3	1.5	2.6	2.5	2.4	1.6	2.6	3.5	2.2	3.2	2.5	0.5
Finland	2.0	0.5	0.5	2.8	0.5	0.8	0.4	0.3	0.8	0.3	0.1	0.0
France	7.1	11.0	10.8	5.9	19.0	9.9	17.4	14.6	6.9	14.4	13.4	1.1
Italy	1.2	0.9	1.1	1.1	1.8	1.2	2.0	1.5	0.9	1.4	0.9	2.3
Netherlands	3.6	6.1	5.7	3.0	3.3	4.3	4.7	4.2	5.4	8.5	4.6	0.4
Poland	4.9	3.3	2.5	4.7	10.9	3.3	3.4	4.7	7.0	3.5	15.8	0.5
UK	21.3	13.7	30.4	26.1	8.4	7.5	5.9	6.9	12.0	4.4	6.2	14.1
Sweden	1.2	0.5	0.4	1.6	0.4	1.1	0.7	0.5	0.9	0.3	0.3	0.2
Norway	0.9	0.3	0.4	1.2	0.6	0.4	0.6	0.6	0.3	0.1	0.1	0.1

Source: Own calculations based on data from Statistical Office (2023).

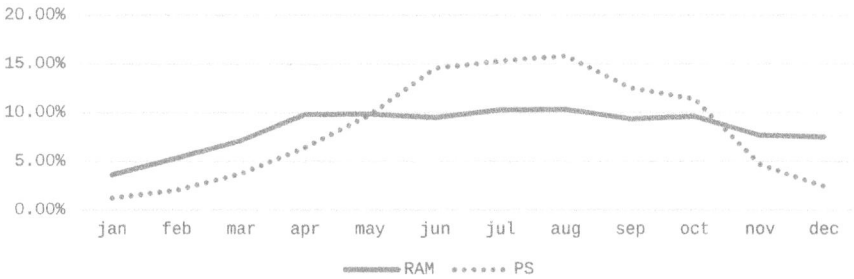

Figure 4.4 Guest per month in Madeira (RAM) and Porto Santo (PS)
Source: Own calculations based on data released by Statistical Office (DREM, 2023c).

In economic terms, Porto Santo exhibits a purchasing power (a proxy for the standard of living) higher than the average for the region. The local value added is generated mainly by the service sector; and by public administration in particular. Data on employment suggests that unemployment in Porto Santo is slightly higher than the region (12.5% vs. 12.1%) but there is a slightly better activity rate (54.6% vs. 52.3%). The total labour force in Porto Santo comprises around 2,000 persons. The primary sector has fallen from 700 persons in 1960 to just 24 in 2021; and those working in the secondary sector increased somewhat from 124 in 1960 to 341 in 2011 and then declined to 279 in 2021. The tertiary sector, meanwhile, increased from 198 individuals in 1960 to 1944 in 2011 but fell to 877 in 2021, during the coronavirus pandemic lockdown (DREM, 2023d).

In terms of the impact of the tourism sector in the local economy, a useful approximation can be made by calculating the share of the total expenditure attributable to Porto Santo based on the region's Tourism Satellite Account, which contains data for 2019 (DREM, 2022). According to this, tourism expenditure in Madeira totalled €1.26 billion. The share of Porto Santo in terms of the total of overnights was in the range of 5.8%. Taking the 2019 share of overnights recorded in the hotel establishments in Porto Santo as a basis, the impact of total expenditure on the municipality can be estimated at €73.5 million. In terms of accommodation capacity, data for 2022 shows 19 establishments out of a total of 401 (4.7%), and 1,247 rooms (out of a total of 16,902). The data on the combined capacity suggest that Porto Santo has 3,172 tourist accommodation units out of the total of 37,667 for the region of Madeira (DREM, 2023e).

Dynamics of tourism development in Porto Santo

Tourism in Porto Santo has been historically reliant on the local and national markets, those being Madeira and mainland Portugal respectively. This has resulted in Porto Santo often being found in a 'double marginalised' position: first by Portugal along with the main island of Madeira, and second by the island of Madeira itself. Data concerning 2002, when the first statistics at the county level started to be released by the Statistical Office, suggest that the Portuguese market then accounted for 71% of the total. The Portuguese market still accounted for 60% in 2022. This suggests that a small-but-significant geographical diversification of origin markets has taken place in the last 20 years, offering prospects for the island to take its own place on the world tourism stage, and join Madeira Island as an equal partner rather than as a subordinate. The success of this transition will, however, depend on the adoption of suitable product development and destination marketing strategies: government functions that are held at the regional rather than municipality level (and hence by Madeira Island rather than Porto Santo).

The region's destination marketing authorities are, indeed, beginning to attempt to stress the contrasting geography and tourism offer of Porto Santo (in relation to Madeira Island) in their promotional campaigns. Porto Santo is increasingly being portrayed in the international level as a complement to Madeira, offering tourists a place to wind down and relax. Tourist impressions that the smallness of the island means there are few facilities and little to do are being countered with images of tranquillity, intimacy, escapism and transcendence, and prospects of freedom for those who need time to think, relax, rebuild, and disconnect from everyday life (Baldacchino, 2012; Baum, 1997). The Porto Santo beach was elected by "European Best Destinations" as the best beach in Europe in 2022 (Visit Madeira, 2023). Apart from UNESCO's World Network of Biosphere Reserves as well as Natura 2000 Network areas, visitors can enjoy horse-riding, golf, whale-watching, diving, stand-up paddle-boarding, windsurfing, kayaking, walking the forest paths, and wellness-related therapies in spas (Visit Madeira, 2023). These are activities that are not well-developed on the main island.

As with much destination marketing, there is an element of hyperbole to the promotion of Porto Santo as a tourism resort. The Visit Madeira website advertises Porto Santo as a 'veritable golden oasis suspended in the Atlantic Ocean'. The website of one of the local hotel groups (Barceló Hotel Group, 2023), meanwhile, claims that:

> [T]he fine sand found on the beaches ... has special properties and is even used as a treatment for rheumatic and mobility conditions. The waters around the island contain iodine, magnesium, calcium and more – elements that are superb in aiding recovery from periods of stress.
>
> (Visit Madeira, 2023)

Porto Santo also has an interesting historical heritage to offer, related to its links to Christopher Columbus and the 'discovery' of the 'New World'. Populated 600 years ago by settlers from Madeira, the island has a rich history that links back both to the historical development of Madeira and local events related to major changes in the landscape and ecosystems, changes in agricultural practices and pirate raids.

Tourism development in Porto Santo, with its 5,000 inhabitants, means that residents, albeit unwillingly, must accept that the local population increases threefold in the summer period. For residents, tourism provides a few hundred permanent jobs, substantial earnings for a small number of local entrepreneurs in the areas of local lodging and restaurants, as well as some additional income for local farmers. Given that the local number of inhabitants accounts for just 2.1% of the total population of the region, it should not come as a surprise that the tourist dynamics of Porto Santo are controlled largely by Madeiran entrepreneurs, who own the key local industry assets (DREM, 2023d, 2023e). Some residents may resent the way that Madeira's key players adopt a patronising attitude, and 'exploit' their island by buying plots of land to build second homes as a status symbol. Nevertheless, a brief analysis of news items about Porto Santo in the local newspapers suggests that most residents accommodate to the cultural, social, and economic dominance practised by their larger island neighbour. In general, people from Porto Santo regard tourism positively because they have no other alternative, except for a small chance of getting a well-paying job in public administration on the island. Nor does there seem any great surprise or concern that the most recent regional development plan asserts that "Porto Santo is a constituent part of Madeira, as a touristic destination" (Assembleia Legislativa, 2020, p. 65).

The tourism sector of Madeira has often been analysed for the purpose of diagnostic analysis and strategic planning (ACIF, 2015; Publituris, 2021). The most recent tourism development plan, 'The Madeira Tourism Strategy 2022–2027' (Publituris, 2021), identifies several strategic pillars for the development of Porto Santo based on high value and sustainable market niches. The guiding drivers of the strategic plan, which apply to the region as a whole, include: (1) strengthening the destination's management, improving knowledge and monitoring the performance of the tourism sector; (2) reinforcing the degree of diversity, differentiation and sophistication of the tourist offer; (3) investing in a reinforced image abroad and destination's visibility; (4) attracting, qualifying and valuing the human resources of the sector; (5) stimulating private investment; and (6) ensuring a high degree of sustainability of the destination (this being cultural, environmental, economic, and social). These goals give some assurance to the people of Porto Santo that their island is a valued constituent part of Madeira's tourism offer. The moot question, however, is who exactly will benefit from this subordinate positioning.

Indeed, Porto Santo is only a municipality from an administrative point of view: the interests of the Madeira's regional government are represented in Porto Santo by the 'Public Administration Coordinator's Office in Porto Santo', which clearly indicates that the policymaking and instruments that have an impact on Porto Santo are determined by politicians in the capital city. Moreover, the 47 MPs in the Madeiran regional parliament are elected in a single constituency via proportional representation, based on the D'Hondt method. Since September 2023, only one MP from Porto Santo (affiliated to the ruling party) is represented in this regional assembly. In terms of the total town council expenditure, the share of Porto Santo averages 3.1% over the period 2009–2019. The corresponding figures for the share of Porto Santo in terms of geographical area and population of the archipelago are 5.4% and 2.1%, respectively.

The Plan for Economic and Social Development (PESD) of the Autonomous Region of Madeira 2020–2030 acknowledges that Porto Santo should be given special attention at this stage of development, owing to the problem of 'double insularity' (Assembleia Legislativa, 2020). The PESD acknowledged that island's limited size constraints a wide range of economic activities, which is to say that the island's status in terms of welfare is not currently economically sustainable without financial support from Funchal, Lisbon, and the EU. The tourism sector is considered the only economically competitive one. The PESD document assigns Porto Santo just a few clearly identifiable roles, those being: (1) the promotion of the sun-sand-sea product as a complementary segment to the archipelago main product; (2) the main centre for a contingency and recovery plan covering emergency and abnormal situations that may occur at the Funchal airport in Madeira owing to disruptive weather, based on a reinforced link by ferry between both islands to transport tourists, (3) a smart fossil-fuel-free island where the renewable energy sector can be developed; and (4) to harness the island's natural resources through the development of the 'blue' and 'circular' economies.

Conclusion

As on other island territories, tourism is accorded a high priority by the regional government of Madeira. This is in recognition of its vital role as a strategic engine of economic growth, in the generation of employment, as a driver of infrastructural development and in the creation of new firms. The region's degree of dependence on tourism is high: in terms of expenditure, the tourism sector accounts for 25% of its GDP, 17% of its employment, and 16% of its GVA. Porto Santo's high reliance on Madeira – as a source of both summer tourists as well as of strategic planning and marketing – suggests a situation in which the former is in a position of 'double dependency' on the latter. Such a relationship makes it difficult for the tourism potential of Porto Santo to be fully achieved. Indeed, the development of tourism in Porto Santo is more likely to be shaped by the needs of Madeira than its own.

An important question to be considered is whether this relationship between the two islands is an immutable one and, if so, whether the key stakeholders in Porto Santo should be concerned about this. The regional government wants to reinforce the sustainable and high-quality pillar of tourism development in Madeira and Porto Santo; to spread the benefits of tourism away from Funchal and its neighbourhood, and towards the north coast, the rural areas, and Porto Santo. This would help sustain a high combined rate of economic growth, based on better and well-paid jobs and a greater appreciation of the identity, culture, and history of the islands. Even in a disadvantaged position, or resigned sub-ordination, in comparison to other territories, it can be argued that most residents of Porto Santo will welcome further development on this basis.

The other option is to bring Porto Santo into the strategy as a full partner, harnessing the advantages of its double insularity to attract tourists not only from Madeira and the mainland of Portugal but also from other international markets. The growing diversity in the origins of its inbound tourists suggest that this is possible. If Porto Santo can find a way to better marshal its distinctiveness to its advantage, it may be able to form a more equal partnership with the island of Madeira. This will require an intentional strategy of product development and destination marketing. Indeed, Porto Santo already has a very different tourism offer to that of the island of Madeira. It will be important for Porto Santo to maintain and build on this difference, rather than to turn into a smaller, still dependent microcosm of Madeira. The difficulty is, of course, that Porto Santo is politically a mere municipality of the region of Madeira, with tourism strategy and policy being set in Funchal, primarily with the interests of the region as a whole in mind. Porto Santo represents something of a footnote in most strategic thinking.

Porto Santo thus remains at a fork in the road in terms of the development of its true tourism potential. Unlike many smaller members in island pairs, it has the unique tourism offer that would allow it to come out from the shadow of its larger sibling. These attractions are, perhaps somewhat ironically, to some extent the consequences of its double insularity. Escaping its doubly marginalised situation is, however, largely at the discretion of its regional government, where, because of its small size and underdeveloped character, Porto Santo has limited influence. The small island's future depends on whether the authorities in Funchal recognise Porto Santo's potential and are willing to adopt an intentional strategy to harness it to best effect.

References

ACIF (2015). *Documento Estratégico para o Turismo na RAM (2015–2020)*. www.turismodeportugal.pt/SiteCollectionDocuments/estrategia/Estrategias-Regionais-Madeira/Documento-Estrategico-Turismo-Madeira-2015-2020.pdf.

Agius, K. and Chaperon, S. (2023). The dependency-autonomy paradox: A core-periphery analysis of tourism development in Mediterranean archipelagos. *International Journal of Tourism Research*, 25(5), 506–516.

Almeida, A. and Garrod, B. (2022). Determinants of visitors' expenditure across a portfolio of events, *Tourism Economics*, 28(8), 2099–2125.

Amoamo, M. (2018). More thoughts on core-periphery and tourism: Brexit and the UK Overseas Territories, *Tourism Recreation Research*, 43(3), 289–304, doi:10.1080/02508281.2018.1455015.

Assembleia Legislativa (2020). Plano de Desenvolvimento Económico e Social da Região Autónoma da Madeira 2030.

Avelar, D., Garrett, P., Ulm, F., Hobson, P., and Penha-Lopes, G. (2020). Ecological complexity effects on thermal signature of different Madeira island ecosystems, *Ecological Complexity*, 43, 100837, doi:10.1016/j.ecocom.2020.100837.

Baldacchino, G. (2012). The lure of the island: A spatial analysis of power relations. *Journal of Marine and Island Cultures*, 1(2), 55–62.

Baldacchino, G. (2014). Small island states: Vulnerable, resilient, doggedly perseverant or cleverly opportunistic? *Études Caribbéennes*, 27–28. https://journals.openedition. org/etudescaribeennes/6984.

Barceló Hotel Group (2023). *Porto Santo: Madeira's finest beached.* www.barcelo.com/ guia-turismo/en/portugal/madeira/things-to-do/porto-santo/.

Barton, E. (2001). *Canary and Portugal currents.* https://digital.csic.es/bitstream/10261/ 26951/1/ms0360_final.pdf.

Baum, T. (1997). The fascination of islands: A tourist perspective. In D. G. Lockhart and D. Drakakis-Smith (Eds), *Island tourism: Trends and prospects* (pp. 21–35). London: Pinter.

Bianchi, R. V. (2002). Towards a new political economy of global tourism. In R. Sharpley and D. Telfer (Eds) *Tourism and development: Concepts and issues* (pp. 265–299). Bristol: Channel View Publications.

Boissevain, J. (1979). The impact of tourism on a dependent island: Gozo, Malta. *Annals of Tourism Research*, 6(1), 76–90. doi:10.1016/0160-7383(79)90096-3.

Butler, R. (2017). Thoughts on core–periphery and small island tourism. *Tourism Recreation Research*, 42(4), 537–539. doi:10.1080/02508281.2017.1359958.

Chaperon, S. and Bramwell, B. (2023). Dependency and agency in peripheral tourism development. *Annals of Tourism Research*, 40, 132–154. doi:10.1026/j. annals.2012.08.003.

DREM (2022), Conta satélite do turismo 2019. https://estatistica.madeira.gov.pt/ download-now/economica/contaseconomicas-pt/contaseconomicas-cst-pt/contasec onomicas-cst-quadros-pt.html.

DREM (2023a). Estatísticas dos transportes da Região Autónoma da Madeira, 2022. https://estatistica.madeira.gov.pt/.

DREM (2023b). Análise dos principais resultados definitivos. https://estatistica.madeira. gov.pt/.

DREM (2023c). Estimativa rápida – Em setembro de 2023, o alojamento turístico na Região Autónoma da Madeira registou um crescimento homólogo de 9.4% nas dormidas. https://estatistica.madeira.gov.pt/.

DREM (2023d). Série retrospetiva trimestral do inquérito ao emprego da Região Autónoma da Madeira, 2011–2023 (série 2021 revista). https://estatistica.madeira. gov.pt/.

DREM (2023e). Série retrospetiva das estatísticas do turismo (1976–2022). https://esta tistica.madeira.gov.pt/download-now/economica/turismo-pt/turismo-serie-pt/turism o-series-longas-pt.html.

DREM (2023f). População Residente nos Recenseamentos de 1864 a 2021, por Freguesia. https://estatistica.madeira.gov.pt/download-now/social/popcondsoc-p t/popcondsoc-censos-pt/popcondsoc-censos-serie-pt.html.

Duić, N. and Carvalho, M., (2004). Increasing renewable energy sources in island energy supply: Case study Porto Santo. *Renewable and Sustainable Energy Reviews*, 8(4), 383–399. doi:10.1016/j.rser.2003.11.004.

Fróis, L. H. C. (2019). *The Madeira tip-jet and variability of extratropical water vapor pumping in the Atlantic basin*, Master's Thesis. https://repositorio.ul.pt/bitstream/ 10451/40472/1/ulfc125161_tm_Lu%C3%ADs_Fr%C3%B3is.pdf.

Fragoso, M., Trigo, R. M., Pinto, J. G., Lopes, S., Lopes, A., Ulbrich, S., and Magro, C. (2012). The 20 February 2010 Madeira flash-floods: Synoptic analysis and extreme rainfall assessment. *Natura. Hazards & Earth System Sciences*, 12, 715–730. doi:10.5194/nhess-12-715-2012.

Harrigan, N. (1974). The legacy of Caribbean history and tourism. *Annals of Tourism Research*, 2(1), 13–25.

Ismeri Europa (2011). *Growth factors in the Outermost Regions*, Final Report Vol. II, CONTRACT N°2009.CE.16.0.AT.101. www3.gobiernodecanarias.org/aciisi/ris3/docum entos/otros/rup/13-estudio-de-los-factores-de-crecimiento-en-las-rup-vol-1-ingles/file.

Leal, M., Fragoso, M., Lopes, S., and Reis, E. (2020). Material damage caused by high-magnitude rainfall based on insurance data: Comparing two flooding events in the Lisbon Metropolitan Area and Madeira Island, Portugal, *International Journal of Disaster Risk Reduction*, 51, 101806. doi:10.1016/j.ijdrr.2020.101806.

McLeod, M., Lewis, E., and Spencer, A. (2017). Re-inventing, revolutionizing and transforming Caribbean tourism: Multi-country regional institutions and a research agenda. *Journal of Destination Marketing & Management*, 6(1), 1–4. doi:10.1016/j.jdmm.2016.08.009.

Opção Turismo (2020). *Madeirenses não deixam cair turismo no Porto Santo.* [Madeirans do not let tourism drop in Porto Santo.] https://opcaoturismo.pt/wp/ma deirenses-nao-deixam-cair-o-turismo-no-porto-santo/.

Oppermann, M. (1993). Tourism space in developing countries, *Annals of Tourism Research*, 20(3), 535–556. https://doi.org/10.1016/0160-7383(93)90008-Q.

Osagie, I. and Buzinde, C., (2011). Culture and postcolonial resistance: Antigua in Kincaid's A Small Place, *Annals of Tourism Research*, 38(1), 210–230. https://doi. org/10.1016/j.annals.2010.08.004.

Pleasant, T. and Spalding, A. (2021). Development and dependency in the periphery: From bananas to tourism in Bocas del Toro, Panama, *World Development Perspectives*, 24, 100363. doi:10.1016/j.wdp.2021.100363.

Publituris (2021). *Madeira coloca estratégia para o turismo 2022–2027 em consulta pública.* www.publituris.pt/2021/12/29/madeira-coloca-estrategia-para-o-turismo-2022-2027-em-consulta-publica.

Rodrigues, J. (2017). *Land and urban management of Madeira island: the relevance of Funchal in this process.* Estudo Prévio 14. Lisboa: Centro de Estudos de Arquitetura, Cidade e Território da Universidade Autónoma de Lisboa. www.estudoprevio.net.

Sharpley, R. (2002). Tourism: A vehicle for development. In R. Sharpley and D. Telfer (Eds), *Tourism and development: Concepts and issues* (pp. 11–34). Bristol: Channel View Publications.

Sultana, T. (2018). *Residents' perceptions and attitudes towards tourism in Malta*, Master's degree dissertation. Malta: University of Malta. www.um.edu.mt/library/oar/handle/123456789/39254.

Turner, L. and Ash, J. (1975). *The golden hordes: International tourism and the pleasure periphery.* New York: St. Martin's Press.

UNESCO (2023). *Porto Santo Biosphere Reserve.* https://en.unesco.org/biosphere/eu-na/porto-santo#:~:text=The%20biosphere%20reserve%2C%20located%20in,are %20exclusive%20to%20Porto%20Santo.

Visit Madeira (2023). *Madeira welcomes you.* www.visitmadeira.com/.

Weaver, D. B. (1998). Peripheries of the periphery: Tourism in Tobago and Barbuda, *Annals of Tourism Research*, 25(2), 292–313. doi:10.1016/S0160-7383(97)00094-7.

Weaver, D. B. (2017). Core–periphery relationships and the sustainability paradox of small island tourism. *Tourism Recreation Research*, 42(1), 11–21.

WHC (2023). *Laurisilva*, Madeira. UNESCO World Heritage Convention. https://whc. unesco.org/en/list/934/.

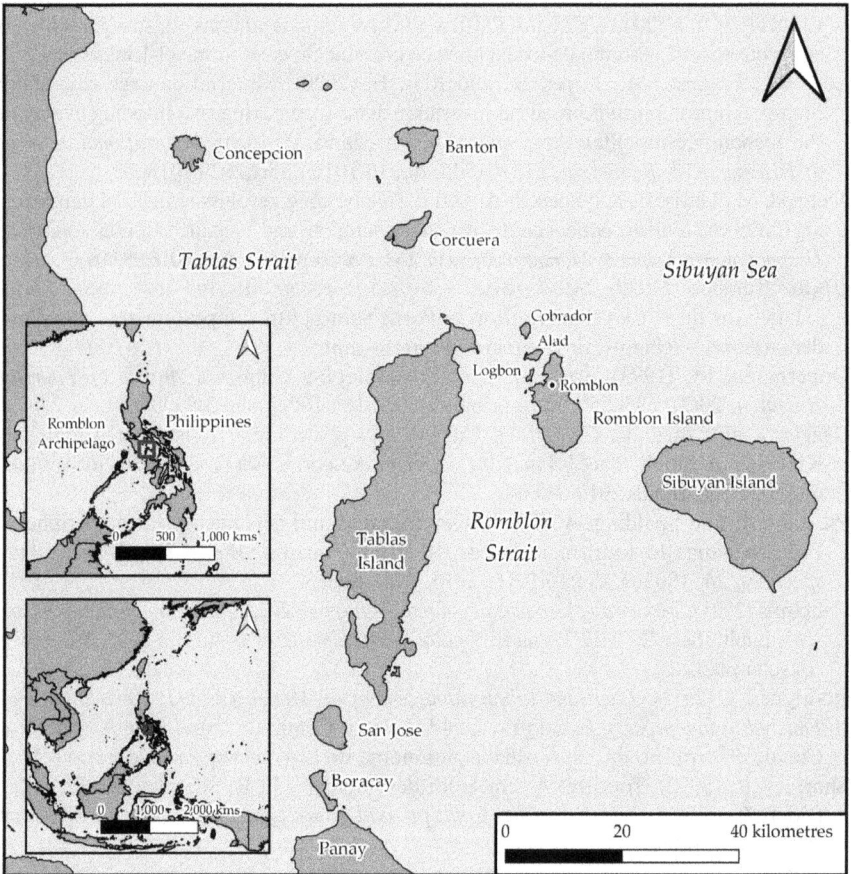

Figure 5.1 Romblon archipelago (including Tablas, Sibuyan, San Jose), The Philippines

5 Tranquillity and exclusivity

Archipelago tourism after the pandemic in Romblon, The Philippines

Joefe B. Santarita

Introduction

The Philippines is a Southeast Asian nation with a rapidly growing population and is the world's third largest archipelago country (after Indonesia and Japan). Tourism is a development option that has been actively pursued in the Philippines since the 1970s and is now a pillar of the national economy, responsible for some 5 million jobs. International travellers, keen on beach tourism, have been increasing; but domestic tourism continues to outpace international arrivals. The country has a huge archipelagic diversity – it officially has 7,641 islands, broadly categorised in three main geographical divisions from north to south: Luzon, Visayas and Mindanao (Santarita, 2020). And yet, The Philippines has yet to reach its potential as an international tourism destination (Henderson 2011; Maguigad, 2013).

The Philippines was not spared from the Covid-19 virus. The pandemic claimed 66,660 lives and over four million Filipinos were officially infected over almost three years (Department of Health, August 2023). All forms of social and economic life were disrupted, including the tourism industry.

The Philippines started the shutdown of all airports on the main island of Luzon on 17 March 2020 in an effort to slow down the spread of Covid-19. Outbound passengers were given 72 hours to fly out of Luzon, including Metro Manila. Inbound international passengers in transit were allowed entry subject to applicable quarantine procedures if coming from countries with existing travel restrictions. Over the following nervous months, the government gradually re-opened its ports and airports to travellers and Filipino workers returning from overseas. This move, however, was not enough to revive the declining number of travellers in the country. As the pandemic weeks dragged into months and years, the dire situation placed the Philippine economy in a serious quandary.

Romblon is geographically situated in the heart of the Philippine archipelago; but that does not make it central. With its wide expanses of rainforest, the province, and Sibuyan Island in particular, is one of the few places in the Philippines with a well-preserved natural environment. It is also home to the country's cleanest inland body of water, the Cantigas River, as well as 34

DOI: 10.4324/9781003451037-7

waterfalls. Sibuyan is domestically referred to as 'the Galapagos of Asia' because of its endemic flora and fauna (Licap, 2013). Apart from its rich biodiversity, the island includes: Cresta de Gallo (literally, the comb or crest on a chicken's head) a remote islet with a stunning sandbar; Mount Guiting-Guiting, with its challenging climbs; the clean Cantingas River; and the exhilarating Busay Falls (San Fernando Tourism Office, 2023). It is claimed that it is Cresta de Gallo's remoteness that makes the vibe quite unique to an ordinary beach (Groves, 2023).

In the case of Romblon, the number of pre-pandemic tourists began to decline even as early as 2019. The seven-month, temporary closure of Boracay Island – located to its immediate south – for six months may have greatly affected travellers' intent to visit the province (Canoy et al., 2020). The final blow, however, was delivered by the pandemic: the lowest number of visitors ever recorded was in 2020, with less than 1,000 combined foreign and domestic travellers. The number of tourists recovered somewhat in 2021 and amazingly grew almost ten-fold the following year (see Table 5.1).

The increase reflects the 'bouncing back' phenomenon in the tourism industry of the whole country. It may have been supported and enhanced by the launch of the "Love the Philippines" campaign by the Department of Tourism (DoT) of the Philippines in early 2023. Such a slogan reflects the serious intentions of the Philippine government to quickly recover from the tourism hiatus after more than two years of unforeseen quarantine measures and community lockdowns. The campaign is addressed at the 'changed traveller' who is looking not just for fun but for memorable and unique experiences (Doctolero, 2023). 'Love', after all, can capture multiple sensations, including country, history, people, beaches, cuisine, and the creative arts (Rocamora, 2023).

The Philippine tourism industry's stakeholders have taken various initiatives to help secure the industry's immediate resuscitation. One of these is a welcome diversification of tourist flows. The Philippine Travel Agencies Association has emphasised the need to actively and deliberately advertise less-well-known domestic destinations other than 'the usual suspects' such as Manila, Baguio, Cebu and Davao, in order to cater for all types of travellers, regardless of their preferences and budget. And, with a population of over

Table 5.1 Travellers to Romblon, 2017–2022

Year	Foreign Travellers	Overseas Filipinos	Domestic Travellers	Total
2017	7,347	2,780	53,009	63,136
2018	1,158	2	43,983	45,143
2019	800	0	3,335	4,135
2020	488	0	479	967
2021	15	0	1,326	1,341
2022	426	0	10,834	11,260

Source: Department of Tourism, Philippines (2022).

110 million, domestic tourism is a powerful economic stimulus and cannot be ignored: 'word of mouth advertising' was also encouraged, to persuade kith and kin to travel to one's preferred tourist destinations in their own country (Ignacio, 2023).

These policies and strategies of recovery aligned nicely with those of travellers trying to 'make up for lost time' during the pandemic by organising trips to family reunions, revisiting favourite vacation spots, embarking on splurge vacations or simply spending time away from home. Industry players are hoping that the re-opening of the economy and the 'revenge travel' phenomenon will lead tourism to a great rebound (Manalang, 2023). Such hope was also expressed by President Ferdinand Marcos Jr in his second 'State of the Nation Address' (SONA) in July 2023.

In addition, several transportation companies relaunched their company's services by implementing fleet expansion and service enhancements to both domestic and international routes. Philippine Airlines, for instance, purchased nine Airbus A350–1000 long-range jetliners and restored flights to both 32 domestic destinations as well as non-stop options to North America, Japan, West Asia, and Australia (Philippine Airlines, 2023). On top of this, there are new, small players in the airline industry whose intention is to capitalise on the needs of those destinations that are not normally serviced by the larger airline companies. One of them is AirSWIFT, a regional boutique airline, which services nine domestic scheduled destinations including Tugdan Airport (IATA code: TBH), on Tablas Island, and the only commercial airport in Romblon province.

All these initiatives (recovery, revenge, and relaunch) have contributed to the 'bouncing back' of tourism lately all over the country. And yet, despite such initiatives, the very geography, politics, and logistics that relate to Romblon have contributed to the maintenance of tranquil and uncrowded beaches and tourist sites in the province. These factors have somehow shaped the decision of various stakeholders to forego short-term but fickle gains and instead embrace and capitalise on tranquillity and seclusion as a distinct branding of Romblon in the post-pandemic era.

Romblon: The underrated neighbour

Romblon as a province is unfairly seen in many instances as a free-rider to the popularity of Boracay; an underrated neighbour of sorts. While that is partly true in the case of Carabao Island on its southern tip, and the immediate neighbour to Boracay, Romblon has its own distinct charms that appeal greatly to a particular type of traveller. Tranquillity and seclusion nowadays have evolved as brand components in their own right. The visibility of Romblon to travellers has been enhanced during the pandemic where its isolation, uncrowded beaches and communities, and largely pristine nature made it a more desirable locale to escape to, away from the virus, from the anxiety of numerous quarantine restrictions and lockdowns, and from the malaise of urban modernity generally.

Romblon is an archipelagic province located in the southwestern Tagalog region. The province comprises of 17 municipalities and 219 barangays. The main islands include Tablas (the largest, which covers nine municipalities), Sibuyan with its three towns, and the smaller island municipalities of Corcuera, Banton, Concepcion, San Jose, as well as Romblon, the provincial capital (MIMAROPA-DILG, 2023). It has a total land area of some 1,540 km^2. Most of its residents still depend on agriculture (mainly rice and copra) and fishing for their livelihoods; while marble quarrying and processing are the major economic activities in the province (MIMAROPA-NEDA, 2023). Romblon is also home to interesting mountain ranges, impeccable beaches, appealing coral reefs, amazing historical sites, as well as natural stone formations and landscapes which qualify the area as a prime ecological tourism (eco-tourism) destination (Ramilo, 2022).

Romblon lies at a considerable distance from the national capital, Manila; and, in spite of its beguiling geographical centrality in the country, it remains relatively isolated in all directions: it lies west of Mindoro, south of Marinduque, east of Masbate, and north of Panay. As a result, its people face perennial problems that might not be common in the other provinces on the larger islands. One of these relates to inter-island transportation and communication issues (Esquejo, 2018), about which more below.

At just over 300,000, the population of the province is relatively small compared to most of the other 81 provinces in the country; and this has political and developmental implications for Romblon. In addition, several residents have moved and are also working in Luzon, Panay, or abroad, looking for greener pastures, which subsequently lessens the number of voters during elections. Tablas island has almost 200,000 residents; while the rest are scattered in much smaller communities (Table 5.2). With such a small constituency and electorate, politicians at the national level tend not to give much attention to the development of hard and soft infrastructures of the province, and much less so to the islands with small voting populations. If there are aspirations for monumental projects such as the construction and maintenance of ports, airports, and other forms of connectivity, these tend to be allocated meagre funding by the national government, leading to their shelving and postponement.

Table 5.2 Population of islands in the province of Romblon, 2020

Tablas	*174,447*
Sibuyan	62,815
Romblon	40,554
San Jose	11,759
Corcuera	10,112
Concepcion	3,561
Banton	5,737
TOTAL	**308,985**

Source: Data generated from www.philatlas.com.

Demography has implications on the decisions of politicians, the distribution of developmental projects, and even on the plan to shift the capital from Romblon to Tablas Island, where the only commercial airport in the province is located (as has already been attempted in the 1970s). From the administrative point of view, the municipality of Romblon is 'the mainland' in relation to the other islands; and it serves as the religious and political capital of the province. The regional offices of various national agencies are also located here. However, Odiongan has also developed into an urban zone on nearby Tablas Island, where it is also the site of the main campus of Romblon State University, Romblon Provincial Hospital, and the Philippine Science High School-MIMAROPA campus. Odiongan is also the base of the provincial extension office (Office of the Governor, Health Office) as well as of the regional and satellite offices of Tourism, Social Security, Health, Social Welfare and Development, Technical Education Skills and Development, Agriculture, Environment and Natural Resources, Disaster Risk Reduction and Management, Land Transportation, Public Works and Highways, Science and Technology, Labour and Employment, and many more (Villegas, personal communication). Thus, residents as well as visitors to Tablas and to other islands that are logistically connected to it can transact business with various government offices in Odiongan without the hassle of several hour ship rides to the provincial capital in Romblon Island. In case it is inevitable to go to the capital, the satellite office can fast track the processing of papers and consequently shorten one's stay in Romblon. Economically, it is more advanced than the capital town, having more banks, a Jollibee fast food branch and its own main port with ferry services to Panay, Mindoro, and Batangas (Esquejo, personal communication). Its demographic and economic heft as well as its connectivity options have made Tablas the provincial 'mainland', marginalising smaller Romblon Island.

Another factor in preserving tranquillity and seclusion of the island is the lack of appropriate avenues of connectivity. With just one airport, no inter-island commercial travel is possible by air within the province. The only option is ferry transport; and here, Tablas and Romblon islands act as the respective hubs to the rest of the archipelago, and the selective transport links a spatial representation of core-periphery relations within Ramblon province, including nested hierarchies of insularity: see Figure 5.2, inspired by Baldacchino and Ferreira (2013). It takes over eight hours by ferries (Montenegro, Starlite Ferry, 2Go) to get directly to Romblon from the port of Batangas, itself a two-hour drive in dense traffic from Manila. Note also that there are various uninhabited islands (marked in grey in Figure 5.2), aside from 132 other unnamed islets and rock formations, that are visited by tourists due to their uncrowded beaches and additional attractions. A good case in point is the installation of a zipline between San Agustin, Tablas, to Biaringan island, an islet just off its north-east tip.

Tablas Island is equipped with several ports that allow for multi-island connectivity, using a combination of sea and land transports. An air transport service was launched in 2022 between the island and Manila. In contrast,

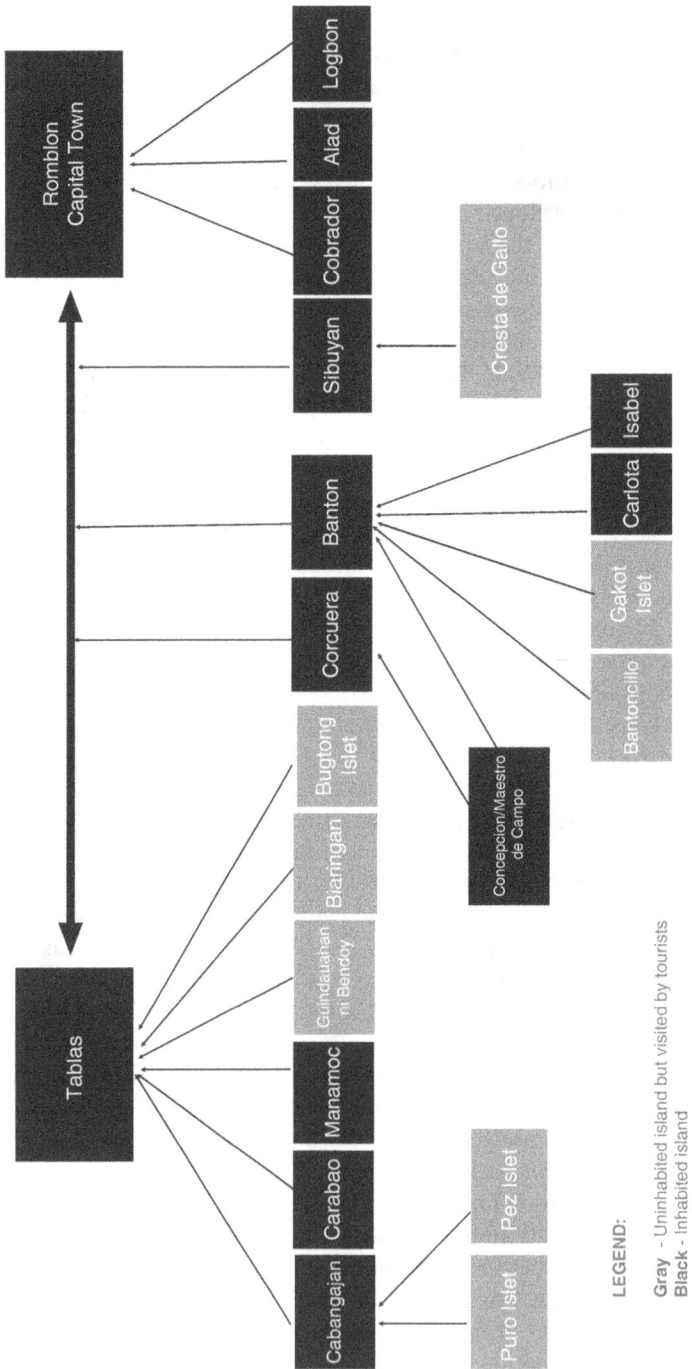

Figure 5.2 The hierarchy of islands of the Romblon archipelago in logistical terms

LEGEND:

Gray - Uninhabited island but visited by tourists
Black - Inhabited island

travelling to the capital town of Romblon and to Sibuyan remains time-consuming, tiring, and costly. Speeding up the journey requires a combination of air, water and land links.

Politically, all roads must lead to the capital town of Romblon in the island of Romblon. However, the logistics followed by tourism flows suggest different routes. Since Romblon Island is much smaller in land area than nearby Tablas, the airport was constructed on the latter. Its construction was political by design. The project was initiated by Florencio 'Pensoy' Moreno. Initially, the site of the airport was meant to be in Bonbon, San Agustin municipality, because of its suitable location. Pensoy's father, however, did not want to sell his vast tracts of land to the government (Esquejo, 2018). This led to the transfer of the building site to its current location in Tugdan.

People and cargo mostly go to Tablas and/or stop at Romblon Island first if originating from Sibuyan. Weather permitting, Sibuyan can be reached through Cajidiocan port from Caticlan, Aklan; Masbate City; and from Romblon city, Romblon Island. However, there are additional routes. For instance, Carabao Island also welcomes people and products from the popular and well-connected island of Boracay. Same for Tablas, which has a direct shipping service from Batangas in Luzon via Starlite Ferries to San Augustin. Romblon Island, on the other hand, has two routes. The first route requires a stopover first in Odiongan Port in Tablas before making its way to Batangas on the island of Luzon; and the other route is taking the daily ferry service directly to Batangas, which takes nine to 12 hours.

If leaving from Sibuyan, there are two stopovers before reaching Batangas: via Romblon or Tablas islands using Starlite Ferries. At the same time, Sibuyan can be reached via Marinduque. Since September 2022, however, the province has been serviced by AirSWIFT. The trip from Manila to Tablas takes just one hour and uses a 50-seater Avions de Transport Régional (ATR) 42–600 turboprop (WanderGeneration Inc, 2022); but there are only two scheduled flights per week. Cebu Pacific used to operate the ATR 72–600 aircraft between Manila and Tablas four times a week prior to the Covid-19 outbreak. This service, however, was not revived post-pandemic.

Passengers from Manila visiting Romblon and Sibuyan islands as final destinations by air from Manila have to travel further to complete the final leg. One has to ride a jeepney or van for an hour to catch a ferry in San Agustin port to reach Romblon Island or a five-hour ship ride to Magdiwang port in Sibuyan.

While there is a direct flight that servicing people to go to Romblon province in contemporary period courtesy of AirSWIFT, the passenger capacity and the scope of travel are still limited. Unless, travellers have the serious intention of reaching the farthest islands, tourists would normally go to places proximate to their airport of landing to save money and time. Thus, it is not surprising that Odiongan and Ferrol, both on Tablas, received the highest number of tourists in 2022 (a total of 4,338) out of the registered travellers that entered the province using Ferrol and Odiongan as ports of entry (11,260) (Table 5.3).

Table 5.3 Travellers to Romblon, 2022

Port	Foreign travellers	Domestic travellers	Total
Ferrol, Tablas Island	3	40	43
Odiongan, Tablas Island	89	4,206	4,295
Romblon, Romblon Island	104	3,291	3,395
San Fernando, Sibuyan Island	0	464	464
San Jose, Carabao Island	230	2,833	3,063
TOTAL	**426**	**10,834**	**11,260**

Now here is the paradox. What are challenging and harrowing transport logistics for the locals of Romblon are now being hailed as parts of the infrastructure that protect and lock the exclusive and tranquil nature of the province. The twin concepts of tranquillity and exclusivity have been appropriated by tourism stakeholders as a strategic tourism approach in the post-pandemic epoch. According to a tourism practitioner, such an approach is preferred, on the premise that 'the more secluded the destination, the more expensive it is', replicating the cases of Amanpulo in Palawan or Balesin in Polillo. That is why several hotels in the province are discouraging walk-ins and require reservations prior to date of arrival. Moreover, the source also believed that the "lack of better connectivity can be addressed by chartered trips (e.g. helicopter, fastcraft) for those who can afford it" (Anonymous, personal communication). If this is indeed the new discourse, and this is being intoned with the connivance of the national government, then Romblon province can expect an immediate future with hardly any developments in connectivity.

Capitalising on tranquillity and seclusion

Being an underrated neighbour of world famous Boracay, next door Carabao Island in Romblon Province has benefitted from the tourist flows. In fact, a good number of Carabao Island residents work at the Aklan's Boracay (ABS-CBN, 2018). This advantage was brought to a halt when Boracay was closed for six months in 2018–2019. Nonetheless, some islands in Romblon province became incidental recipients of tourism in the post-pandemic period where their marginality turned the tables to their advantage as the hubs of exclusive, tranquil and more authentic destinations in the archipelago. Romblon offers Boracay-like beaches without Boracay-like crowds (Fos, 2019). In the blogs of people who have visited the area, the beauty of the powdery white sands and crystal-clear waters similar to Boracay's but minus the crowd and the parties is vividly highlighted. Travellers can find peace and solace in the beaches located in Carabao, Talipasak, Tiamban, Bonbon, and Tablas islands (Miniano, 2022). Even then DoT's regional director, Danilo Intong, has acknowledged that Romblon offers a total package of discovery, recreation and leisure especially "for those who are craving for something sedate" (Explora.ph, 2017).

The words 'tranquillity and exclusivity' have been employed by the DoT in describing the nature of Romblon, particularly the islands of Sibuyan and Cresta de Gallo. In the department's social media, Sibuyan, Romblon is likened to a "perfect paradise where you can experience tranquillity and exclusivity" (Department of Tourism FB, 2022). Even the names of prominent beaches and tourist destinations in Romblon province witness this 'off the beaten path' appeal: Binucot Beach, in Ferrol, Tablas Island, means 'hidden'; while Tinagong Dagat, a salt lake carved into the limestone rock of Kabibihan Point, in the north of Tablas Island, means "hidden sea" (Explore Romblon, 2013). The narrative consistently peddles the notion that Romblon is in a natural state:

> Romblon is still largely untouched by mass tourism which allows guests to immerse themselves in the local culture and way of life. In addition, the province boasts some of the most breath-taking scenery in the Philippines, with crystal clear waters, hidden lagoons, and picturesque beaches.
>
> (TourGuidePh, 2023)

The visit by the *Silver Shadow* luxury cruise ship to Romblon right after the pandemic brought renewed attention to the province as "[it] is fast becoming the destination of choice for travellers seeking an authentic, off-the-beaten-path experience" in the Philippines (TourGuidePh, 2023).

Many travellers to Boracay take day tours to Carabao or stay there for a couple of days to enjoy its more serene vibe before heading back to party in Boracay. The pandemic did trigger a sober re-evaluation of the use of space and the desirability to market various 'second tier' destinations in the Philippines as explicit alternatives to mass tourism, while being more attractive to visitors because of lower chances of infection during the virus scare. International cruise ships started including Cresta de Gallo as part of their itineraries in 2023. The Local Government Unit (LGU) of San Fernando welcomed this development and assisted the companies in their plans. For instance, Silver Shadow of Silversea Cruises with 330 guests visited the island in February 2023 (Fos, 2023a). It was followed by the luxury cruise ship MS *Europa*, owned and operated by Hapag-Lloyd, which arrived in April, with 205 tourist passengers and 283 crew on board. Two months later, in June 2023, the cruise ship *Heritage Adventurer* made its maiden voyage and cruise call in San Fernando, Sibuyan Island, with 81 crew and 81 passengers/guests on board. For more than three decades, Heritage Expeditions (2023) has pioneered conservation-driven, small ship expedition cruises to some of the wildest, least-explored and biologically rich regions on the planet including the Southeast Asian islands. The guests on board – composed of retired scientists/professionals, the ship owner and family, and cruise staff – also visited several local tourist attractions (San Fernando Tourism Office, 2023).

To maintain the tranquillity and exclusivity in some islands and to protect the environment as well, the LGU wanted to make islands such as Cresta de

Gallo an off-beat destination where travellers would partake in day tours only. In case they wanted to stay overnight, tourists would need to bring their own camping gear and provisions since there is no single lodge or restaurant to be found on the island. Peace and seclusion are clearly complementary to the LGU's plan of making the islands and the whole province collectively recognised as eco-tourism destinations. Travellers are encouraged to observe and appreciate nature as well as to partake in the traditional cultures prevailing in the region. By doing so, negative impacts upon the natural and socio-cultural environments are minimised and at the same time valuable economic benefits are generated for local communities, institutions and local conservation-driven authorities (UN World Tourism Organization, 2002). Pushing for the serious implementation of eco-tourism projects would cumulatively maintain the exclusivity of some tourist destinations in the province, while thwarting the onset of mass tourism in Romblon.

Conclusion

The conceptualisation and marketing of Romblon islands as spaces of seclusion and tranquillity to travellers is shaped by their geography, geology, politics, and transport logistics, all of which manifest a clear, nested 'core-periphery' relationship: between Manila and the province; and, within the province, the rivalry for primacy between Tablas and Romblon, and the multiple insularity experienced by the rest of the archipelago. The visibility of these islands has been amplified by the pandemic and attracted the attention of various travellers who are participating in the 'revenge tourism' of the post-pandemic era (Wassler and Fan, 2021). However, this increased interest notwithstanding, sea and air connectivity has not improved that much to open the floodgates; with only 100 airplane seats available weekly from Manila to Tablas, only the stronghearted and budget-constrained individuals will get to Romblon by sea.

One shift in worker and employer preference, seen during and after the pandemic, is the feasibility of working from home. The Philippines government should prepare suitable regulatory frameworks and explore targeting foreign nationals who are 'digital nomads' and may wish to settle in such pristine locations as Romblon while continuing with their work (e.g. Woldoff and Litchfield, 2021). But such aspirations sound rather farfetched in a context of poor connectivity. It is hoped that the 2021 initiative by the Philippine Long Distance Telephone Company (PLDT) to instal submarine fibre cables in Romblon, Sibuyan, Tablas, and Carabao islands will hasten their broadband connectivity. This facility enables the offering of 5G network to address the businesses, educational institutions, and household connectivity requirements in the province (Smart Communications, 2021). As of 2023, the Department of Information and Communications Technology (DICT) has installed free Wifi in 21 strategic sites located in the provincial capital, hospital, and national high schools and all campuses of Romblon State

University across the province (Fos, 2023b). According to Gerard Montojo, mayor of the town of Romblon, the "fibre connectivity in our town will not only enhance the internet experience of our residents but will also boost our town's tourism goals and our LGU's digitalization initiatives" (Amojelar, 2022).

Tourism numbers in Romblon have been low and are likely to remain so, now even more with a declared policy that privileges isolation, seclusion and tranquillity. Only eight places of accommodation and one dining place in the province were accredited by the Department of Tourism in 2019 (Maestro and Dumlao, 2019, p. 246). The 'Love the Philippines' touted in the country's most recent marketing campaign might be understood to include subtle hints that Romblon residents need to love their country and so embrace their marginality because this is what is going to lure the eco-tourists and their spending.

References

Amojelar, D. (2022). PLDT rolls out fibre Internet in Romblon. *Manila Standard*. https://ma nilastandard.net/business/it-telecom/314284465/pldt-rolls-out-fiber-internet-in-rom blon.html.

Baldacchino, G. and Ferreira, E. C. D. (2013). Competing notions of diversity in archipelago tourism: Transport logistics, official rhetoric and inter-island rivalry in the Azores. *Island Studies Journal*, 8(1), 84–104.

Barile, C. (2017). Administrator Tiangco welcomes 2017. Manila: *National Mapping and Resource Information Authority*. www.namria.gov.ph/list.php?id=1032&alias=a dministrator-tiangco-welcomes-2017&Archive=1.

Canoy, N. A., Roxas, G. K. T., Robles, A. M. Q., Alingasa, A. P. T., and Ceperiano, A. M. (2020). From cesspool to fortified paradise: Analyzing news media territorial assemblages of rehabilitating Boracay Island, Western Philippines. *Journal of Sustainable Tourism*, 28(8), 1138–1157. https://doi.org/10.1080/09669582.2020.1726934.

Department of Tourism, Philippines. (2022). *Facebook page.* www.facebook.com/Depa rtmentOfTourism/posts/cresta-de-gallos-stunning-sandbar-and-clear-turquoise-waters-make-this-hidden-is/454052506766885/.

Doctolero, J. (2023). What to expect from the 'Love the Philippines' campaign? *Philippine Information Agency.* https://pia.gov.ph/features/2023/06/29/what-to-expect-from -the-love-the-philippines-campaign#:~:text=The%20Department%20of%20Tourism% 20(DOT,heart%20of%20every%20single%20Filipino.

Esquejo, K.R. (2018). Family and politics in an archipelagic province: The Moreno dynasty in postwar Romblon, 1949–1969. *Kasarinlan: Philippine Journal of Third World Studies*, 33(2), 89–124. www.journals.upd.edu.ph/index.php/kasarinlan/a rticle/view/7069/6154.

Explora.ph (2017). MIMAROPA Magazine: Romblon.

Fos, P. (2019). Romblon: The place to go for white sand and uncrowded beaches, marine sanctuaries, and marble. www.gmanetwork.com/news/lifestyle/travel/689522/ romblon-the-place-to-go-for-white-sand-and-uncrowded-beaches-marine-sanctuaries-a nd-marble/story/.

Fos, P. (2023a). Cruise ship with 205 tourists arrives in Cresta de Gallo. Manila: Philippine Information Agency. https://pia.gov.ph/news/2023/04/11/cruise-ship-with-205-tourists-arrived-in-cresta-de-gallo.

Fos, P. (2023b). *Free Wi-Fi sites launched in 21 locations within Romblon province.* Manila: Philippine Information Agency. https://pia.gov.ph/news/2023/02/03/free-wi-fi-sites-launched-in-21-locations-within-romblon-province.

Fuentes, A. (2018). The search for the next Boracay: Carabao Island, Romblon. *ABS-CBN.* https://news.abs-cbn.com/life/multimedia/slideshow/05/29/18/the-search-for-the-next-boracay-carabao-island-romblon.

Groves, J. (2023). *Cresta de Gallo island: Itinerary and travel guide.* www.journeyera.com/cresta-de-gallo-island-romblon/.

Henderson, J. (2011). Tourism development and politics in the Philippines. *Tourismos,* 6(2), 159–173.

Heritage Expeditions (2023). www.heritage-expeditions.com/.

Ignacio, C. (2023). Bolstering Philippine tourism as it bounces back. *Business World.* www.bworldonline.com/special-features/2023/05/09/521712/bolstering-philippine-tourism-as-it-bounces-back/.

Licap, E. (2013, April 16). The Galapagos of Asia: Sibuyan island. *Choose Philippines.* www.choosephilippines.com/go/islands-and-beaches/46/galapagos-asia-sibuyan-island/.

Maestro, N. B. and Dumlao, M. F. (2019). Romblon islands into a smart tourism destination through point of interest recommender, augmented reality and near field communication: A proposal. *International Journal of Innovative Technology and Exploring Engineering (IJITEE),* 8(2), 242–248.

Maguigad, V. M. (2013). Tourism planning in archipelagic Philippines: A case review. *Tourism Management Perspectives,* 7(1), 25–33. https://doi.org/10.1016/j.tmp.2013.03.003.

Manalang, M. (2023). Making up for lost time through revenge travel. *Philippine News Agency.* www.pna.gov.ph/articles/1195479.

MIMAROPA Regional Development Plan 2023–2028 (2023). National Economic and Development Authority. MIMAROPA Region. https://mimaropa.neda.gov.ph/regions-profile-2/.

MIMAROPA-DILG (2023). *DILG Romblon.* Department of Interior and Local Government. http://mimaropa.dilg.gov.ph/dilg-romblon/.

Miniano, M. (2022). Eight reasons why you should visit Romblon: Boracay's under-rated neighbour. www.tripzilla.ph/romblon-tourist-spot/36103/.

Permanent Mission of the Republic of the Philippines to the United Nations (n.d.). The Philippines at a Glance. www.un.int/philippines/philippines/philippines-glance.

Philippine Airlines (2023). PAL to invest in fleet, services amid Q2 net income growth. www.philippineairlines.com/en/newsevent-listingpage/press-releases-statements/pal-to-invest-in-fleet-services-amid-q2-net-income-growth.

Rocamora, J. (2023, August 10). Support for 'Love the Philippines' campaign grows: DOT chief. *Philippines News Agency.* www.pna.gov.ph/articles/1207593.

Ramilo, S. (2021). Potential of San Fernando, Province of Romblon, as ecotourism destination. *Turkish Online Journal of Qualitative Inquiry,* 12(7). www.tojqi.net/index.php/journal/article/view/4100/2811.

San Fernando Tourism Office (2023). www.facebook.com/sanfernandotourismoffice.gov.ph.

Santarita, J. (2020). The saga of a maritime nation sans maritime doctrine. *Timon: The Proceedings of the Philippine Maritime Heritage Forum*: Vol. 1. Pasay: Asian Institute of Maritime Studies and Museo Maritime, pp. 84–99. www.aims.edu.ph/files/conference%20proceedings/AIMS_Museo_Maritimo_Conference_Proceedings_Volume_1.pdf.

Smart Communications (2021). Fiber, 5G finally arrive in Romblon islands. https://smart.com.ph/About/newsroom/full-news/2021/11/10/fiber-5g-finally-arrive-in-romblon-islands.

TourGuidePh (2023, February 14). Silversea's Silver Shadow makes historic arrival in Romblon. www.tourguideph.com/2023/02/Silverseas-Silver-Shadow-Makes-Historic-Arrival-in-Romblon.html.

United Nations World Tourism Organization (2002). Ecotourism and protected areas. www.unwto.org/sustainable-development/ecotourism-and-protected-areas.

WanderGeneration Inc (2022). Airswift now flies to Romblon. www.thepoortraveler.net/airswift-romblon-airport/.

Wassler, P. and Fan, D. X. (2021). A tale of four futures: Tourism management and Covid-19. *Tourism Management Perspectives*, 38, 100818. https://doi.org/10.1016/j.tmp.2021.100818.

Woldoff, R. A. and Litchfield, R. C. (2021). *Digital nomads: In search of freedom, community, and meaningful work in the new economy*. Oxford: Oxford University Press.

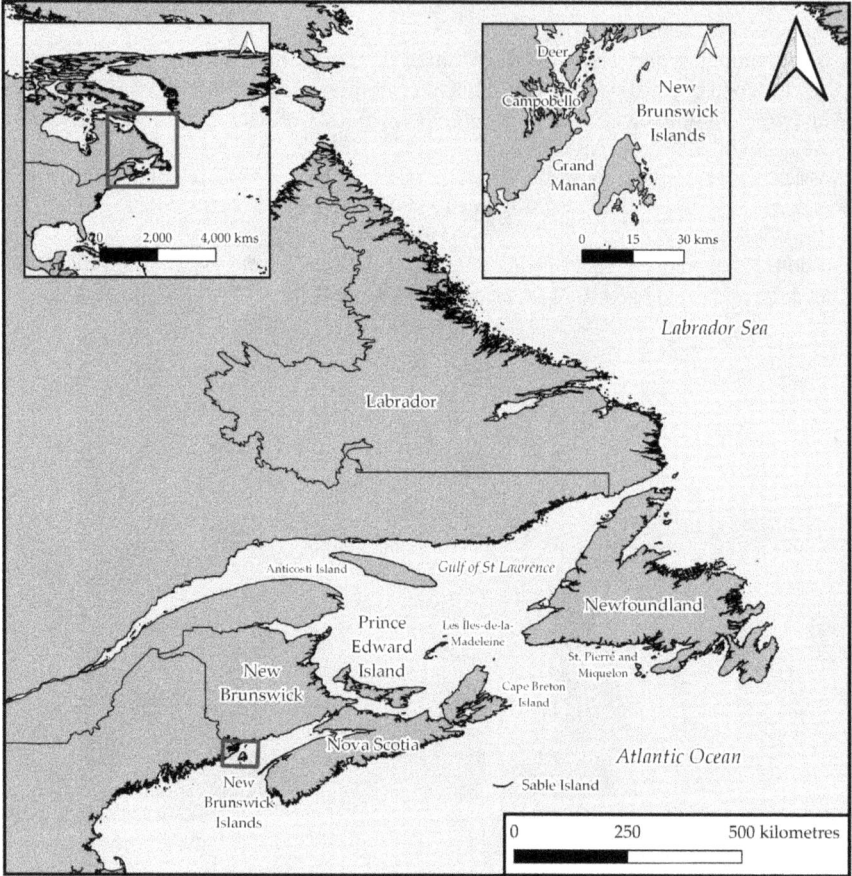

Figure 6.1 The Atlantic Bubble, Canada

6 Beyond 'the Atlantic Bubble'

Considering archipelagic tourism on Canada's east coast

Laurie Brinklow, Louise Campbell, Andrew Halliday and Isabel MacDougall

Introduction

In the summer of 2020, while Covid-19 and the hardening of internal borders brought tourism nearly to a standstill, the Atlantic provinces of Canada banded together to create 'the Atlantic Bubble', promoting non-quarantine 'stay-cation' travel for residents within the region. With their active Covid-19 cases nearly nonexistent, the provinces of Nova Scotia (which includes Cape Breton Island), New Brunswick, Prince Edward Island, and Newfoundland and Labrador (particularly the island portion of Newfoundland) dropped some of their individual restrictions and isolation rules, allowing preregistered intra-regional travel while maintaining strict isolation and entry controls to all non-residents, including other Canadians. The Atlantic Bubble provided a semblance of a peak summer season, averting near disaster for tourism operators, and maintained the sanity of many Atlantic Canadians suffering from cabin fever who seized the opportunity to vacation within the region.

Reflecting on how well the Atlantic Bubble worked in the summer of 2020 and to some extent in 2021, it was easy to envision Atlantic Canada as an archipelago where 'island-hopping' might have some tourism potential. It follows on a novel conceptualisation of Canada as an archipelago (Suthren, 2009), whereby thousands of islands off Canada's Atlantic, Arctic, and Pacific coasts, as well as through the central Great Lakes, encircle the mainland. Just as all these islands interact with the main 'island' of Canada, so do Atlantic Canada's islands have relations with the main 'island' of Atlantic Canada – some of which are actual islands or, in the case of Nova Scotia, separated from New Brunswick by the 24-km-wide Isthmus of Chignecto, making it an 'almost island'. Historically, the Atlantic region – physically, linguistically, and culturally cut off from the rest of Canada by the province of Quebec – has often been "habitually derided … as a bunch of aging hard-luck cases reliant on federal transfer payments" (O'Connor, 2021, n.p.). Though the country's major air carriers cut flights to the region during the pandemic, further isolating Atlantic Canada, doing so offered an opportunity for "the castaways in New Brunswick, Newfoundland and Labrador, Nova Scotia and Prince Edward Island … to forge a different future" (ibid.) by taking

DOI: 10.4324/9781003451037-8

advantage of their islanded status and working as an archipelagic unit to become more innovative and entrepreneurial – including doing tourism in the Atlantic Bubble.

The idea of 'thinking with the archipelago' follows on groundbreaking work by Stratford et al. (2011) which presents two dominant topological relations: of land and sea, and of island and continent/mainland. As the authors explain:

> We seek to understand archipelagos: to ask how those who inhabit them or contemplate their spatialities and topological forms might view, represent, talk and write about, or otherwise experience disjuncture, connection and entanglement between and among islands.
>
> (ibid., p. 114)

They, too, posit Canada as an archipelago, considering "Canada not as a unitary land mass but as a series of multiple assemblages of coastal, oceanic and insular identities" (ibid., p. 121). An Atlantic Canada archipelago, consisting as it does of mainland and islands, and the concomitant entanglements, is the Canadian archipelago writ small.

In exploring this new idea, the authors – who are island studies scholars from the cognate disciplines of history and political science, literature and human geography, and tourism studies – draw upon their knowledge and networks to gather historical and modern-day transportation statistics and logistics, the history of Maritime Union and Atlantica, tourism statistics, and tourism marketing efforts within the Atlantic region. Through the lens of regionalism, island and archipelago studies, and Covid-19, they synthesise an argument that considers historical factors including: traditional 'north-south' cultural linkages between Atlantic Canada and 'the Boston States'; the concept of regionalism and how it can be considered an archipelago within the Gulf of St. Lawrence; transportation infrastructure (or the lack thereof) that makes island-hopping in the region possible (or not); and how the Atlantic archipelago is – and might be better – marketed in Canada, the US, and beyond – all in relation to proposing archipelagic tourism for Atlantic Canada. It considers what Baldacchino (2015, p. 1) has termed "a centrifugal tendency" or "domination and subordination, liminality or layering" within the archipelago. Finally, the chapter identifies some of the hidden island gems that might be shined up to become part of a new 'Atlantic Canada archipelago' going forward.

History's long shadow

Before the railway transformed continental North America by 'opening up the west', and the TransCanada highway eventually made it possible to traverse Canada coast to coast, the traditional transportation linkages in British North America were by water; hence the significance of major navigational infrastructures such as the Erie and Rideau Canals (Rodrigue, 2020); thence, east to England; south to the Eastern Seaboard of the United States; and beyond to the Caribbean. Continentally, the 'north-south' bonds held greater sway than the 'west-east' ones of nation building, including the construction of the St. Lawrence Seaway, a joint post-WWII project of the Canadian and US Governments, completed in 1959 (ibid.). Many would argue that this N-S axis remains true to this day.

One illustrative example of the historically strong transportation linkages within the region is that of the Dominion Atlantic Railway (DAR), incorporated in 1895. DAR transported Annapolis Valley produce and passengers throughout western Nova Scotia and as far south as Boston, with a strong focus on the tourist market (Library and Archives Canada, 2003). Their strongly branded 'Land of Evangeline' route built upon Longfellow's so-named heroine in his 1847 poem and the area's Acadian history (ibid.). A 1900 advertising brochure announced the Boston to Halifax route as 'just' a 22-hour journey and implored potential customers to "think what the difference means in time, comfort and money" (Dominion Atlantic Railway, 1900, n.p.). From Halifax, possible onward connections were highlighted to Newfoundland, Liverpool, or the West Indies (see also map at Geographicus.com, 2023, n.p.).

The subject of ferry linkages in the region is intricately tied to the railways and the federal government. It is also important as both Prince Edward Island (1873) and Newfoundland (1949) joined the Canadian federation upon the terms of having the federal government finance and maintain their respective linkages to the mainland (Bruce, 1977). These linkages were ferry connections in both instances up until 1997 with the opening of the 12.9-km-long Confederation Bridge, linking Prince Edward Island to New Brunswick and the rest of mainland Canada. Marine Atlantic remains in operation with the North Sydney to Newfoundland service while the other regional ferry routes were turned over to Northumberland Ferries Limited (founded in 1941) which is responsible for the Wood Islands (PEI) to Caribou (NS) route, and its subsidiary Bay Ferries Limited.

Regionalism in Canada and the Atlantic provinces

In thinking of the geography of Atlantic Canada, it is important to consider regionalism, which is a factor of the Canadian federation. It is also a contested term that is "highly ambiguous" (Tomblin, 1995, p. 5). Canadian regionalism has been attributed to factors such as geographic size and isolation, population distribution and historic patterns of settlement, a long history of province-building, variation in economic prosperity, and perhaps the most potent factor: linguistic identity (Savoie, 2000). Conrad (2002, p. 171) suggests that the "peculiarities of the Canadian political system" have shaped region-building more so than the actions taken within the regions. Indeed, "[w]ithin Canada, regional disparities are as old as the country itself" (Tomblin, 1991, p. 102). Conrad (2002, p. 161) argues that, notwithstanding regional frustrations throughout the country with national policy, "few areas of the country have such deep scars [as Atlantic Canada] to prove their case".

The geographic concept of 'Atlantic Canada' is prickly, as it "is not a region in any academic sense of the term" (Conrad, 2002, p. 161). McKay (2000, p. 96) suggests that the term has become naturalised, "impos[ing] a top-down homogenization on historically distinct experiences". It is worth noting that the boundaries of and within the region have not remained fixed in the landscape, but rather have been fluid throughout the expansionist stages of Canadian Confederation (McKay, 2000). It has also been suggested that from the perspective of individual provinces, their borders (except the inland Labrador/Quebec border) actually pre-date Confederation, making for "deeply-rooted people" (Conrad, 2002, p. 168). The establishment and entrenchment of these subnational borders means "premiers are resistant to outside interference with local powers or a provincial sense of identity" (Tomblin, 1995, p. 4).

Concerning the perception of Atlantic Canada as a region, McKay (2000, p. 96) reviews a "structural-functionalist" batch of approaches with similar characteristics: "*periphery* to a *centre*, a *hinterland* to a *heartland*, an *internal colony* to a *metropole*". In this lens, Atlantic Canada is defined by what it lacks: a metropolis, local pools of capital, a well-developed industrial base, and a clearly defined class structure (McKay, 2000). Put more bluntly by Conrad (2002, p. 162), Atlantic Canada may be characterised as "[b]ackward, conservative, and juiced up on handouts ... a region blighted by location, culture and identity."

The region does have a historical precedent however, even if just an aspirational one. In the years where the secession of the province of Quebec from Canada was a distinct possibility, the four Atlantic provinces were bracing for a bleak future as the marginalised backwater of a new, diminished country. The idea of Atlantica – not to be confused with the mythical Atlantis – as a region and two new countries, was peddled by individuals, organisations and think-tanks, including the vocal Atlantic Institute of Market Studies (AIMS), emphasising its "geography, shared history, difficult economic challenges [and] long neglect by two national governments" [In this model, the NE region of the US – Maine, New Hampshire, Vermont and upstate New York – were equally disadvantaged and should contemplate their own separate country: AIMS (2005)].

Given what the literature tells us, then, the application of Atlantic Canadian regionalism in the context of Covid-19 and the Atlantic Bubble is key. It would seem to be an acceptance of the regional brand by the four small subnational provinces, whose collective population (around 2.4 million) was – and still is – less than that of the city of Montreal. Further, the allure of this conceptual creation means that in this exact context, Atlantic Canada became a place of envy and want owing to its exclusive status as a Covid archipelagic safe haven. From an internal vantage point, the region was now being defined as what it did have: a seemingly safe idyllic space where subnational inter-provincial travel was permitted to residents; a place where the 'new normal' was being put into practice.

The experience of the Atlantic Bubble

In the aftermath of the initial pandemic wave to hit the region, a temporary *Covid-19 archipelago* was formed. This 'Atlantic Bubble' endured for five months. It was announced on 24 June 2020, via the Council of Atlantic Premiers and came into effect on 3 July 2020. It burst on 23 November 2020, when the Canadian island province of PEI (connected by the Confederation Bridge to NB, and by seasonal ferry to NS) and the coastal-island province of Newfoundland and Labrador (connected to NS by year-round ferry) suspended their participation, owing to an alarming rise of case numbers of Covid-19 in NB and NS.

To grasp the Atlantic Bubble, one must understand the external and internal borders which delineated it. Externally, the Atlantic Bubble (via NB) shared a 513-km terrestrial border with the US state of Maine as well as an international maritime border. A small international maritime boundary was also shared with France, through its overseas territory of St. Pierre and Miquelon. Moreover, two separate portions of land border, covering some 4,000 km, and a maritime border involving a vehicle passenger ferry (to les Îles de la Madeleine) connected the 'bubble' with the Canadian province of Quebec (via NB, NL, and PEI respectively).

The Atlantic Bubble involved a targeted and geographically limited easing of travel restrictions for residents within the region's hardened internal subnational borders, initially established by their respective subnational governments, in concert with measures taken by the federal government to harden international borders to limit virus transmission and spread. Residents of the four provinces were afforded the opportunity to travel intra-region without any quarantine regime, while any approved visitors/travellers from outside (including other Canadians) were subject to a mandatory 14-day quarantine period. The Council of Atlantic Premiers, the regional institution which created the Atlantic Bubble, is unique in Canada for being the only such permanent, resourced, and established body in the country.

Transportation infrastructure and logistics in the Atlantic Canadian archipelago: Layering, domination, and multiple peripherality

There is a clear subnational hierarchy centred on layers of jurisdiction and autonomy in Atlantic Canada. The regional construct rests upon the foundation of the four subnational provinces which formed the Atlantic Bubble: Nova Scotia, New Brunswick, Prince Edward Island, and Newfoundland and Labrador. As provincial jurisdictions, each is an equal member of the Canadian federation and has its own government with the associated fiscal and policy instruments, including the purview of tourism marketing and branding which is discussed later in this chapter. This status provides a platform of domination for the four provinces in comparison to smaller subnational islands without that level of jurisdiction or autonomy. However, given the central role of the federal government in Ottawa, this relationship, itself a manifestation of the core-periphery power relation, requires more detailed examination. This is especially pertinent with respect to the funding of infrastructure, particularly the investment and ownership of ferries, the transport infrastructure, and logistics in the region.

Highways, bridges, and causeways

Within the tighter identity and geographic unit of the three Maritime provinces, there are significant highway linkages for both Prince Edward Island (Confederation Bridge to mainland NB) and Cape Breton Island in Nova Scotia (the Canso Causeway to mainland NS). The Confederation Bridge, which opened in 1997, the result of a Can$1 billion public-private partnership, was one of the largest infrastructure projects in recent Canadian history. While PEI also has a seasonal ferry link maintained to Pictou County, NS, the significance of permanent 'fixed links' – bridges, tunnels, causeways – cannot be understated (Baldacchino, 2008). Smaller highway-linked coastal 'near islands' include Lameque and Miscou Islands (NB), and Lennox and Panmure Islands (PEI). Thus, the provinces of NB, NS, and PEI form a nucleus of the Atlantic Canada archipelago, entirely interconnected by permanent highway infrastructure.

Ferry services

As an archipelagic construct, what of the current state of the ferry infrastructure and routes serviced? As noted above, the interprovincial ferries were extensions of the railway network funded and operated by the federal government. To this day, these important intra-regional connections are entirely dependent upon the attentiveness and willingness of the Canadian national government in Ottawa to maintain and invest in vessels and port infrastructure.

Ferries plying routes along additional coastal and inland waters are owned and operated by respective provincial governments. As such, discretionary decision-making around the procurement, staffing, and frequency of the services offered is left to the relevant provincial departments. In the absence of overland highway connections, the island of Newfoundland uses coastal ferries to serve communities and outports on the southern coast as well as on the Labrador coast, primarily with cargo and foot passengers.

The imposition of international travel restrictions and the formulation of the Atlantic Bubble impacted certain ferry routes in the region. For instance, the St. Pierre and Miquelon service to Newfoundland was suspended for 17 months, with service resuming only in August 2021 (CBC News, 2021).

Intra-region air connections

The pandemic saw the removal of significant air linkages in the region, with commercial service disappearing completely from Sydney, NS, and Saint

John, NB, in 2021, and being significantly reduced in all other airports (Edwards, 2021). As of 2023, service had resumed, but several flights which were removed by national carrier Air Canada were never reinstated. Moreover, its main national competitor, Westjet, reacted similarly and mainly exited the region, leaving only reduced services from primarily Ontario's Toronto Pearson Airport (IATA Code: YYZ) to Halifax, NS, and St. John's, NL (Atlantic Canada Airports Association, 2020); some of the suspended routes have returned while others have not. This hollowing out and erosion of intra-regional air service had a significant local impact on regional transportation logistics. However, in the last year, smaller airlines, namely Porter Airlines and the Newfoundland-based Provincial AirLines (PAL), have moved in to fill some of the void, with intra-regional flights; that is, flights that do not require a connection via the hubs of Montreal or Toronto. In July 2023, Atlantic Premiers and federal Cabinet Ministers from the region announced a new Atlantic Working Group on Regional Air Travel, seeking to address "regional air access challenges", with an implementation plan requested by 1 October 2023 (Atlantic Canada Opportunities Agency, 2023).

The Atlantic Bubble and tourism

Tourism numbers show just how drastically visitation fell during Covid-19: The region witnessed a contraction of almost 60%: from around 5 million tourists visiting in 2019 to 1.5 million in 2020 and not much more in 2021. This translated into an approximately Can$3.3 billion loss of revenue (Atlantic Economic Council, 2020). But without the interprovincial travel afforded by the Atlantic Bubble in 2020, the numbers would have been even worse (Department of Tourism, Culture, Arts and Recreation [TCAR], Government of Newfoundland and Labrador, 2021, 2022; Province of PEI, 2022–2023; Tourism New Brunswick, 2020, 2021, 2022; Tourism Nova Scotia, 2021a, 2021b, 2022).

Beyond the 'Bubble'

And yet, visitation numbers post-pandemic have rebounded in Atlantic Canada and were on track to match or surpass pre-pandemic levels in 2023. With slowdowns in transportation connectivity within Canada and this being one of the key tourism trends impacting post-pandemic recovery internationally (Destination Canada, 2021), we are at an opportune time to build on the Atlantic Bubble concept. As Destination Canada (2021, p. 22) notes, "Reduced air access and limited ground transportation will directly impact both the cost and ease of travel from one destination to the next". And, as the cost of living – and especially the cost of transportation – increases globally, and people become more aware of their environmental footprint during what is now deemed a climate crisis, tourism marketers could build on the bubble concept to promote Atlantic Canada as an archipelagic unit of islands that are located relatively close together and thus physically and conceptually easier to navigate. Visitors would become part of a slower or more sustainable tourism movement, staying longer in one area and contributing more to the region's tourism economy.

Our conceptual Atlantic Canadian regional archipelago goes beyond the four Canadian provinces of NS, NB, PEI, and NL to encompass other regional islands governed by other Canadian provinces – Quebec's Anticosti Island and the Îles-de-la-Madeleine archipelago – as well as beyond Canadian jurisdiction: France's Overseas Territory of St. Pierre and Miquelon.

The sprawling Atlantic archipelago would be geographically bound by Parks Canada-administered Sable Island, NS, in the south; NB and its coastal islands of Deer, Campobello, and Grand Manan in the west; Quebec's Anticosti Island at the mouth of the St. Lawrence River towards the north of the Gulf of St. Lawrence; and Labrador and Newfoundland Island to the extreme north and east. Of note for potential future collaborative research is the striking similarity of these boundaries with the Mi'kma'ki Territories, home to Atlantic Canada's First Nations (Native Land Digital, 2023), as well as ideas for a potential Vinland-Mi'kma'ki tourist trail (Traustason, 2017).

Branding and marketing the Atlantic Canada archipelago

The Covid-19 isolation requirements forced Atlantic Canadians to explore closer to home. Non-active types who were looking for safe activities outside their homes were suddenly scoping out hiking trails; while bike shops could not get enough stock to meet the surging demand. This trend started within each province and then expanded to the rest of the Atlantic provinces once they "bubbled". People who may have once eschewed travel within the region for farther-flung locales were suddenly delighted to be able to move beyond their borders and seek fun and adventure in neighbouring provinces. While the tourism industry took a major hit, the captive audience of Atlantic Canadian travellers helped, to a degree, to recoup some of the losses. More importantly, though, pandemic travel opened people's eyes to what Atlantic Canada has to offer and the low number of Covid-19 cases made the region extremely attractive to those in the rest of Canada, the United States, and overseas.

Destination Canada plays an important role in helping the Canadian tourism industry reach domestic and international markets (Destination Canada, 2022). Its 2022 Forecast Highlights report showed that the Canadian tourism sector was on a recovery trajectory: domestic tourism will lead first, with international tourism taking over by 2024–2025. Domestic tourism reached almost 100% of 2019 levels in 2022; and was expected to fully recover in and by 2023. Visits from the US should reach 2019 levels in 2024; while overseas travel from key international markets in Asia-Pacific and Europe should recover fully by 2026. This projected recovery presents an opportunity to capitalise on the attraction of Atlantic Canada that was engendered during

Covid-19. But what is the best way to take advantage of Canadians' pent-up demand for travel, as found in Destination Canada's Global Tourism Watch (2021)? How can the Atlantic provinces attract international travellers as they once again hit the seas and skies to seek out adventures and experiences?

The individual Atlantic provinces have been strategising on how to best entice travellers post-pandemic. While these provincial strategies and visions are sure to boost tourism visitations, there is definite value in marketing Atlantic Canada as a whole. The Atlantic Bubble, which helped offset tourism losses during the pandemic, saw the provinces band together to increase visitation and spending in the region. If it worked then, why not now? The Atlantic provinces have a collective wealth of natural, cultural, heritage, and artistic treasures robust enough to compete for domestic and international travel. With big and small islands as well as mainland areas, the region can be thought of as an archipelago. There are, of course, two subnational island jurisdictions (SNIJs) – namely Prince Edward Island and Newfoundland and Labrador – but there is also a collection of 17 non-provincial subnational island jurisdictions (see Table 6.1), 13 of which belong to one of the four Atlantic provinces, with the remaining four belonging to Quebec, Maine, and France. These hidden island gems further the notion of an archipelagic entity for tourism purposes. Some of these provincial and non-provincial islands market themselves as tourism destinations in their own right.

Tourism marketing of Atlantic Canada as a whole is not a new idea. Since 2016, the Atlantic Canada Agreement on Tourism (ACAT) has been growing the industry in the region (ACAT, n.d.). Comprised of the Atlantic Canada Opportunities Agency (ACOA), the department responsible for tourism in each of the four provinces, and the four provincial tourism industry associations, ACAT drives growth via research-driven marketing campaigns and activities in key international markets as well as in select Canadian markets. Marketing strategies, which include major consumer advertising campaigns, travel trade programs, and media relations activities, allow the four Atlantic provinces to penetrate markets which are otherwise largely inaccessible; intensify international tourism marketing efforts; extend the reach of provincial and tourism brands in the United States; and benefit from the promotion of the regional Atlantic Canada brand into priority overseas markets.

ACAT's four-province marketing strategy is a good foundation for promoting an Atlantic archipelago. In particular, ACAT's strategy of targeting potential visitors from the United Kingdom could be expanded to include the lesser-known island gems, thereby highlighting more of the region's beauty, culture, and relationship to the sea. The UK marketing campaign slogan, *Awaken to the Rhythm of the Sea* (ACAT, 2023), would work as well for archipelagic marketing as for the current promotion; as would the invitation: 'Embark on an enchanting seacoast holiday in Atlantic Canada.' The campaign could be expanded to domestic as well as other international markets and indeed within the Atlantic region itself.

Table 6.1 Non-provincial subnational island jurisdictions

Island	Jurisdiction	Tourism marketing	Website
Campobello	New Brunswick	X	www.visitcampobello.com
Grand Manan	New Brunswick	X	www.grandmanannb.com
Miscou	New Brunswick		https://tourismnewbrunswick.ca/listing/miscou-island
Bell	Newfoundland	X	www.tourismbellisland.com
Fogo and Change Islands	Newfoundland		www.newfoundlandlabrador.com/top-destinations/fogo-and-change-islands
New World	Newfoundland	X	https://newworldisland.com
Big Tancook and Small Tancook	Nova Scotia		www.novascotia.com/search?query=+big+tancook
Brier and Long Islands	Nova Scotia		www.novascotia.com/trip-ideas/stories/brier-island
Cape Breton	Nova Scotia	X	www.cbisland.com
Isle Madame	Nova Scotia		www.cbisland.com/search/?q=isle+madame
Sable	Nova Scotia		https://parks.canada.ca/pn-np/ns/sable/visit
Hog	Prince Edward Island		www.tourismpei.com
Lennox	Prince Edward Island	X	https://experiencelennoxisland.com
St. Pierre and Miquelon	France	X	https://en.spm-tourisme.fr
Machias Seal	Maine		https://downeastacadia.com/story/machias-seal-island-light
Anticosti	Quebec	X	www.sepaq.com/sepaq-anticosti/vacances-ete
Îles-de-la-Madeleine (Magdalen Islands)	Quebec	X	www.tourismeilesdelamadeleine.com/en/

Like most non-tropical areas, Atlantic Canada struggles with the seasonality of its tourism product. Agius and Briguglio (2021) suggest that ecotourism can play a role in mitigating seasonality patterns in the Aegadian archipelago off Sicily, Italy, largely because ecotourism activities can be conducted all year-round. Marine ecotourism, a subcategory, is defined by Sakellariadou (2014) as any kind of responsible travel to coastal and marine settings, considering environmental conservation, elimination of environmental impacts, and improvement of the well-being of local communities with respect to culture and Indigenous people. Moreover, it should be nature-based and educational. With whale watching, an abundance of seabirds, diving and snorkelling, deep-sea fishing, coastal and beach walking, and boat tours in the mix, an Atlantic archipelago could capitalise on marine ecotourism in the summer months as well as in the shoulder seasons and support the movement in the individual provinces toward year-round tourism.

Complementary relationships between islands in an archipelago produce synergies (Cannas and Giudici, 2015). In the case of an Atlantic Canada archipelago, those synergies would be between islands as well as between islands and their respective mainland components. The concept of 'coopetition' is useful in the development and marketing of neighbouring tourism destinations (Almeida and Moreno-Gil, 2018). The concept, stemming from the tendency, even obligation, to both cooperate and compete (Brandenburger and Stuart, 1996), is especially relevant to Atlantic Canada as the provinces attempt to attract tourists from the same target market. Joint branding of the region, factoring in the non-provincial SNIJs, can make a greater impact in the tourism realm. In the spirit of 'coopetiting', Almeida and Moreno-Gil (2018) suggest that islands with complementary brands would benefit from having a joint presence at tourism fairs and in various promotional and marketing media. They also suggest that "complementary island brands could carry out promotional strategies at the airports of other islands ... seeking to attract island tourists for their future holidays" (ibid., p. 86).

Conclusion

Atlantic Canada is a uniquely positioned archipelago with regards to tourism due to the concentration of islands within the region and their close proximity to the rest of the 'mainland' of North America. Because of this, even before the 'bubble', a large portion of tourism in the region was interprovincial travel: nearly 40% of total visitations to Nova Scotia are by other Atlantic Canadians (Tourism Nova Scotia, 2021a). Atlantic Canadian travellers have traditionally represented the largest group of visitors to PEI: 59% in 2019 and 80% in 2020 (Tourism PEI, 2023a, 2023b). The intra-regional tourist is often a repeat visitor and is visiting friends or family; of the 1,108,400 Atlantic Canadian visitors to Nova Scotia in 2019, only 2% were first-time visitors (Tourism Nova Scotia, 2021b, p. 9). This concept of a repeat visitor brings a different set of values and attachment to place (Smaldone, 2007). Arguably, the 'short-haul' visitor comes with a lighter carbon footprint, even though their stays may be shorter. The downside is that the Atlantic Canadian visitor spends less money, tends to stay with friends and family, comes for day trips, does not participate in tours, and does not purchase typical tourism products or souvenirs. And yet, measuring value in dollars (tourism expenditure) is no longer the best way to gauge success. The environmental impact of travel and tourism is becoming increasingly significant as are the cultural and social impacts. At the same time, spreading out tourism to extend over the year instead of peak summer months has been top-of-mind for years. Short-haul Atlantic Canadian tourists could be the cornerstone of winter tourism. Visitors may not travel across the world for the region's winter tourism; but visitors will drive a few hours to escape and experience something different in the winter months.

'Thinking with the archipelago' (Stratford, 2013; Pugh, 2013) is one way we can contribute to a more holistic way of doing tourism in Atlantic Canada. As Pugh (2013, p. 10) argues, "archipelagic thinking denaturalizes the conceptual basis of space and place" and allows for more fluidity between and among its constituent pieces – the mainlands of Nova Scotia and New Brunswick and their surrounding islands – reiterating the ideas of boundedness and connectedness that are the hallmarks of island living (Brinklow, 2013). We call upon the region's tourism marketers – individually and collectively – to take advantage of physical, historical, and cultural linkages and networks to promote the region as an assemblage of islands, each nuanced in their own special way but always as part of a distinctive Atlantic Canadian culture that sets the region apart from the rest of Canada. Improving transportation infrastructure – through highways, ferries and bridges, and air links – would go a long way to building a more sustainable tourism product. Connectivity matters.

References

Agius, K. and Briguglio, M. (2021). Mitigating seasonality patterns in an archipelago: The role of ecotourism. *Maritime Studies*, 20(4), 409–421. https://doi.org/10.1007/s40152-021-00238-x.

AIMS (2005). *Atlantica: Two countries, one region*. Newsletter. Atlantic Institute for Market Studies. November. www.aims.ca/site/media/aims/Atlantica110305.pdf.

Almeida, A. and Moreno-Gil, S. (2018). Effective island brand architecture: Promoting island tourism in the Canary Islands and other archipelagos. *Island Studies Journal*, 13(10), 71–92. https://doi.org/10.24043/isj.45..

Atlantic Canada Agreement on Tourism [ACAT]. (n.d.). *Atlantic Canada Agreement on Tourism: Growing tourism together*. http://acat-etra.ca.

Atlantic Canada Agreement on Tourism [ACAT]. (2023). *Atlantic Canada: Awaken to the Rhythm of the Sea*. https://atlanticcanadaholiday.co.uk.

Atlantic Canada Airports Association. (2020). *WestJet service cancellations another blow to already hard-hit Atlantic Canada Airports* [Press Release], 14 October. https://acairports.ca/news/westjet-service-cancellations-another-blow-to-already-hard-hit-atlantic-canada-airports/.

Atlantic Canada Opportunities Agency. (2023). *Premiers and federal ministers agree to renew Atlantic Growth Strategy* [Press Release], 18 July. www.newswire.ca/news-releases/premiers-and-federal-ministers-agree-to-renew-atlantic-growth-strategy-826863599.html.

Atlantic Economic Council. (2020). *COVID-19: Key issues for Atlantic Canada's economy*. https://atlanticeconomiccouncil.ca/page/Covid19KeyIssuesAug20?&hhsearchterms=%22covid%22.

Baldacchino, G. (2015). *Archipelago tourism: Policies and practices.* Farnham: Ashgate.

Baldacchino, G. (2008). *Bridging islands: The impact of fixed links.* Charlottetown: Acorn Press.

Brandenburger, A. M. and Stuart, H. W. (1996). Value-based business strategy. *Journal of Economics & Management Strategy,* 5(1), 5–24. https://doi.org/10.1111/j.1430-9134.1996.00005.x

Brinklow, L. (2013). Stepping-stones to the edge: Artistic expressions of islandness in an ocean of islands. *Island Studies Journal,* 8(1), 39–54. https://islandstudiesjournal.org/files/ISJ-8-1-2013-Brinklow.pdf.

Bruce, H. (1977). *Lifeline: The story of the Atlantic ferries and coastal boats.* Toronto: Macmillan.

Cannas, R. and Giudici, E. (2015). Tourism relationships between Sardinia and its islands: Collaborative or conflicting? In G. Baldacchino (Ed.), *Archipelago tourism: Policies and practices* (pp. 67–81). Farnham: Ashgate.

CBC News. (2021, 12 August). *Smiles and tears of joy as ferry route between Newfoundland and Saint-Pierre reopens.* CBC News Newfoundland and Labrador. www.cbc.ca/news/canada/newfoundland-labrador/st-pierre-miquelon-fortune-ferry-1.6138866.

Conrad, M. (2002). Mistaken identities? Newfoundland and Labrador in the Atlantic region. *Newfoundland Studies,* 18(2), 159–174. https://id.erudit.org/iderudit/nflds18_2art02.

Department of Tourism, Culture, Arts and Recreation, Government of Newfoundland and Labrador [TCAR]. (2022). *Tourism highlights: 2022.* www.gov.nl.ca/tcar/tourism-division/visitor-and-market-insights/.

Department of Tourism, Culture, Arts and Recreation, Government of Newfoundland and Labrador. [TCAR]. (2021). *Tourism highlights: 2021.* www.gov.nl.ca/tcar/files/Monthly-Performance-Report_YTD-December-2021_FINAL_JAN28_2022.pdf.

Department of Tourism, Culture, Arts and Recreation, Government of Newfoundland and Labrador. [TCAR]. (2020). *Tourism highlights: 2020.* www.gov.nl.ca/tcar/files/Tour_perf_YTD_DEC_2020.pdf.

Destination Canada. (2022). *Tourism outlook: Forecast highlights.* Fall 2022. www.destinationcanada.com/sites/default/files/archive/1696-Tourism%20Outlook%20-%20Fall%202022/Tourism%20Outlook%20-%20Forecast%20Hightlights%20-%20Fall%202022%20-%20EN_1.pdf.

Destination Canada. (2021). *2021 Global Tourism Watch Highlights Report.* www.destinationcanada.com/sites/default/files/archive/1710-Global%20Tourism%20Watch%20-%20Canada%20-%202021/2021%20GTW%20Highlights%20Report_Canada_Final_EN.pdf.

Dominion Atlantic Railway. (1900). *Boston and the Maritime Provinces* [Brochure]. https://qspace.library.queensu.ca/bitstream/handle/1974/11842/bostonmaritimepr00domi.pdf?sequence=1.

Edwards, D. (2021, January 11). Airports in N.B., N.S. without commercial service after Air Canada cuts. *CTV News Atlantic.* https://atlantic.ctvnews.ca/airports-in-n-b-n-s-without-commercial-service-after-air-canada-cuts-1.5262251.

Geographicus.com. (2023). *Nova Scotia Dominion Atlantic Railways and Connections.* (Poole Brothers). www.geographicus.com/P/AntiqueMap/dominionatlantic-poolebros-1918.

Library and Archives Canada. (2003). *Nova Scotia - Transcontinental Tour – Canada, by Train.* https://web.archive.org/web/20080429195939/http://www.collectionscanada.gc.ca/trains/h30-2100-e.html#b.

McKay, I. (2000). A note on 'region' in writing the history of Atlantic Canada. *Acadiensis*, 29(2), 89–101.

Native Land Digital. (2023). *Mi'kma'ki Territories*. https://native-land.ca/maps/terri tories/mikmaq/.

O'Connor, J. (2021, February 5). Atlantic Canada is the next land of opportunity (even if the rest of Canada doesn't know it yet). *The Financial Post*. https://financia lpost.com/news/economy/atlantic-canada-is-the-next-land-of-opportunity-even-if-th e-rest-of-canada-doesnt-know-it-yet.

Province of New Brunswick. (2022). *Strategic vision: The invitation*. www2.gnb.ca/ content/dam/gnb/Corporate/Promo/l-the-invitation/tourism-strategy.pdf.

Province of Prince Edward Island. (2022–2023). *Tourism strategy: Charting the course of tourism in PEI for the next two years*. www.princeedwardisland.ca/en/publication/ tourism-strategy-2022-2023.

Pugh, J. (2013). Island movements: Thinking with the archipelago. *Island Studies Journal*, 8(1), 9–24. https://islandstudiesjournal.org/files/ISJ-8-1-2013-Pugh_0.pdf.

Rodrigue, J. P. (2020). *The geography of transport systems*. London: Routledge.

Savoie, D. J. (2000). All things Canadian are now regional. *Journal of Canadian Studies*, 35(1), 203–217.

Sakellariadou F. (2014). The concept of marine ecotourism: Case study in a Mediterranean Island. *International Journal of Climate Change: Impacts & Responses*, 6 (1), 33–39. www.cabidigitallibrary.org/doi/full/10.5555/20143371204.

Smaldone, D. (2007). The role of time in place attachment. In R. Burns and K. Robinson (comps.) *Proceedings of the 2006 Northeastern Recreation Research Symposium*. U.S. Department of Agriculture, Forest Service, Northern Research Station. www.nrs.fs.usda.gov/pubs/gtr/gtr_nrs-p-14/7-smaldone-p-14.pdf.

Stratford, E. (2013). Guest editorial introduction. The idea of the archipelago: contemplating island relations. *Island Studies Journal*, 8(1), 3–8. https://doi.org/10. 24043/isj.272.

Stratford, E., Baldacchino, G., McMahon, E., Farbotko, C., and Harwood, A. (2011). Envisioning the archipelago. *Island Studies Journal*, 6(2), 113–130. https://doi.org/ 10.24043/isj.253.

Suthren, V. (2009). *The island of Canada: How three oceans shaped our nation*. Toronto: Dundurn Press.

Tomblin, S. G. (1991). The Council of Maritime Premiers and the battle for territorial integrity. *Journal of Canadian Studies*, 26(1), 100–119.

Tomblin, S. G. (1995). *Shifting boundaries and regional integration. Ottawa and the outer provinces: The challenge of regional integration in Canada*. Toronto: Lorimer.

Tourism New Brunswick. (2023). *2022 Tourism indicators / Indicateurs du tourisme 2022*. NB: Government of New Brunswick. www2.gnb.ca/content/dam/gnb/Depa rtments/thc-tpc/pdf/RSP/Indicators_Indicateurs/IndicateursDuTourisme2022Tour ismIndicators.pdf.

Tourism New Brunswick. (2022). *2021 Tourism indicators / Indicateurs du tourisme 2021*. NB: Government of New Brunswick. www2.gnb.ca/content/dam/gnb/Depa rtments/thc-tpc/pdf/RSP/Indicators_Indicateurs/IndicateursDuTourisme2021Tour ismIndicators.pdf.

Tourism New Brunswick. (2021). *2020 Tourism indicators / Indicateurs du tourisme 2020*. NB: Government of New Brunswick. www2.gnb.ca/content/dam/gnb/Depa rtments/thc-tpc/pdf/RSP/Indicators_Indicateurs/IndicateursDuTourisme2022Tour ismIndicators.pdf.

Tourism New Brunswick. (2020). *2019 Tourism indicators / Indicateurs du tourisme 2019*. NB: Government of New Brunswick. www2.gnb.ca/content/dam/gnb/Depa rtments/thc-tpc/pdf/RSP/Indicators_Indicateurs/IndicateursDuTourisme2019Tour ismIndicators.pdf.

Tourism Nova Scotia. (2021a). *Nova Scotia visitor origin*. NS: Government of Nova Scotia. https://tourismns.ca/sites/default/files/2022-05/2021_visitorsByOrigin_YE-Fina l.pdf.

Tourism Nova Scotia. (2021b). *2019 Nova Scotia visitor exit survey*. NS: Government of Nova Scotia. https://tourismns.ca/sites/default/files/2021-01/2019%20VES%20Full %20Year%20Report.pdf.

Tourism Nova Scotia. (2022). *Nova Scotia visitation indicators*. Government of Nova Scotia. https://tourismns.ca/visitation-statistics.

Tourism PEI. (2023a). *Tourism PEI visitor indicators*. www.tourismpei.com/industry/ research/tourism-indicator-dashboards/visitor-volume-value.

Tourism PEI. (2023b). *Visitor volume model: Visitors to PEI*. www.tourismpei.com/ industry/research/tourism-indicator-dashboards/visitor-volume-value.

Traustason, S. R. (2017). *Developing a new and innovative tourism product in Atlantic Canada by deciphering ancient Icelandic sagas*. Master's dissertation. Western Norway University of Applied Sciences, Norway. www.academia.edu/35542971/ VISIT_VINLAND_Developing_a_New_and_Innovative_Tourism_Product_in_Atla ntic_Canada_by_Deciphering_Ancient_Icelandic_Sagas.

Vannini, P., Baldacchino, G., Guay, L., Royle, S. A., and Steinberg, P. E. (2009). Recontinentalizing Canada: Arctic ice's liquid modernity and the imagining of a Canadian archipelago. *Island Studies Journal*, 4(2), 121–138. https://doi.org/10. 24043/isj.231.

Part II

Complete, sovereign, archipelagic states

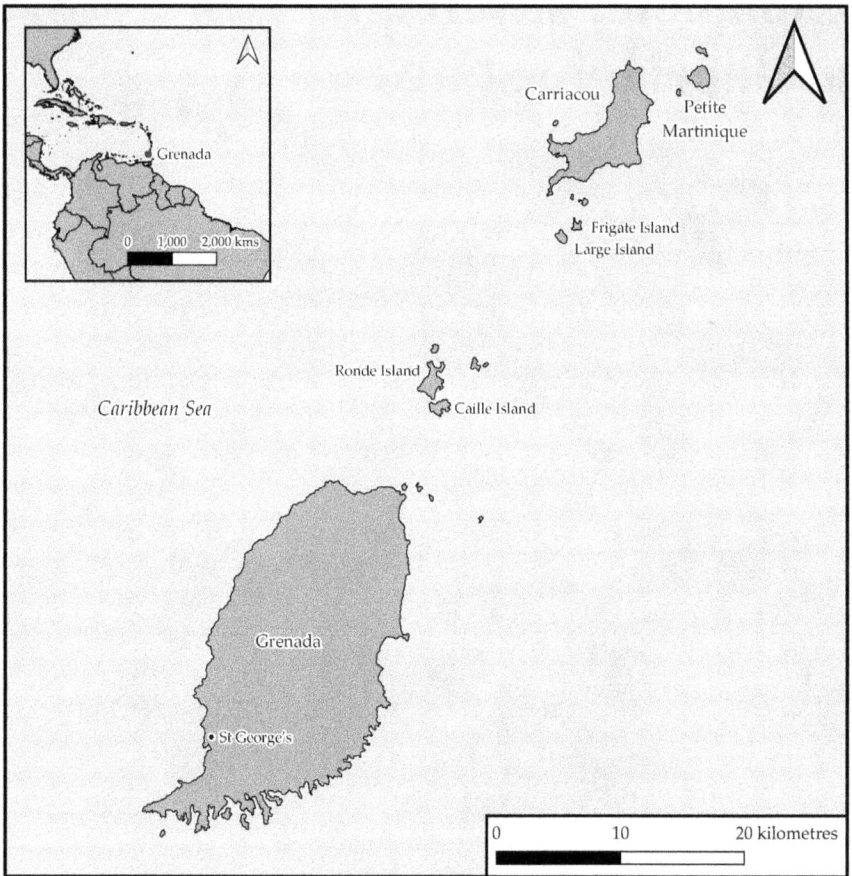

Figure 7.1 Grenada (with the Lesser Grenadines, including Carriacou and Petite Martinique)

7 Archipelago tourism in the tri-island state Grenada

John N. Telesford and Godfrey Baldacchino

Introduction

The Caribbean consists of many small island jurisdictions, many of which consist of multiple island units. The Bahamas comes on top, with some 700 islands, of which 30 are inhabited. At the other end, island states and territories, like Barbados, Saba, and St Lucia, boast single populated islands, and can only count a few islets or rocks beyond the mainland. In between are archipelagic jurisdictions, whose geographical nature may also be reflected in their names: Antigua and Barbuda, St Vincent and the Grenadines, Trinidad and Tobago, St Kitts-Nevis, and Turks and Caicos Islands. Multiple islands are inhabited in all of these. Grenada is another Caribbean archipelagic state; yet its multi-island nature is not reflected in its name. And, like many island configurations, the name of the jurisdiction is the same as the name of the mainland (or main island).

This chapter looks at the nature of archipelagic relations in the tri-island state of Grenada, using the five dimensions identified by Baldacchino (2015) as its conceptual framework. The information on which the chapter is drawn is based on: the insider knowledge of JNT, a Grenadian by birth and now a lecturer at the country's Community College, its highest education institution; and the outsider knowledge of GB, an island studies scholar, who – assisted by JNT – visited Grenada for a few weeks in early 2017 while his wife was conducting research leading to her doctoral thesis. This chapter is another example of collaborative work by the two co-authors, who have known each other over many years.

The Island of Spice, or the tri-island state?

Grenada is a 'small island developing state' (SIDS), with a total population of around 110,000. Although not immediately self-evident, a closer look reveals that Grenada – the country, known as "The Island of Spice" – is more than just Grenada (the main island; population: around 104,000). Of the 12 suggestions as to the 'Best things to do in Grenada', however, Tripadvisor recommends only one activity that does not involve the main island of

DOI: 10.4324/9781003451037-10

Grenada: a visit to small 'Sandy Island' which almost disappeared when Hurricane Ivan struck Grenada in 2004. There is no encouragement to consider taking the short flight with SVG Air or 90- minute trip with Osprey Ferries to the island of Carriacou (or C'cou; population: around 4,600; land area: 31 km^2) whose name allegedly means 'the isle of reefs': a place ideal for snorkelling, and suitable for a tranquil getaway (Tripadvisor, 2023; Carriacou.biz, 2023; but see Martin, 2022, p. 42). "The tiny island has some of the best dive spots in the world with clear waters, pristine coral reefs, and magical drifts" (Pure Grenada, 2023a). Also absent from this list of suggestions is any landmark on the third and final inhabited island component to the country: Petite Martinique (population: around 900; land area: 2.37 km^2). Indeed, most of the residents of Petite Martinique "derive their income from boat building and fishing" (and not from tourism) (Pure Grenada, 2023b). We are told – perhaps as an act of exoticising these islanders – that the residents of C'cou and Petite Martinique are fiercely traditional, welcoming visitors to discover traditional customs of African ancestry or maritime sport at annual events like the 'Kayak Mas' Street Carnival on Carriacou and the Petite Martinique Whitsuntide Regatta (Ins and Outs, 2023a, 2023b). There are various other islets in the country, as well as Kick 'em Jenny, an active submarine volcano or seamount; all are uninhabited. From Petite Martinique, it is just 800m of water to Petit St Vincent, but this islet is part of another jurisdiction – St Vincent and the Grenadines – as well as privately owned; so it is off limits to day-trippers from Grenada.

The implications are self-evident: on websites like Tripadvisor, Grenada is showcased as a mainly single island tourism destination. Carriacou and Petit Martinique miss out on most of the action.

Tourism in Grenada has rebounded fast to pre-pandemic levels. The country welcomed some 528,000 visitors in 2018 (latest year for which complete statistics are available); 65% were cruise ship travellers, spending a few hours on the island and often venturing only as far as the nearest downtown shopping mall or the spectacular Grande Anse beach, a 10-minute taxi or bus ride from the harbour. Additionally, many of these tourists embark on pre-booked tours to the Grand Etang rain forest or on a tour around the island, with very short stops, if any. Numbers then plummeted in 2020 (217,000) and even more in 2021 (72,000); but started picking up again in 2022 (335,000) (Tourism Analytics, 2023). (See Figures 7.2 and 7.3.)

In any case, Carriacou is hardly touched by tourism; and Petite Martinique even less so. Around 4,000 tourists are reported to have visited Carriacou annually on average between 2014 and 2019; with the numbers dipping to just around 1,000 in 2020 and 2021. This may explain why the islands that are a ferry ride or short flight away from the 'mainland' of Grenada are presented as less spoilt and more traditional. Petite Martinique does not have an airport; while Lauriston airport in Carriacou has an 800 m runway and only operates a few short-haul flights. Discussions are underway to install night landing facilities there which, Minister Andrews (see below) hopes, may attract a "major hotel brand here for the first time" (Loop News, 2023).

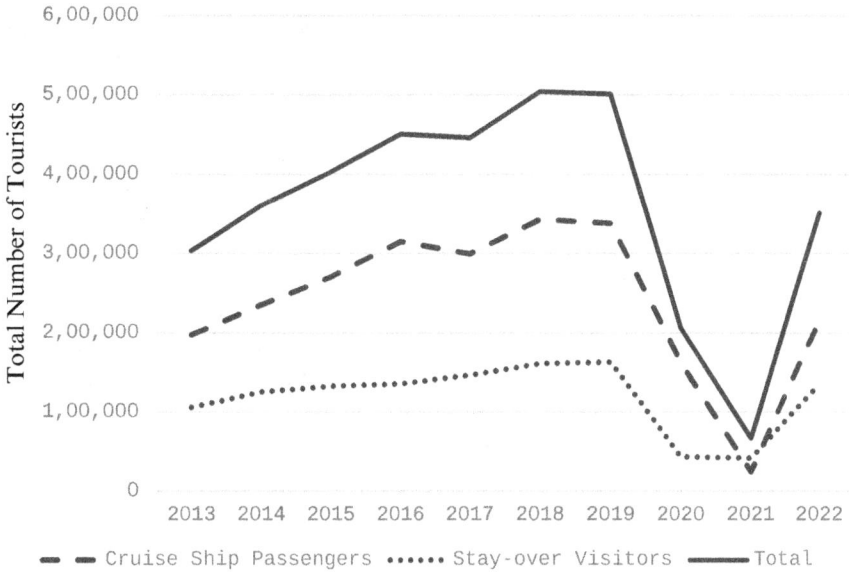

Figure 7.2 Tourist arrivals in Grenada (2013–2022)
Source: Eastern Caribbean Central Bank and Central Statistical Office, Grenada.

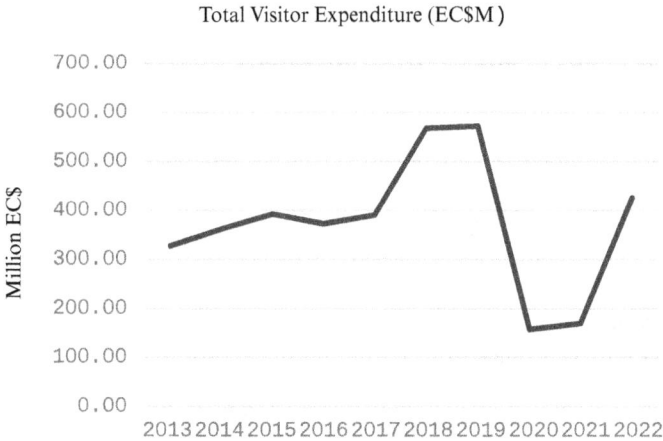

Figure 7.3 Tourist expenditure in Grenada (EC$: 2013–2022)
Source: Eastern Caribbean Central Bank and Central Statistical Office, Grenada.

There is no sense of such core-periphery relations in the official imaginary of the tri-island state. In contrast, one could be excused for thinking that the two smaller islands are very much at the centre of things. The six stars in the red border on the country's flag – designed by Anthony C. George and adopted upon the country's independence from the United Kingdom in

1974 – represent the six parishes that are found on the main island; while the larger middle star, encircled by a red disk, allegedly represents Carriacou and Petite Martinique (Brittanica, 2023), which are counted together as the island's seventh parish. In spite of its small resident population (in both relative and absolute terms), Carriacou has birthed two of Grenada's heads of state: Chief Minister Herbert Blaize (who also served as prime minister once Grenada became independent) and Prime Minister Sir Nicholas Brathwaite. Carriacou and Petite Martinique form a single constituency within the tri-island state and the current sole representative in the Houses of Parliament is Tavin Andrews, from the National Democratic Congress (NDC), who also serves as the Minister for Carriacou and Petite Martinique Affairs since June 2022. He replaced Kindra Mathurin Stewart of the New National Party (NNP) who served in that capacity from 2008.

The locals call their archipelago the 'tri-islands'. While still in early childhood centres and kindergartens, Grenadian toddlers and children are already being reminded that they live in an archipelago, where some islands are smaller or bigger than others. A short, cute poem is written on a wall poster in various early childhood settings. It doubles as an English language lesson in how to use comparative and superlative adjectives:

> Somewhere in the blue Caribbean Sea, lies a chain of islands three.
> The first one said: I am big, you see.
> The second one said: I am bigger than you.
> The third one said: I am the biggest of the three.
> Grenada, Carriacou and Petite Martinique.
> They are three islands in the sea.

Island identities matter, irrespective of size and scale. In the Caribbean, secessionist movements are "ubiquitous in virtually all multi-island states" (Bishop et al., 2022, p. 538). The self-evident geography of an island promotes feelings of distinction in oneself and the Othering of anyone else; it exaggerates and reinforces a self-image of difference. Regional identities in Carriacou and Petite Martinique are palpable (Montero, 2015). As Payne (2008, p. xxxiii) has noted in relation to Carriacou, not only is it "typically conscious of its own individuality in relation to Grenada, but … the people of its dependency, the even tinier Petit Martinique, identify scarcely at all with the people of Carriacou". Caribbean sub-national island jurisdictions, such as Anguilla and Saint-Barthélemy, featured in another chapter in this volume, are living examples of fission from larger political units. The People's Revolutionary Government, which governed Grenada during 1979–1983, had however suggested that the people of both Carriacou and Petite Martinique should "run these islands in their own assemblies" (Miller, 2007, p. 36); but this policy was never enacted. Nor have the citizens of these two islands clamoured for such autonomy. However, then newly-elected MP and Minister Tavin Andrews noted "that the implementation of local government would allow for the people of Carriacou and Petite Martinique to have greater say in their development" (Now Grenada, 2023). Implementation is scheduled before the end of 2023, as the Ministry of Legal Affairs drafting team prepares the legislation (ibid.)

As with Grenada, the rest of the world follows, as this chapter sets out to argue. Our "world of islands" (Baldacchino, 2018) is the plurality for a complex and often unacknowledged interplay of centrality and peripherality, status upgrades and downgrades, visibility and negligence *between* islands, enacted by politico-economic elites, citizens and visitors. In this endeavour, references to identity and identity politics are rife, possibly culminating in threats or initiatives in favour of that ultimate schism: breaking away. After all, an island is the quintessential single jurisdiction, seemingly thus pre-ordained by nature: its geography suggests a unitary polity. Indeed, islands that are shared by or between more than one country tend to have uneasy relationships (Baldacchino, 2015).

Having (at least) one elected representative in Parliament and a minister representing 'the outer islands' within the Cabinet is a workable arrangement that also seeks to mitigate secessionist tendencies. It is a situation that exists in various jurisdictions, including Malta (in relation to Gozo), and the Bahamas (in relation to the 'Family Islands'), apart from Grenada. The condition is described by Watts (2000, p. 24) as a political 'union', whereby it maximises shared rule and cohesion, but substantially at the expense of the constituent islands or their communities. Grenada is, like Malta, a centralised state and is an almost pure, two-party democracy; since 1999, only candidates from the NNP and the NDC have been elected to Parliament. The internal tensions and rivalries between the two parties and their supporters is easily lost to outsiders, including tourists.

Things are not always what they seem

If what appear to be single islands are mostly and actually archipelagos, then physical separation does not just separate them from mainlands but also from each other. Thus, inter-island crossings between (especially inhabited) islands, often at levels of demand that preclude profitability, will be hot items on the local political agenda. Cultural differences between islanders within the same island cluster may be enhanced and exaggerated, even invented, in order to appeal to specific tourist niches or political status; a stretch of water can make a world of difference.

In an archipelago, the ability to attract tourists may need to stretch to include the ability to share and spread the tourist dollar to the far-flung corners of the island group. And an archipelago's political establishment would need to balance, on one hand, the need to offer a sharp, strong, consistent, and clearly identifiable brand of the country as a single destination with, on the other hand, the need to allow each of the islands within its territory to develop and express its own voice and identity. Such an articulation of a harmonious complementarity of a plurality of voices and interests, often in acute competition with each other for scarce fiscal and spatial resources, will not align with notions of an alluring island paradise; but most tourists would be oblivious to this cacophonic and complex power play.

Centrifugal tendencies

In a short but seminal contribution, La Flamme (1983, p. 361) offers a nuanced definition of the archipelago state, by proposing 'four major attributes'. These involve: (1) dealing with a 'large number of islands'; (2) where component islands are small and underdeveloped; (3) where the surrounding waters are considered within the assemblage's boundaries and an integral part of its heritage and territory; and (4) which will experience a centrifugal tendency: "in an archipelago, the temptation is always great, at worst to secede and at best to disregard the political jurisdiction of the centre" (ibid.).

This inter-island and intra-archipelago dynamic strikes right at the heart of invariably uneven, 'island-to-island' politics. How do multi-island jurisdictions, especially if run by democratically elected politicians, balance the wishes of their various island publics and constituencies with the rationale of hub-and-spoke transport logistics (versus costly repetitive infrastructure), tourism differentiation (versus repetition), and complementary (rather than similar) and cooperative (rather than competitive) economic development trajectories? Most islanders implicitly know that they experience tense relations with their island neighbours. Often made fun of in popular idiom, these tensions and rivalries may find expression in discriminatory practices of various kinds, official or otherwise. These strains become more likely if specific islands claim, or are represented as claiming, a linguistic, racial/ethnic, religious, historical, occupational, and/or economic status and identity that are distinct from that pertaining to other islands. Such pseudo-ethnic distinctions can be easily fanned and contrived by the desire to strike a different, albeit complementary, island brand (Baldacchino and Ferreira, 2015).

In archipelagos, politicians and their publics often anchor their aspirations, affiliations and identities to specific island geographies, culminating in expressions of statehood-nationhood which are ultimately sealed via independence. Thus, there may be a Caribbean region – as the world's best known and branded tourism playground, for example – but the region is articulated as comprising Antiguans, Barbadians, Dominicans, Trinidadians and so many others – and this is just in the Anglo-speaking world (there are also US, Dutch and French island territories in the region). Serious attempts at federation have not waned: the Organisation of Eastern Caribbean States (OECS) – of which Grenada is a founding member – provides one successful example of how small island states combine resources and usurp the limitations of scale by crafting *confederal* institutions of shared governance, which therefore do not threaten the national basis of power politics (Roberts, Telesford, and Barrow, 2015). The OECS slate of institutions includes a common currency, central bank, regional air space and court of highest appeal. Such 'bottom up' trans-territorial approaches to governance and policy – rather than federal and 'top down', as in the case of the doomed West Indies Federation – are probably critical for long-term economic survival and institutional sustainability.

Island studies is part of the spatial turn in the social sciences that rejects the non- and anti-materialism that dominate the wave of post-modernity (Pugh, 2013). Geography still matters: ask Mexicans wanting to enter the US; or Eritrean migrants wanting to migrate to Europe. Of course, one should be careful not to essentialise islands, seeing only insular dynamics when domestic island politics – like any other politics – are riddled with their own rich tensions and divisive practices (e.g. Richards, 1982). Yet, archipelagic movements are one of the clear examples of the richly tumultuous ways in which island units converge and diverge (Stratford, 2013).

Considerations

In a landmark text, Lewis and Wigen (1997, p. 200) issued a powerful challenge, contesting and rebuking a dominant continent-oriented model of human geography and culture. Instead, they advocated the merits of:

> devising a *creative cartographic vision* capable of effectively grasping unconventional regional forms … [A]nalysing contemporary human geography requires a different vocabulary. Instead of assuming contiguity, we need a way to visualize discontinuous 'regions' that might take the spatial form of lattices, archipelagos, hollow rings, or patchworks.
>
> (emphasis in original)

The 'archipelagic turn', identified above, helps to assert the significance of inter-island dynamism and island-based nationalism, both of which remain relevant towards a comprehensive assessment of Caribbean development, including tourism. Deploying an archipelagic imagination illuminates and transforms identity theory beyond the nation-state framework.

Therefore, once the unitary identity of the nation-state is recognised for its archipelagic material discontinuity, it is easier to discern and evaluate the core-periphery dynamics underway. This involves an understanding of both urban-rural and main island – outer island considerations. Dealing with a small state does reduce the distance between settlements, but not the significance of these analyses.

Grenada's developmental and touristic predicament is impacted by both these vectors. First: the archipelagic. The country is aligned along a SW-NE axis, with its main Maurice Bishop International Airport at Point Salines, with a 2740 m runway, in the extreme south-west tip of the main island (IATA code: GND). A more central airport, Pearls, with a 1600 m runway, was decommissioned in 1984 (after the opening of Point Salines) and is now a construction site and drag racing strip. The longer but cheaper way to get to

Carriacou from Grenada, a distance of 73 km, is to first land at GND, drive to St George's (the capital), and then catch the ferry, which makes daily crossings; the 56 km journey – which, in open water, can be rough – costs US \$43 and takes two hours. The faster and more expensive option is a flight on an 18-seater plane from GND to Carriacou, which takes 20 minutes and costs US\$136 return (Carriacou.biz, 2023). Getting to Petite Martinique using public transportation is only possible via the daily ferry crossing from St George's. With these logistic challenges, one should not be surprised to note that some 98% of arrivals to both smaller islands are by private yacht.

Second, is the urban-rural divide which, as with many other island states, places St George's, the capital city, in an enviously advantageous position. With a population of 37,000, it is (by far) the largest settlement and commercial and administrative hub of the island. It also has the closest thing to a natural harbour in the country: so it boasts the main sea port, the ferry terminal, and lies a 15-minute drive from the airport. Its waterfront area, known as the Carenage, is "perhaps the most recognizable feature of Grenada" (Nelson, 2005, p. 138). With over half of all tourists travelling to Grenada arriving via cruise ship – and the Covid-19 pandemic hiatus has not made much difference – St George's becomes the main 'zone of exchange' between hosts and guests, and the area in the country where most tourist shopping takes place (Nelson, 2005). In a twist to the official narrative, some interpret the central star on the Grenada flag to actually represent St George's, rather than the two outer islands (Flags World, 2023).

The more naturalistic character attributed to the islands and residents of the South Grenadine twin islands of Carriacou and Petite Martinique is an oblique but not-too-subtle reminder of how tourism has not much affected these islands. If mainland Grenada needs to be sullied and rendered modern to make the comparison with the smaller island duo even starker and more explicit, then so be it:

> Carriacou is one of the last places, if not the last place, in the Caribbean that has retained the charm of yesteryear. It is often deemed the Caribbean's *last uncut diamond*. Even mainland Grenada has lost some of its charms, especially the capital, which is now choking with cars and heaving with entitled tourists (Our Carriacou, 2023; emphasis in original).

From endless possibilities to restricted movements, and much in between

The official website of the Grenada Tourism Authority wears its tri-island status on its sleeve:

> Grenada, Carriacou and Petite Martinique: With three islands at your fingertips, the possibilities are endless … Three Caribbean islands; one unique gem.

> (Pure Grenada, 2023c)

And immediately following is a map of the archipelago which shows only the three populated islands, with the various air and sea connections between them. Comparing this map to a Google map rendition of the same configuration shows that the map on the GTA website actually makes for a good example of "tweaked representation" (Baldacchino, 2015). It shows a distorted geographical representation of the state of Grenada, placing the three main islands closer to each other than they really are; and stylising the points from which the air and ferry connections operate (ibid.).

But the reality on the ground is somewhat different, and core-periphery relations come to play. Grenada has come late to tourism, and the service industry only gained prominence after the collapse of the banana regime and its preferential market access to Europe in the late twentieth century (Bernal, 2020). Carriacou and Petite Martinique have a lot of catching up to do; together, they attract only around 1% of tourist arrivals to Grenada. Tight political control by a central government reduces the space for political action from the islanders living in the single-seat, two-island constituency (Baldacchino, 2020). The situation has not changed during or after the onset of Covid-19.

The spatial dimensions of an archipelago offered some immunity to Grenada against Covid-19 in the early stages of the pandemic, as it did to other island jurisdictions. Geo-physical boundedness, aligned with the jurisdictional powers of a sovereign state, allowed Grenada a thorough and comprehensive control over travel via air and sea, including restrictions on inter-island travel. This would have initially prevented high caseloads on the tri-islands; however, border closures and controls can only be temporary to a small state that depends on various imports (including tourism) for survival. And so, creating an 'island bubble' was not a realistic prospect in the longer term (Telesford, 2021).

As the first coronavirus case was confirmed on the island of Grenada in March 2020, the residents of Carriacou and Petite Martinique became very guarded of their shores, insisting that Grenadians suspected of infection as well as foreign visitors should stay away. However, it soon became apparent that it was necessary for the residents of Carriacou and Petite Martinique to travel to Grenada, also for citizens requiring treatment and ventilator support in the local hospital. So the call for limited passenger travel between the islands soon emerged. Attempts to keep Covid-19 away from Carriacou and Petite Martinique while maintaining a connection with 'mainland' Grenada was part of the archipelagic juggle and intrigue of 'central island, smaller island' dependency (Telesford, 2021).

Conclusion: Renaming Grenada?

Interestingly, the New National Party government (2013–2022) had actually proposed renaming the country as 'Grenada, Carriacou and Petite Martinique', in line with various other Caribbean and Pacific archipelago states and territories. This was one of seven propositions – including setting a three-term limit for the prime minister, establishing fixed dates for elections, and reforming the electoral authority and the body that sets constituency boundaries – that were put to a referendum vote in November 2016, the first in Grenada's history. This particular amendment was intended to respect the three-populated-island nature of the jurisdiction, while specifying constitutionally the areas that have formed part of Grenada since its independence in 1974; just in case someone had second thoughts (Now Grenada, 2016). The measure, along with the other six propositions, was however summarily rejected by an apathetic and antagonistic public: only 43.7% voted in favour, with a voter turnout of just 32.5%. It appears that the Grenadian voting public is not entertaining a more explicit (and more unwieldy) recognition of its tri-island, archipelagic jurisdiction just yet.

References

Baldacchino, G (Ed.) (2013). *The political economy of divided islands: Unified geographies, multiple polities.* New York: Palgrave Macmillan.

Baldacchino, G. (Ed.). (2015). *Archipelago tourism: Policies and practices.* Farnham: Ashgate.

Baldacchino, G. (Ed.). (2018). *The Routledge international handbook of island studies: A world of islands.* London: Routledge.

Baldacchino, G. (2020). 'Together, but not together, together': The politics of identity in island archipelagos. In Y. Martínez-San Miguel and M. Stephens (Eds), *Contemporary archipelagic thinking: Towards new comparative methodologies and disciplinary formations* (pp. 365–382). Lanham, MD: Rowman & Littlefield International.

Baldacchino, G., and Ferreira, E. C. D. (2015). Contrived complementarity: Transport logistics, official rhetoric and inter-island rivalry in the Azorean archipelago. in G. Baldacchino (Ed.), *Archipelago tourism: Policies and practices* (pp. 85–102). Farnham: Ashgate.

Bernal, R. L. (2020). The importance of bananas in the Caribbean. In *Corporate versus national interest in US trade policy: Chiquita and Caribbean bananas* (pp. 41–61). New York: Macmillan.

Bishop, M., Byron-Reid, J., Corbett, J., and Veenendaal, W. (2022). Secession, territorial integrity and (non)-sovereignty: Why do some separatist movements in the Caribbean succeed and others fail?, *Ethnopolitics*, 21(5), 538–560. doi:10.1080/17449057.2021.1975414.

Brittanica (2023). *Flag of Grenada.* www.britannica.com/topic/flag-of-Grenada.

Carriacou.biz (2023). *Grenada Carriacou and the Grenadines: SVG flights to Carriacou - Grenada.* Osprey ferry schedule. https://carriacou.biz/transport/.

Flags World (2023). *Flag of Grenada: colours and meaning.* https://flags-world.com/en/flag-of-grenada.

Ins and Outs of Grenada, Carriacou and Petite Martinique (2023a). www.insandoutsgrenada.com/events/carriacou-carnival.

Ins and Outs of Grenada, Carriacou and Petite Martinique (2023b). www.insandoutsgrenada.com/events/petite-martinique-whitsuntide-regatta.

La Flamme, A. G. (1983). The archipelago state as a societal subtype. *Current Anthropology*, 24(3), 361–362.

Lewis, M. W. and Wigen, K. (1997). *The myth of continents: A critique of metageography*. San Francisco CA: University of California Press.

Loop News (2023, 22 May). Carriacou: Lauriston Airport wants to accommodate night landings. https://caribbean.loopnews.com/content/carriacou-lauriston-airport-exploring-options-night-landings.

Martin, J. A. (2022). *The A-Z of Grenada heritage*. St George's: Gully Press.

Miller, R. S. (2007). *Carriacou string band serenade: Performing identity in the eastern Caribbean*. Baltimore, MD: Wesleyan University Press.

Montero, C. G. (2015). Tourism, cultural heritage and regional identities in the Isle of Spice. *Journal of Tourism and Cultural Change*, 13(1), 1–21, doi:10.1080/14766825.2013.877473.

Nelson, V. (2005). Representation and images of people, place and nature in Grenada's tourism. *Geografiska Annaler: Series B, Human Geography*, 87(2), 131–143. https://doi.org/10.1111/j.0435-3684.2005.00187.x.

Now Grenada (2023, May 16). *Local government top priority for Minister Andrews*. https://nowgrenada.com/2023/05/local-government-top-priority-for-minister-andrews/.

Now Grenada (2016, September 19). *Grenada constitutional reform*. http://nowgrenada.com/2016/09/fact-sheet-grenada-constitution-reform/.

Our Carriacou (2023). *All about Carriacou*. https://ourcarriacou.com/all-about-carriacou/.

Payne, A. (2008). *The political history of CARICOM*. Mona, Jamaica: Ian Randle Publishers.

Pugh, J. (2013). Island movements: Thinking with the archipelago. *Island Studies Journal*, 8(1), 9–24. http://dx.doi.org/10.24043/isj.273.

Pure Grenada (2023a). *About Carriacou*. Grenada Tourism Authority. www.puregrenada.com/about-carriacou/.

Pure Grenada (2023b). *About Petite Martinique*. Grenada Tourism Authority. www.puregrenada.com/about-petite-martinique/.

Pure Grenada (2023c). *Landing Page*. Grenada Tourism Authority. www.puregrenada.com/.

Richards, J. (1982). Politics in small independent communities: conflict or consensus?. *Journal of Commonwealth & Comparative Politics*, 20(2), 155–171.

Roberts, S., Telesford, J. N., and Barrow, J. V. (2015). Navigating the Caribbean archipelago: An examination of regional transportation issues. In G. Baldacchino (Ed.), *Archipelago tourism: Policies and practices* (pp. 147–162). Farnham: Ashgate.

Stratford, E. (2013). The idea of the archipelago: Contemplating island relations. *Island Studies Journal*, 8(1), 3–8. https://doi.org/10.24043/isj.272.

Telesford, J. N. (2021). Critiquing 'islandness' as immunity to COVID-19: A case exploration of the Grenada, Carriacou and Petite Martinique archipelago in the Caribbean region. *Island Studies Journal*, 16(1), 308–324.

Tourism Analytics (2023). *Grenada*. https://tourismanalytics.com/grenada.html.

Trip Advisor (2023). *Top attractions in Grenada*. www.tripadvisor.com/Attractions-g147295-Activities-Grenada.html.

Watts, R. W. (2000). Islands in comparative constitutional perspective. In G. Baldacchino and D. Milne (Eds), *Lessons from the political economy of small islands: The resourcefulness of jurisdiction* (pp. 17–37). New York: Macmillan.

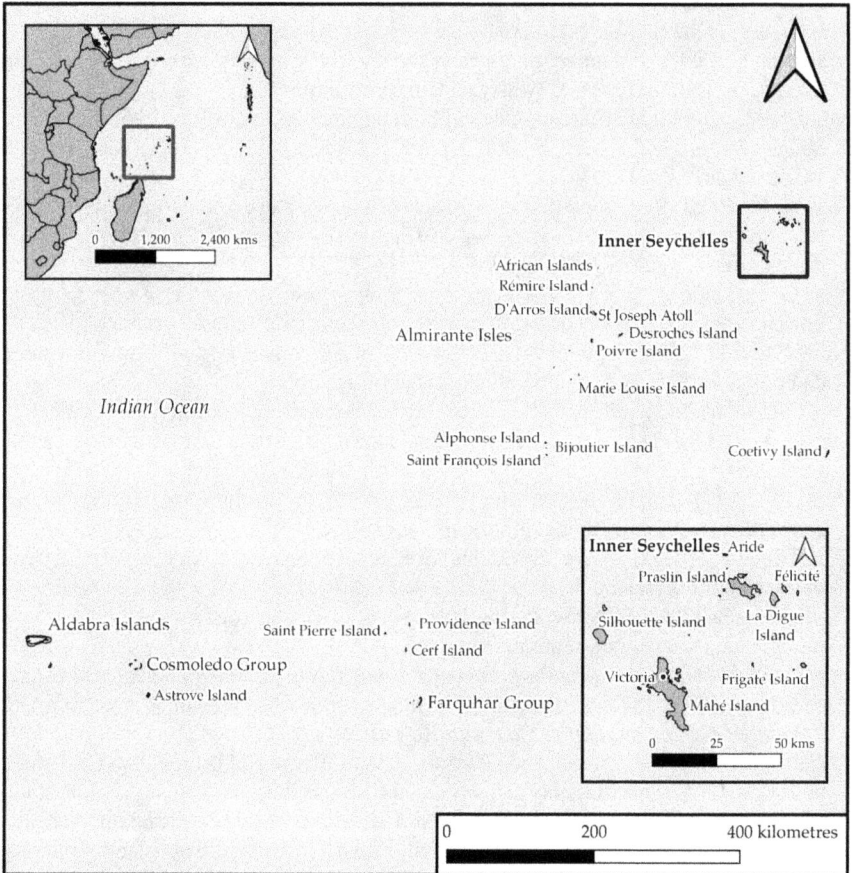

Figure 8.1 Seychelles

8 Archipelago tourism in Seychelles

Journeys through space, time, authenticity, and exclusivity

Godfrey Baldacchino and Hervé Atayi

Introduction

Assomption (land area: 11.6 km^2) is a coral islet in the outer islands of Seychelles. It made the international news when, in 2018, Seychelles and India signed an agreement to build and operate a joint military facility on this small island. A sustained wave of public protests followed in Seychelles, with many locals voicing concerns about their small country being dragged into the geopolitical 'Sino-Indian competition' simmering in the Indian Ocean (McDougall and Taneja, 2020). The agreement was declared 'dead' by the then Opposition Linyon Demokratik Seselwa (LDS) Party, now in government (Seychelles News Agency, 2018a).

This opening sets the stage for a more rounded understanding of Seychelles as something beyond the single identity (Massey, 1994) pedalled by the international media: that of a tropical, all-year-round, pleasure periphery for Western tourists. The small island state has a population just shy of 100,000 (making it the world's twelfth smallest country by population), a total land area of 458 km^2 (making it the fifteenth smallest country in the world by land area), but with an exclusive economic zone (EEZ) of 1.4 million km^2, aptly qualifying the country with the label of a 'large ocean state', with the world's 25th largest EEZ by country (GEF, 2016). The total contribution of the tourism industry to GDP has been steady around 60%, the third highest in the world by country; and the contribution of tourism to total employment in Seychelles is also a staggering 60% (McEwen and Bennett, 2010; Solarin and Lean, 2014; Gabbay and Ghosh, 2017).

These credentials are possible because Seychelles is a multi-island archipelago. Endowed with hundreds of islets and offshore rocks, the country officially recognises 115 named islands, of which Assomption is one. The island of Mahé (population: 95,000) is the country's 'mainland' and the site of Victoria, which is the country's capital city, seat of government, main port, and main urban settlement, with a population of around 35,000. Mahé is also the site of the country's only international airport (IATA Code: SEZ).

DOI: 10.4324/9781003451037-11

This chapter

This chapter explores how the fascination of this small island state as an Indian Ocean tourist 'paradise' is enabled, differentiated, and rendered more sustainable and exceptional by virtue of the country's multi-island geographic configuration, or "archipelagicity" (Wöll et al., 2023). 'Remoteness' is a readily understood characteristic of islands, as well as of island tourism (Dodds and Graci, 2012; Sharpley, 2012, p. 167); and, like Assomption, this condition also makes islands tempting candidates for military outposts. But the features and implications of the *nested* nature of *archipelago* tourism have often been neglected. Which is why this chapter argues that the inter-island distance – measured by the time it takes to get from one place to another – is a crucial differentiating factor in the making of an archipelago (including the Seychelles), and in the socio-spatial construction of its remoteness (Ronström, 2021). It also argues that the brush with the coronavirus and ensuing pandemic (2020–2022) has hardly changed the tourism paradigm in this small Indian Ocean island state.

This chapter unfolds as follows. The general context to the development challenges of Seychelles is first presented. The chapter then introduces the 'idea' of the archipelago, and its usefulness in addressing tourism in the Seychelles: not simply as a geographic term for island fragmentation and dispersion, but of the forces and tensions at work to craft a rich and diverse, multi-island tourism brand while operating as a unified state with a history of central planning. In the third section, we argue that 'Seychelles' stands for an archipelago which, for tourism purposes, goes beyond the spatial: it offers a tacit invitation to multiple journeys across affordability, authenticity, and time itself. A discussion unfolds in the fourth section, including an assessment of tourism practice just before, during and after the onset of Covid-19 in the country. The fifth and final section concludes.

This chapter is also an attempt to address the paucity of scholarly literature about Seychelles by Seychellois. All too often, as happens with many small (and often island and warm water) jurisdictions, spaces are represented from the lens of outsiders; while the voice of the insider is unheard or belittled. Insiders are privileged interlocutors of their own predicament; but, on the other hand, they may be too close and too imbricated in what they do and see. They may also believe, or have an obligation to peddle, the romantic stereotypes associated with their country, especially the 'paradise island' tropes that accompany their local tourism industry's messaging and aspirations (e.g. Connell, 2003), and which may form the bedrock to their economic appeal. We have tried to address these valid concerns by developing a chapter that is itself an outcome of a conversation between two scholars: an outsider/ visitor (GB) and an insider/local (HA) respectively. The references below to specific tourist perspectives and experiences in Seychelles are to be credited to the local author. The outsider has a longstanding academic interest in the Seychelles and has visited the country a number of times, mainly in connection with the University of Seychelles. Otherwise, this chapter is based essentially on secondary data, and is supported by a systematic literature review. Specific data on the characteristics of each of Seychelles' 115 islands – such as land area and population (if any) and how to get there – has been collected from official sources and has informed the ensuing discussion.

The state of development of Seychelles

The state of development of Seychelles includes some palpable contradictions. The islands claim a strong commitment to environmental conservation and have placed around half their terrestrial surface area under protection: no other country has done as much proportionately (Shah, 2002; Gössling and Wall, 2007). Yet, a short stay in Mahé reveals traffic regularly snarled along the single, main, north-south thoroughfare, and with a public fleet of Tata buses that are heavy air polluters. While tourists congregate along the 65 beaches and adjacent turquoise waters around the main island – tourism is almost exclusively marine, which adds to the vulnerability of the industry – most of the locals live in town or up in the hills: flat land is rare and the airport runway was built on reclaimed land, as late as the 1970s. Most items, including food, are imported and are very expensive. This is perhaps not of much concern to the wealthy upscale clientele that the islands have targeted. But many islanders try to beat the exorbitant cost of living by growing their own food – especially fruit and vegetables – or catching their own fish. Indeed, "soaring land prices, rapid inflation, increasing volumes of exports, increasing poverty and malnutrition and neglect of agriculture" have long been recognised as part of the opportunity costs of the Seychelles tourism 'business model' (Wilson, 1979). Additionally, "most hotels and resorts are owned by foreign transnational hotel chains with a propensity to import personnel, technologies, food and drink" (Giampiccoli, Mtapuri, and Nauright, 2021, p. 444). Apart from the island trio of Mahé, Praslin and La Digue – the largest in terms of population and permanently inhabited – other islands host luxury resorts, private homes, farming facilities, and conservation programmes. The small island state encourages those visitors who have the time, and deeper pockets, to visit the 'outer islands'.

The islands are grouped within six different geographical clusters. Closest to the main three islands, and referred to as 'Inner Islands', are Silhouette, North, St Anne, Moyenne, Round, Long Island, Therese, Conception, Aride, Curieuse, Cousin, Cousine, Sisters Islands, Félicité, Marianne, Cerf, Bird, and Denis. In the 'Amirantes Group' are Remire, St Joseph, Desroches, Marie Louise, and Desnoeoufs. Platte, on which the first seven-star facility in Seychelles is under construction, is located in the 'Southern Group', along with Coetivy, which is allocated to farming. The 'Alphonse Group' hosts Alphonse, St Francois, and Bijoutier. Towards the West, lies the 'Farquhar Group', with St Pierre, Providence, and Farquhar Atoll. Finally, close to the African coast, lies the 'Aldabra Group', composed of Aldabra Atoll, Assomption, Cosmoledo Atoll, and Astove (see Figure 8.1).

Over 380,000 visitors overstayed on Seychelles in 2019, an all-time high. But the country then experienced serious declines in its tourism numbers during the Covid-19 pandemic, with the international airport closed for four months and domestic tourism practically inexistent. The number of visitors plunged to 124,000 in 2020, was marginally better in 2021 with 182,000 arrivals, and had still not reached 2019 levels in 2022, with 332,000 arrivals (National Bureau of Statistics, 2023). Visitors tend to be high spenders; such numbers are enough to maintain the national economy. The country has the highest gross domestic product (GDP) as well as the highest greenhouse gas emitter per capita in Africa, since international tourism is the main revenue generator (Ahmed et al., 2021, p. 4; Saddington, 2023).

The other notable revenue and employment generator is tuna processing: Seychelles hosts one of the world's largest tuna canning operations. Tuna comprises 95% of Seychelles' tangible exports by value. Fishery-related industries resisted the economic shock resulting from Covid-19 and become more prominent; a foreign-owned tuna fleet supplies the local canning plant, which remains the main provider of private sector jobs in the country (Guillotreau et al., 2023).

Enter the archipelago

We say 'island' but we usually mean 'archipelago': the relationality of multiple islands is a common occurrence in geography and politics. Indeed, the existence of a country as a single island unit – like Nauru – is a rare occurrence. A conceptualisation of the archipelago was proposed by La Flamme (1983) and then relaunched more recently within island studies (Stratford et al., 2011; Pugh, 2013). Interest in the usefulness of this notion has grown and diversified (e.g. Roberts and Stephens, 2017; Martínez-San Miguel and Stephens, 2020), including within tourism research (e.g. Baldacchino, 2015b; Baldacchino and Ferreira, 2013). Here, issues of concern have included transport logistics, official rhetoric, and inter-island rivalry leading to tensions and forces of fission or fusion, all triggered by competing economic and political elites from different islands within the same jurisdiction, wanting to cash in on lucrative economic opportunities, and notably tourism. La Flamme (1983, p. 361) also claims that every archipelago has a centrifugal tendency: "in an archipelago, the temptation is always great, at worst to secede and at best to disregard the political jurisdiction of the centre".

Such an argument does *not* apply as much to Seychelles. The archipelago has a post-independence history of rigid central planning and control, set up and finessed during decades of one-party authoritarian rule. None of its islands, singly or collectively, has any powers of self-determination – unlike, say, in the neighbouring Republic of Mauritius, where Rodrigues, the second most populated island, is a recognised subnational jurisdiction since 2002 and has its own House of Assembly (Toulouse and Vithilingem, 2007). And so, in Seychelles, and in spite of its many islands, only the island of Mahé boasts an international airport. Of the other inhabited islands, Praslin, the largest, has a resident population of just around 8,800. Three out of the country's 27 electoral districts are located on the islands – two of which on Praslin, and one on La Digue (population: 2,973) and which includes other small islands in its constituency (combined population: 583) – but these islands have never spawned their own political party, which would have been a clear sign of a distinct political consciousness, island-based proto-ethnicity and mobilisation (Baldacchino, 2013). This centralised model of governance is reflected in tourism advertising and branding, where individual islands and their offers are always integrated and encapsulated within a single, solid, national narrative. Hence, tourists visiting and spending time on some of the inner or outer islands would still refer to 'Seychelles' as their destination. This is a rare example of a jurisdiction where the name of the main island (Mahé) is not identical to the name of the country or territory as a whole (Seychelles) (Baldacchino, 2015a).

Instead, with its 115 islands, from Mahé to Aldabra, 'Seychelles' stands for an archipelago which, for tourism purposes, is a tacit invitation to a repertoire of multiple journeys across not just space but also time, authenticity and affordability. In this chapter, we propose a nuanced understanding of the archipelago as an instrument that enables a particular tourist experience which is paralleled by the physical journey/s away from Mahé to increasingly more distant and/or less accessible and visited places within the country.

For most tourists, getting to Seychelles is already expensive and typically involves long haul flights. Based on 2022 data, 73% of tourists hailed from Western Europe: here, Germany, France, UK, Italy, Switzerland, and Austria are the top source countries, in that order. An economy return ticket from Mauritius to SEZ – a relatively short and direct flight of 2.5 hours with Air Seychelles – cost €400 / US$430 in March 2022 (along with the cost of two PCR Covid-19 related tests; plus a €10 electronic and non-refundable application fee to enter the country). Once in the country, accommodation and food is expensive; and this is just the start: a 20-minute return flight from Mahé to Praslin costs around €180–250. Other islands, further away, are accessible via chartered boat or plane, or a combination of the two. In some cases, islands can only be visited after transiting through another island, a logistic challenge that is common in other archipelagic regions, such as Greece (Karampela, Kizos, and Spilanis, 2014)., For example, Curieuse Island (land area: 2.9 km^2; population: 7) can be visited as a one-day trip by boat from Praslin.

The journey beyond the spatial

The journey from Mahé to any of the other islands is, however, more than a logistic consideration. In accordance with other explanations of island tourism, the journey to and from the main island is itself part of the tourist experience: the obligatory transfer over water, by plane or boat, is a physical and spatial representation of the distinct island experience that awaits the visitor, entertaining archetypal notions of pilgrimage, transformation, catharsis and redemption (e.g. Péron, 2004): just what weary and stressed travellers from the metropolitan heartland of the West need and desire.

When the multi-island and archipelagic features of Seychelles are added to this rich canvas, we realise that other dimensions are added and overlaid onto the spatial. These constitute increasing levels of quality, *as if* the further away from the main island, and the greater the challenge of access, the stronger the essence or intensity of what is being sought. We use the clause *as if* because the proposed correlation is more likely fictional, and is not necessarily actually there; however, there is a strong imaginary and suggestive understanding that it is real in its consequences.

We recognise at least three such additional dimensions, although there may be more, and they tend to be inter-related: affordability, authenticity, and time. These are reviewed in turn below.

Affordability

Travel to small oceanic islands, located far from their source countries, is expensive. Travel to even smaller islands is even more so. Wide-bodied jets give way to small turbo-prop planes, ferryboats or catamarans. Travel itineraries are more flexible and irregular. The number of passengers per journey drops; so costs per passenger invariably increase.

Despite the use of wide-bodied jets by international carriers such as Air France, Condor, Emirates, Etihad, Ethiopian, Kenyan, and Qatar Airways, the cost of getting to Seychelles is still high compared to other destinations. For example, an economy return flight from London (UK) to Seychelles is no less than €1,000/ US$1085, depending on the season. Residents rely on frequent promotions by the big airlines to make long haul journeys, unless they are urgent and necessary. The high cost of getting into Seychelles is not limited to airfares. Once at the destination, tourists face unexpected costs relating to internal transportation. There is a lack of transparency in the pricing strategy of taxi operators: one tourist claimed paying €56 for a trip from the airport to his hotel, and then €30 for the same trip back to the airport (Atayi, 2022). Further irrationality in transportation pricing on the islands is the discrimination between residents and visitors, with the latter paying more than the former; thus visitors ultimately subsidise the residents. Residents pay some €40 for a return trip to Praslin (a 20-minute flight with a small plane on Air Seychelles); visitors pay between €180–250. The ferry to Praslin from Mahé applies a similar policy, charging around €140 return for a visitor while the resident is charged €35.

The affordability of Seychelles goes beyond transportation. In spite of its multitudinous and diverse islands, Seychelles hardly produces anything. Everything is imported, agricultural land is scarce and the accessibility to the outer islands is not within reach of most local inhabitants or entrepreneurs.

Basic necessities such as food and beverages are also expensive. Here, residents as well as locals face the high price of such basic commodities. A 2018 study by the Seychelles Bureau of Statistics (SBS) suggested that 40% of the population were living under the poverty threshold. Within tourist spaces, the cost of food and beverages is even more exorbitant. This practice is often rationalised with cynicism since the country is a niche market and an exclusive destination. A former Minister of Tourism warned,

> the cost of operation in Seychelles continues to rise, irrespective of the fact that the islands are in competition with other similar tourism destinations. Increases in food and electricity costs are serving to price Seychelles out of the market. Smaller operators are feeling the burden of these increases, which are stifling their variability and potential growth.
>
> (St Ange, 2018)

The affordability factor is seeing by some as a differentiator. When it is expensive it is good; or, as a resident put it, "it is paradise, so you have to spend more to get it" (Atayi, 2020).

Authenticity

For international tourists, the journey across the Seychelles archipelago is also a journey that, given the location of SEZ, must start from the most urbanised, populated, and cosmopolitan island; and therefore clearly the one that is most impacted by civilisation and modernity. With a land area of 158 km^2 and some 90,000 residents, plus being the obligatory point of arrival, transit and departure of all international visitors to the country, Mahé has, by far, the largest population (and traffic) density within the archipelago. Solid waste continues to increase, while landfilling continues to be the main disposal path (Meylan et al., 2018). This state of affairs offers a stark contrast to what awaits visitors to the other islands.

To those who have some knowledge of the word, the mere sound of the name 'Seychelles' can evoke not only the typical association with 'sun, sea and sand' tourism, but also unspoilt and untouched nature, with not one but several paradises or gardens of Eden across the vast array of a sprawling archipelago. Some of the attributes of Seychelles – its remoteness, pristine environment, absence of dangerous species, and pirate narratives – make it a paradise destination (Wilson, 1994). A French tourist confirmed this by exclaiming "le paradis existe" (Dupaquier, 2019). While many may refer to Seychelles as if it were a journey to one single island destination, the reality is that it *can* be a journey through different and diverse small islands with their own unique specificities. From the busy and fast-paced atmosphere of Mahé, the next natural destinations to visitors are Praslin and La Digue.

Praslin is the second largest island, measuring 45km^2 and a short hop by air from Mahé or one hour by ferry. The island is renowned for its exceptional beaches, resorts and the world heritage site Vallée de Mai with its unique *coco de mer* (Lodoicea, commonly known as the sea coconut). It was the place that General Gordon (1899) dubbed the lost 'garden of Eden'. From Praslin, visitors are able to visit other smaller, closer islands such as Curieuse (a former leprosy centre during the colonial era), St Pierre, Aride (a bird reserve), but most importantly, La Digue, 7 km from Praslin.

The only way to get to La Digue is via a 15-minute ferry from Praslin. It is a popular destination for honeymooners because of its surreal pristine environment. It is the island of bicycles because it is the common mode of transportation. One of the world's most beautiful and photographed beaches, Anse Source d'Argent, is situated here. La Digue is also the gateway to its own other smaller islands of Petite Soeur, Félicité, Coco, and Marianne.

As the visitor moves away from the three main islands, the uniqueness of the Seychelles archipelago is revealed as a series of cameos where each episode entices the viewer to watch the following episode. Fregate – home to the endangered Seychelles magpie robin (*Copsychus sechellarum*) – is a private island with luxurious resorts where hundreds of tortoises roam freely. Cousine is a haven for untouched and protected wildlife; an island synonymous with nature where time stands still. And finally, Aldabra is the world heritage site, untouched and fragile. It is the home of some of the largest and oldest turtles in the archipelago. These are only a few of the different identities of the islands which composed Seychelles.

One could argue that such diversity makes Seychelles unique and its authenticity lies in the diversity of its islands. But for those tourists who do not have the means or the time to visit other islands in the archipelago, they tend to be impressed by the sandy beaches, so starkly different from what they are used to in their normal life in the West (Urry, 1990). Those lucky enough to be able to move beyond Mahé, Praslin, and La Digue can continue to gaze on the relatively pristine environment but also to experience the freedom of being cut off from the bustle of modernity and have a 'one to one' with nature in its purest state. It is unlikely that one would view authenticity in Seychelles as staged (MacCannell, 1973). In this context, visiting Seychelles is like going on a pilgrimage in search of an authentic experience, with Nature as the muse. The archipelago defies most theories of fakeness sold to tourists as a front or back stage experience (Goffman, 1990) or as pseudo-event tourism (Boorstin, 1964).

Matryoshka-like, the further the visitor moves away from the inner settled islands towards the deserted outer islands, the signal is that of a more authentically natural encounter (Echtner, 2010). While some tourist brochures manipulate and deliberately exclude local people from scenes or photos, in Seychelles, most islands are naturally devoid of people. Echtner (2010) argued that omitting people in brochures is also a contemporary marketing strategy, showing that empty space means upscale tourism, privacy, seclusion, and exclusiveness. But the remote islands of Seychelles do not need a wilful erasure of humanity; they are the "antithesis to the frenzy or busy bustle of everyday life" (Shellhorn and Perkins, 2004).

That said, the authenticity of Seychelles archipelago has its dark side. Despite the government and the NGOs' efforts towards protection, the concern for a holistic, sustainable, and consistent conservation of the environment remains a challenge. Plastic, glass, and other waste can be disposed of irresponsibly, and steps towards the care and preservation of the unique and endangered species of the country can be slow. A heavy dependency on foreign resources, expertise, and financial capital tends to slow the progress of conservation efforts. Moreover, economic and social issues such as drug and alcohol abuse affect a considerable number of locals and the latter have been transparent and candid about how such issues unmask the paradise myth, at least for the locals (Atayi, 2021). Some 5,000 Seychellois are addicted to heroin: the highest per capita figure in the world (Saigal, 2019). The contrast with the tourist experience can be shocking:

> Even if life in Seychelles is expensive, open your eyes: there are a lot of beauties around. Despite the difficult life, it is surrounded by beauty and moreover, there is no winter, this is paradise. There is no misery in the Seychelles, but there is still poverty. Tourists pay to experience paradise and we are magicians for them.
>
> (Atayi, 2020)

Time

It takes time to travel; and if one is visiting the Seychelles for a short spell, then some of the journeys within the archipelago that require long hours on boats operating on schedules that are liable to change with little notice, would simply have to be abandoned. Seychelles tourism is essentially marine and the journey to and from islands, by boat or plane, is wrapped deftly as part of the tourism experience in the country's marketing and brand messaging (Seychelles Tourist Board, 2022).

But the time dimension goes beyond this condition. Travelling away from Mahé is also a journey *in time*, away from the present, with all its discontents; and a journey towards a past, ideal and primordial 'state of nature'. The Valle de Mai, potential site of the Garden of Eden and a UNESCO World Heritage Site, a reserve with "the vestiges of a natural palm forest preserved in almost its original state" (UNESCO, 2022a), is not on Mahé but on Praslin, requiring a separate journey by boat or plane. And this is just the beginning: venture further afield, as the archipelagic tourism narrative goes, and the human footprint becomes progressively scarcer, while natural elements increasingly take over, subject to the geology of the 41 inner granite islands or the 74 flat, coralline outer islands. Many of the islands are really small, making them circumnavigable, walkable, and discoverable in an hour or less: so the sensation of grasping the essence of a place after even a fleeting visit is less elusive.

Discussion

Seychelles finds itself in the twenty-first century as a recognised quality tourism destination: the quality referring both to its tourist product (especially in terms of natural assets) as well as to the relative affluence of its visitors. The archipelago is not for the faint hearted or shallow pocketed. The country's official clutch of 115 islands spreads out across the South-West Indian Ocean. They are not too close to each other: that would have made them more easily accessible via a small boat or ferry after a short trip, or even via a tunnel or bridge; and probably all with permanent populations, since access to the mainland would not be hard: as in the cases of Gozo and Malta, the Faroe Islands, or the British/ US Virgin Islands (Ankre and Nilsson, 2015). But nor are the islands too far away from each other: making them accessible only via expensive and occasional plane journeys, or long ferry crossings, which makes multi-island hopping less likely: as in the case of Kiribati, French Polynesia or the Federated States of Micronesia.

No wonder, therefore, that the ultimate island destination within Seychelles, we propose, is Aldabra Atoll, inscribed since 1982 as a UNESCO World Heritage Site (UNESCO, 2022b). It is the Seychelles island that is furthest away from Mahé: over 1,140 km, and it has no airstrip. With a land area of 155 km^2, it is reputably the world's second largest raised coral atoll, the largest being Kiritimati (pronounced: Christmas), one of the Line islands of Kiribati, with a land area of 388 km^2 (Wikipedia, 2023). The island of Mahé would fit easily within Aldabra's lagoon. It has no permanent residents. It is also quite complicated to get to: to reach Aldabra from Mahé, one could either charter a private boat or take a boat cruise over the long distance each way. For those tourists who may have only one week in Seychelles, this option is simply not suitable. Otherwise, the only option is to book a 2.5-hour flight from Mahé to Assomption Island with the Islands Development Company (IDC) Aviation, and from there travel another 30 km by boat to Aldabra (IDC, 2023; Tripadvisor, 2021). There are no scheduled flights and details on charters can be obtained from the IDC that manages the islands (Seychelles News Agency, 2018b). Moreover, Aldabra is a Special Reserve under Seychelles Law; and so it is subject to a "limited and strictly controlled tourism policy". All visitors must receive prior authorisation; be accompanied by a Seychelles Islands Foundation (SIF) staff member at all times; and their access to the island is limited to specific areas (SIF, 2022). Getting to Aldabra is like getting to the end of the world, while travelling to another world.

Discussion

This understanding that the Seychelles archipelago is a journey through space, time, authenticity and exclusivity is, like all tourism narratives, partly fictional and aspirational. The islands of the Seychelles are also places where people live; places where people have been exiled, still imprisoned, or undergoing drug rehabilitation; places where charcoal and phosphate have been mined, copra produced, vegetables grown and animals reared; places where native species have been rendered extinct after the introduction of predatory cats and rats: the IUCN Red List includes 22 such, now extinct, species (Mongabay, 2023); and places that lend themselves to strategic or military purposes (as the opening paragraph of this chapter illustrates, with respect to Assomption). This is hardly the stuff of Eden (Stoddart, Benson, and Peake, 1970; IDC, 2022). Tourism idylls must operate alongside these pragmatic considerations which, in all fairness, approximate more typical attempts to render the different components of an archipelago complementary to each other, with some efforts at small island specialisation (e.g. Grydehøj and Casagrande, 2020).

From a theoretical perspective, the analysis presented in this chapter proposes a more nuanced and complex approach to 'archipelagicity' than that proffered by Stratford et al. (2011) in their seminal paper. There are added dimensions and layers to be acknowledged – and not all of which are illustrated in this chapter – corresponding to "alternative cultural geographies and alternative performances, representations and experiences of islands" (ibid., p. 114). There are also the "changing natures" of islands to consider (Benítez-Rojo, 1996): an analytic framework that remains largely underutilised.

The situation has hardly morphed under the impact of Covid-19. Seychelles was early to start inoculating its citizens, quickly becoming proportionately "the most vaccinated nation on earth" by May 2021, using Sinovac vaccines donated by China. However, cases of the virus surged, in spite of high levels of inoculation (CNBC, 2021). The logistic nightmares that gripped so many of the imported food items during the pandemic has impressed on the Seychelles government to consider food security as a policy priority and key strategic concern. Mahé has a serious shortage of agricultural land; and so growing crops and livestock on the other islands is a national economic and security imperative.

Just as the Seychelles government was witnessing the recovery of tourism to almost 2019 levels, a consultancy report published in July 2023 warned that the roads, and the sea and airport functioning on Mahé and La Digue, were "at capacity" and the best beaches were already "very crowded" (Seychelles News Agency, 2023). Is this manifestation of excessive tourism numbers also going to spread to the outer, nested islands? The state authorities are cognisant of the risks, and they are trying to restrict the many applications by Seychellois homeowners for 'change of use', which would allow them to rent private homes to foreigners, Airbnb-style. Such establishments may undercut regular accommodation providers and charge close to the bare minimum, thus eroding the quality and high-spending tourism model that the country has so far managed to preserve and nurture (ibid.).

Conclusion

It is in the analysis of how currents, real and metaphoric, flow and move between its islands that the character of an archipelago reveals itself (Baldacchino, 2015b, p. 8). This chapter has proposed a nuanced understanding of the archipelago as an instrument that enables a particular tourist experience which is paralleled by the physical journey/s away from the capital city to increasingly more distant and/or less accessible and visited places within the country. In Seychelles, the expression and manifestation of differentiation lies the progressive exclusivity provided by distance, and enabled by a plurality of island geographies. These assemblages, different material pieces of the country, "act in concert" (Stratford et al., 2011, p. 122): hence emerges a metanarrative where visitors are transported (pun intended) to more remote island worlds that can be coveted and enjoyed by progressively lower visitor numbers and higher natural content. Such a practice is self-reinforcing, since less numbers implicitly mean a reduced human footprint. In so doing, the journey is also understood and felt as one in time, ushering encounters with increasingly less adulterated, more authentic, and more pristine islandscapes. This begets a narrative of the utopian 'lost Eden': the world as it once was, and as it sadly is no more, other than in these few elusive niches. That such a strong and consistent brand message is at all possible is also attributable to the strong central state in Seychelles, which thwarts the 'centrifugal tendencies' (La Flamme, 1983) visible in many archipelagos around the world.

Islands are generative 'elsewheres': part-real, part-symbolic spaces that are highly susceptible to translatability because they articulate perspectives on the shifting relationship between self and other, centre and periphery (Bonnett, 2020; Stephanides and Bassnett, 2008). In the context of the marine tourism industry of Seychelles, they also present a privileged relationship that imbricates affordability, authenticity and time itself.

References

Ahmed, F., Houessenou, P., Nikiema, A., and Zougmore, R. (2021). *Transforming agriculture in Africa's small island developing states: Lessons learnt and options for climate-smart agriculture investments in Cabo Verde, Guinea-Bissau and Seychelles.* Rome: Food & Agriculture Organisation.

Ankre, R. and Nilsson, P.-A. (2015). Remote yet close: The question of accessibility in the Faroe Islands. In G. Baldacchino (Ed.), *Archipelago tourism: Policies and practices* (pp. 137–146). Farnham: Ashgate.

Atayi, H. (2020). *Unlocking the Garden of Eden: A postcolonial reading of tourists' and locals' image of Seychelles.* PhD thesis. University of Leicester. https://figshare.le.ac.uk/ndownloader/files/23930786.

Atayi, H. (2021). As real as it gets: Authenticity and tourism in Seychelles. *Seychelles Research Journal,* 3(2), 88–106.

Atayi, H. (2022). Seychelles' image as a tourist destination: The good, the bad, and the ugly. *Seychelles Research Journal,* 4(1), 74–98.

Baldacchino, G. (Ed.). (2013). *The political economy of divided islands: Unified geographies, multiple polities.* New York: Springer.

Baldacchino, G. (2015a). More than island tourism. In G. Baldacchino (Ed.), *Archipelago tourism: Policies and practices* (pp. 1–18). Farnham: Ashgate.

Baldacchino, G. (2015b). *Archipelago tourism: Policies and practices.* Farnham: Ashgate.

Baldacchino, G. and Ferreira, E. (2013). Competing notions of diversity in archipelago tourism: transport logistics, official rhetoric and inter-island rivalry in the Azores. *Island Studies Journal*, 8(1), 84–104. http://dx.doi.org/10.24043/isj.278.

Benítez-Rojo, A. (1996). *The repeating island: The Caribbean and the postmodern perspective*. Durham, NC: Duke University Press.

Bonnett, A. (2020). *Elsewhere*. Chicago: University of Chicago Press.

Boorstin, D. (1964). *The image: A guide to pseudo-event in America*. New York: Harper.

CNBC (2021, May 13). The Seychelles is the most vaccinated nation on Earth. But Covid-19 has surged. *CNBC*. www.cnbc.com/2021/05/13/seychelles-most-vaccinated-nation-on-earth-but-covid-19-has-surged.html.

Connell, J. (2003). Island dreaming: The contemplation of Polynesian paradise. *Journal of Historical Geography*, 29(4), 554–581. https://doi.org/10.1006/jhge.2002.0461.

Dodds, R. and Graci, S. (2012). *Sustainable tourism in island destinations*. New York: Routledge.

Dupaquier, J.-F. (2019). *Les Seychelles: L'envers de la carte postale*. Paris: Karthala.

Echtner, C. M. (2010). Paradise without people: Exclusive destination promotion. *Tourism, Culture and Communication*, 10(1), 83–99. http://dx.doi.org/10.3727/109830410X12815527582747.

Gabbay, R. and Ghosh, R. (2017). Tourism in the Seychelles. In R. Ghosh and M. Siddique (Eds), *Tourism and economic development: Case studies from the Indian Ocean region* (pp. 104–127). London: Routledge.

GEF (2016, 26 October). *We are a large ocean state: Environmental innovations for sustainable development of Seychelles*. www.thegef.org/news/we-are-large-ocean-state-environmental-innovations-sustainable-development-seychelles.

Giampiccoli, A., Mtapuri, O., and Nauright, J. (2021). Tourism development in the Seychelles: a proposal for a unique community-based tourism alternative. *Journal of Tourism and Cultural Change*, 19(4), 444–457. https://doi.org/10.1080/14766825.2020.1743297.

Goffman, E. (1990). *The presentation of self in everyday life*. London: Penguin.

Gordon, C. G. (1899). The site of the Garden of Eden. *Strand Magazine: An illustrated monthly*, 17(99), 314–317.

Gössling, S. and Wall, G. (2007). Island tourism. In G. Baldacchino (Ed.), *A world of islands: An island studies reader* (pp. 429–454). Malta and Charlottetown, Canada: Agenda Academic and Institute of Island Studies, University of Prince Edward Island.

Grydehøj, A. and Casagrande, M. (2020). Islands of connectivity: Archipelago relationality and transport infrastructure in Venice Lagoon. *Area*, 52(1), 56–64. https://doi.org/10.1111/area.12529.

Guillotreau, P., Antoine, S., Bistoquet, K., Chassot, E., and Rassool, K. (2023). How fisheries can support a small island economy in pandemic times: the Seychelles case. *Aquatic Living Resources*, 36, 24. https://doi.org/10.1051/alr/2023020.

IDC (2022). *Island Development Company*. www.idcseychelles.com/.

IDC (2023). *Aviation: Travel to Assomption*. www.idcseychelles.com/aviation.html.

Karampela, S., Kizos, T., and Spilanis, I. (2014). Accessibility of islands: Towards a new geography based on transportation modes and choices. *Island Studies Journal*, 9(2), 293–306. http://dx.doi.org/10.24043/isj.307.

La Flamme, A. G. (1983). The archipelago state as a societal subtype. *Current Anthropology*, 24(3), 361–362.

Martínez-San Miguel, Y. and Stephens, M. (Eds). (2020). *Contemporary archipelagic thinking: Towards new comparative methodologies and disciplinary formations*. Lanham, MD: Rowman & Littlefield International.

Massey, D. (1994). *Space, place and gender*. Cambridge: Polity Press.

MacCannell, D. (1973). Staged authenticity: Arrangements of social space in tourist settings. *American Journal of Sociology*, 79(3), 589–603.

McDougall, D. and Taneja, P. (2020). Sino-Indian competition in the Indian Ocean island countries: The scope for small state agency. *Journal of the Indian Ocean Region*, 16(2), 124–145. https://doi.org/10.1080/19480881.2020.1704987.

McEwen, D. and Bennett, O. (2010). *Seychelles tourism value chain analysis*. London: Commonwealth Secretariat.

Meylan, G., Lai, A., Hensley, J., Stauffacher, M., and Krütli, P. (2018). Solid waste management of small island developing states. The case of the Seychelles: a systemic and collaborative study of Swiss and Seychellois students to support policy. *Environmental Science and Pollution Research*, 25(36), 35791–35804. https://doi.org/10.1007/s11356-018-2139-3.

Mongabay (2023). *List of extinct species in Seychelles according to the IUCN Red List*. https://rainforests.mongabay.com/biodiversity/en/seychelles/EX.html.

National Bureau of Statistics (2023). *December 2022 Tourism Data*. www.nbs.gov.sc/downloads/38-economic-statistics/14-monthly-visitors-arrivals/121-2022.

Péron, F. (2004). The contemporary lure of the island. *Tijdschrift voor Economische en Sociale Geografie*, 95(3), 326–339. doi:10.1111/j.1467-9663.2004.00311.x.

Pugh, J. (2013). Island movements: Thinking with the archipelago. *Island Studies Journal*, 8(1), 9–24. http://dx.doi.org/10.24043/isj.273.

Revi, V. (2020, May 23). Seychelles and Assumption Island: Another test for India. *Observer Research Foundation*. www.orfonline.org/expert-speak/seychelles-assumption-island-project-another-test-for-india-66551/.

Roberts, B. R. and Stephens, M. A. (Eds). (2017). *Archipelagic American studies*. Durham, NC: Duke University Press.

Ronström, O. (2021). Remoteness, islands and islandness. *Island Studies Journal*, 16(2), 270–297. https://doi.org/10.24043/isj.162.

Saddington, L. (2023). Geopolitical imaginaries in climate and ocean governance: Seychelles and the Blue Economy. *Geoforum*, 139, 103682. https://doi.org/10.1016/j.geoforum.2023.103682.

Saigal, K. (2019, November 21). Why Seychelles has world's worst heroin problem. *BBC*. www.bbc.com/news/world-africa-50488877.

Seychelles News Agency (2018a, 20 March). *Opposition in Seychelles says it won't support military base deal with India*. www.seychellesnewsagency.com/articles/8864.

Seychelles News Agency (2018b). *Two cool UNESCO World Heritage Sites to visit in Seychelles*. www.seychellesnewsagency.com/articles/8723/+cool+UNESCO+World+Heritage+Sites+to+visit+in+Seychelles.

Seychelles News Agency (2023, 12 July). *Seychelles' two main islands reach carrying capacity - change of use permits on hold*. www.seychellesnewsagency.com/articles/18974/Seychelles%27++main+islands+reach+carrying+capacity+-+change+of+use+permits+on+hold#:~:text=Mobile%20%7C%20Desktop-,Seychelles%202%20main%20islands%20reach%20carrying%20capacity%20%2D%20change,of%20use%20permits%20on%20hold&text=Findings%20from%20the%20latest%20tourism,a%20top%20official%20on%20Wednesday.

Seychelles Tourist Board (2022). *The islands.* www.seychelles.travel/app-planyourvi sit-categories?selectedcat=DESTINFO&cla=ISLANDS&bannerCat=islands.

Shah, N. J. (2002). Bikini and biodiversity: Tourism and conservation on Cousin island, Seychelles. In E. Di Castri and V. Balaji (Eds), *Tourism, biodiversity and information* (pp. 185–196). Amsterdam: Backhuys.

Sharpley, R. (2012). Island tourism or tourism on islands?. *Tourism Recreation Research*, 37(2), 167–172. http://dx.doi.org/10.1080/02508281.2012.11081701.

Shellhorn, M. and Perkins, H. (2004). The stuff of which dreams are made: Representations of the South Sea in German-language tourist brochures. *Current Issues in Tourism*, 7(2), 95–133. http://dx.doi.org/10.1080/13683500408667975.

SIF (2022). *Visitor Information: Aldabra.* Seychelles Islands Foundation. www.sif.sc/a ldabra.

Solarin, S. A. and Lean, H. H. (2014). Nonlinearity convergence of tourism markets in Seychelles. *Current Issues in Tourism*, 17(6), 475–479. doi:10.1080/ 13683500.2013.810612.

St Ange Tourism Report: 2018. www.visitbhutan.org/latest-travel-news/saint-ange-tour ism-report-16th-april-2018/.

Stephanides, S. and Bassnett, S. (2008). Islands, literature, and cultural translatability. *Transtext(e)s Transcultures* 跨文本跨文化. *Journal of Global Cultural Studies* (Special issue), 5–21. https://journals.openedition.org/transtexts/212.

Stoddart, D. R., Benson, C. W., and Peake, J. F. (1970). Ecological change and effects of phosphate mining on Assumption Island. *Atoll Research Bulletin*, 136, 121–145.

Stratford, E., Baldacchino, G., McMahon, E., Farbotko, C., and Harwood, A. (2011). Envisioning the archipelago. *Island Studies Journal*, 6(2), 113–130. http://dx.doi.org/ 10.24043/isj.253.

Toulouse, L. R. and Vithilingem, S. (2007). Rodrigues' autonomy arrangements with the Republic of Mauritius. *Parliamentarian*, 88(1), 33–36.

Trip advisor (2021). *How to get to Aldabra.* www.tripadvisor.com/ShowTop ic-g293738-i9311-k10935209-How_to_get_to_Aldabra-Seychelles.html.

UNESCO (2022a) *Vallée de mai.* https://whc.unesco.org/en/list/261/.

UNESCO (2022b). *Aldabra.* https://whc.unesco.org/en/list/185/.

Urry, J. (2002). *The tourist gaze* (2nd edn). London: Sage.

Wilson, D. (1979). The early effects of tourism in the Seychelles. In E. B. De Kadt (Ed.), *Tourism: Passport to development?* (pp. 205–236). Oxford: Oxford University Press for the World Bank and UNESCO.

Wikipedia (2023). *Kiritimati atoll.* https://en.wikipedia.org/wiki/Kiritimati.

Wilson, D. (1994). Unique by a thousand miles. *Annals of Tourism Research*, 21(1), 20–45. https://doi.org/10.1016/0160-7383(94)90003–90005.

Wöll, S., Gfoellner, B., Pisarz-Ramirez, G., and Ganser, A. (2023). Introduction: Conceptualizing archipelagic mobilities. *Journal of Transnational American Studies*, 14(1). https://doi.org/10.5070/T814160878.

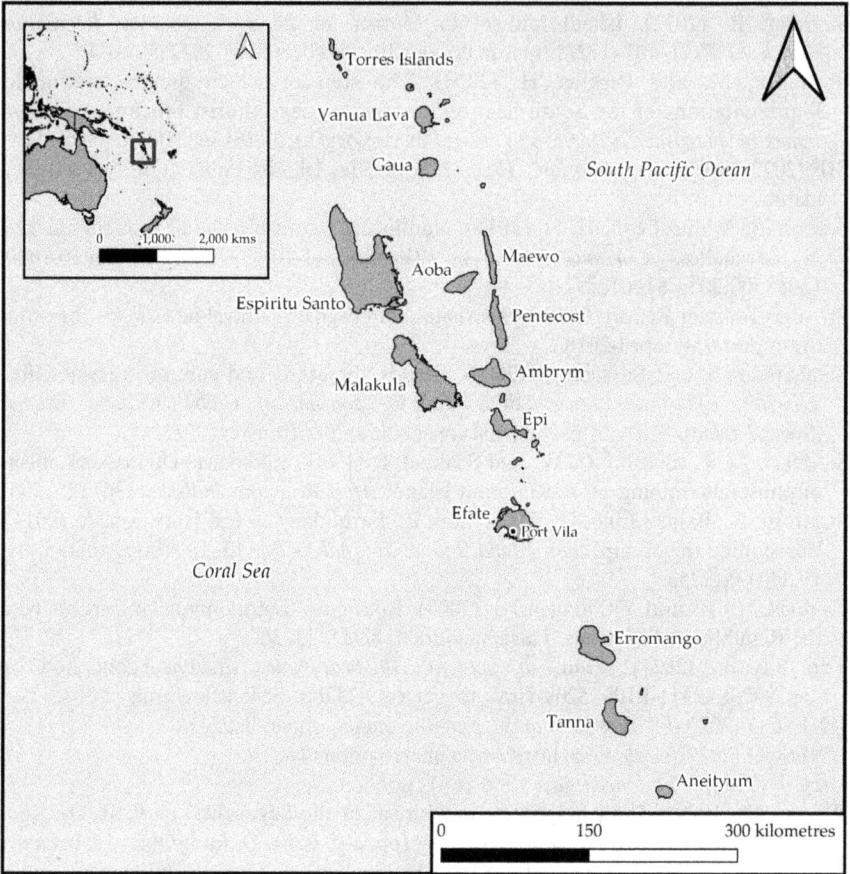

Figure 9.1 Vanuatu

9 Culture, Covid-19, cyclones, and supply chains

The contraction of tourism in Vanuatu

John Connell

Introduction

The Pacific island state of Vanuatu consists of about 83 islands, most of volcanic origin. Of these, 65 are inhabited by a population of about 335,000, and extending 1,300 kilometres between its northern and southern islands. Formerly known as the New Hebrides, a shared colony (condominium) of France and the United Kingdom since 1906, it became independent in 1980, after a brief, localised, but violent struggle. A modern tourist economy was emerging around the same time.

This chapter takes a historical perspective on the recent evolution of tourism in the country, as it seeks to rebound from both Covid-19 and the impacts of several severe cyclones that have affected different parts of the archipelago. Collectively, these problems have contributed to difficulties in re-launching tourism in a low-income state and have affected most parts of the country, including the outer islands (OIs): usually regarded as every island except the trio of Efate, (Espiritu) Santo, and Tanna. Material has been collected by means of a site visit by the author in mid-2023, where various leaflets and webpages were also sourced – and from which the quotes that follow are taken – as well as a review of four standard Australian newspapers collected over a period of 25 years: *The Australian* (A); *The Sun-Herald* (S-H), *Sunday Telegraph* (ST), and the *Sydney Morning Herald* (SMH).

The country's economic development is hindered by: dependence on relatively few commodity exports – mainly fish but otherwise almost all agricultural products (including copra, beef, kava and small quantities of coffee and chocolate) – vulnerability to natural disasters; and considerable distances between islands (with low volumes and high costs) and from main markets and sources of imports. In the post-Covid-19 era, these conditions have contributed to supply chain problems. Vanuatu's trade deficit has worsened since the economic slowdown of the 2020s, resulting from the collapse of tourism during Covid-19 and cyclone damage. Financial services include passport sales, mainly to Chinese business owners, and are increasingly significant revenue generators. With few avenues for economic diversification, Vanuatu has increasingly relied on tourism as the core of the economy and a

DOI: 10.4324/9781003451037-12

key source of employment, and international arrivals expanded steadily until 2020. The reopening of borders in mid-2022 brought the gradual restoration of the tourism industry; although in mid-2023 the number of arrivals was still substantially below that of 2019. Tourism's continued revival is crucial to national development.

Vanuatu is widely regarded as the most hazard prone country in the world. In this century, cyclones (including Cyclone Pam in 2015, Harold in 2020 and two, Kevin and Judy, in a single week in 2023) have delivered devastating blows to livelihoods (and the tourist industry). Being part of the Pacific Ring of Fire, volcanic eruptions and earthquakes (some accompanied by tsunamis) have damaged infrastructure, and required temporary population resettlement. Typically, recovery efforts are still ongoing when the next disaster arrives (Cruz and Passmore, 2023). Cyclones may worsen as climate change proceeds and Vanuatu has taken the lead, through the United Nations, in demanding 'loss and damage' payments.

Port Vila (Vila) has a population of around 50,000 while the second town, Luganville, has less than half of that, although in both places informal settlements overflow beyond urban boundaries. Together they account for about a quarter of the country's total population. Poverty, unemployment, and a lack of food security exist, mainly in informal settlements, and a growing incidence of non-communicable diseases is a source of concern. Cultural diversity is significant, between and within islands, with more than 100 indigenous languages, so averaging barely 3,000 speakers per language, with the lingua franca, Bislama, becoming a creole. Considerable religious diversity and fluidity occur even within villages.

Vanuatu was one of the last countries in the world to have avoided the coronavirus outbreak, recording its first case of Covid-19 in November 2020. By then, the country's borders had been closed and tourism abruptly ended after March 2020 (Naupa et al., 2021; Connell and Taulealo, 2021). The effective closure of the border, and of the internal 'border' between Efate and other islands, kept the incidence of Covid-19 low, so that, when the border was reopened in June 2022, hospitalisations and deaths due to Covid-19 had been few, with 14 recorded fatalities and 10,500 cases. The opening of the border allowed Vanuatu to belatedly rejoin the tourist market, anxious to catch up with other Pacific states, notably Fiji, that had reopened earlier. As it did so, concerns existed that Vanuatu was barely ready to restart, with a lack of agreements with foreign airlines posing a significant problem (overseas airlines being less interested in prioritising Vanuatu, as a small market), and its own national airline, Air Vanuatu, which declared bankruptcy in May 2024 and went into receivership. Such economic issues, and the still partly incomplete reconstruction after Cyclone Pam, impeded restoration of the vital tourist industry.

Genesis of tourism

Until the 1970s, tourism in Vanuatu was effectively unknown, with just a little offshore tourism from Australia (which was to become the country's primary market). Nonetheless, that coincided with a global context where tourism was being seen by scholars and international agencies, such as the World Bank and the UNDP, as not merely a recreational extravagance or cultural concern but a means of achieving some degree of economic and social development in developing countries (de Kadt, 1989). There was however some caution that, while national benefits were more obvious, tourism might not benefit outlying islands in far-flung archipelagos, especially in developing economies, because of tourist preferences for good infrastructure and services (Tisdell and McKee, 1988).

Political unrest around independence reduced even the little tourist activity that then existed. However, Air Vanuatu was established in 1981, and was operated by the Australian Ansett Airlines for the first five years, with tourist numbers reaching 32,180 in 1982. Numbers fell when the national currency, the vatu, gained value against the Australian dollar and civil unrest in neighbouring New Caledonia discouraged tourism (with Vanuatu not always being distinguished from New Caledonia in Australia). The end of Ansett's involvement and cyclone damage to Vila hotels resulted in numbers falling to 14,652 in 1987. Air Vanuatu's purchase of a Boeing 727 in 1989, an advertising campaign in Australia, more flights and, ironically, political crises in Fiji, the main Australian market in the Pacific, boosted numbers. By 2001, numbers of tourists arriving by air had grown to 53,300. In the same year, cruise ship arrivals were as numerous as air arrivals and, by the end of the decade, had surpassed them.

While many twentieth-century tourists did travel beyond Efate (see below), the construction of large international resorts brought a greater focus on Vila. In 2002, 60% of visitor arrivals were from Australia, 15% from New Zealand and 10% from New Caledonia. Two decades later, in the first quarter of 2023, these proportions were almost identical; nothing had changed despite the new presence of China (with 2% of tourists in 2023), which has been welcomed as a profitable market. In 2002, the average length of stay of tourists was 9 days and in 2023 it was 10.5 days, which in theory at least allowed time for tourists to visit OIs. Cruises brought a significant additional dominantly Australian market, but cruise numbers were recorded separately from air arrivals (presumably since they had little economic effect).

The promotion of tourism in Vanuatu never specifically focused on OIs, excluding intermittently Tanna and Santo. Brochures from large travel agents, such as Flight Centre, also excluded all other islands. Only belatedly and briefly were these OIs the centre of a marketing campaign in 2016 ('Discover what matters') that coincided with the establishment of provincial tourism offices (and the Bislama slogan *Turisim hemi bisnis blong evriwan*) and encouragement of local tourism bungalow development. On several islands, local people optimistically constructed tourist bungalows, but most were empty most of the time and the targeted promotion of the OIs has effectively been abandoned. (By the way, the Bislama slogan means: Tourism is everyone's business.)

1980s: Moon and Lonely Planet

Before Vila began to exert a post-independence primacy, tourist facilities were more evenly distributed over the three main islands, but tourist numbers were relatively few. Independence was matched by Vanuatu slowly opening up but tourism was not yet a priority in any vision of economic diversity. Vanuatu as a country was practically unknown internationally; and tourist brochures from the 1980s show that it was being marketed as something of a footnote to New Caledonia (partly to emphasise a distinctive French culture). There were occasional hints of OIs; but few had heard of the new country, let alone its many small islands. Cruises began at around the same time, but were effectively placeless with brochures saying almost nothing about the country and ships usually only called at Vila.

For Vanuatu as a whole, and for the OIs, the partial extension of the predominantly Asian 'hippy trail' into the Pacific, and its encouragement and direction, from two series of independent guidebooks – Moon and Lonely Planet (LP) – cast a focus on more of the archipelago. The age of low-cost discovery and adventure (and security) eventually extended the backpacker world, to the Pacific, redolent of an era of personal exploration, serendipity, and occasional whimsy. The first edition of Moon Publications' *South Pacific Handbook* provided reasonably detailed information on the four islands of Efate, Tanna, Malekula, and Santo, and brief mentions of Erromango, Aneityum (Anatom), Ambrym, and Pentecost. At Erromango,

> It is possible to fly from Vila to Ipota and walk across the island's centre to Dillon's Bay in the west in two days. At Dillon's Bay, camp on the grass beside the river and buy supplies in the village co-op store. Buy fresh mackerel from the fishermen for 10 cents each.
>
> (Dalton and Stanley, 1979, p. 180)

No hint was offered of why that might be worthwhile, whether a track existed, or what might occur if it rained. At Lamap (Malakula), a government rest house could be booked where "the gendarme will show you to your room. Bread and other supplies are available from local stores" (ibid., p. 171). No hints suggested restaurants or cooking facilities. Any lack of electricity or nightlife was not presumed to be an inconvenience. In 1995, to visit Aniwa Island, "make enquiries at the Tafea provincial government secretary's office at Isangel in Tanna" (Harcombe and O'Byrne, 1995, p. 173). Even on Lelepa island (off Efate) in 1979, hospitality was uncertain: "On Lelepa is a cave and good swimming. Perhaps stay the night with the locals" (Dalton and Stanley, 1979, p. 160). Much depended on possible – and presumed – islander hospitality. The OIs were clearly for the intrepid, adventurers rather than tourists, a theme that remains true in some contemporary descriptions. Both Moon and LP relied on tourist contributions and experiences to supplement the authors' own observations and experiences; but, by the end of the century, after a decade that was probably the relative heyday of the OIs, much of Vanuatu remained a blank slate for tourism. Even so, LP (by then in its seventh edition, but now combining Vanuatu with New Caledonia) was still the source of information for many visitors to OIs, at least into the late 2010s (Bennett, 2017).

Outer islands in the media

While even the Moon and LP guidebooks – and there were no others – presented relatively straightforward accounts of Vanuatu, other sources had more idiosyncratic perspectives on the country's tourist potential. Travel supplements in the Australian print media have always intermittently featured Pacific countries and particular islands, thus including Vanuatu and its islands, often and disproportionately emphasising the supposed exoticism of the OIs. Many were sponsored – and therefore positive and colourful – with pictures and headlines alone enough to generate an obvious sense of difference. Rarely did photographs of people not show them wearing grass skirts or *nambas* (penis sheaths) alongside cliched smiling children in floral headdresses.

Tanna was most commonly featured in the press, and entitled 'Custom made' (SMH, 29 March 2002), 'Ring of fire' (A, 25 September 2004), 'Island of fire' (SMH, 23 November 1996), 'Rumbles in the jungle' (S-H, 25 November 2007), 'Yasur, That's my volcano' (A, 28 November 1998), 'Isle of smiles' (SMH, 21 May 2021), and 'Tanna from heaven' (25 June 2023). Extended subtitles were even more expressive: 'life in a remote village fighting the onslaught of "civilisation"', 'right to the edge on a seething peak in Vanuatu', 'the wonders of nature are anything but pacific', 'in love with an unspoilt corner of Vanuatu', 'A pygmy village is the newest tourist attraction in Vanuatu', 'ni-Vanuatu on Tanna Island, among the happiest people on earth' and 'hears strange tales and meets true believers in the jungles of Vanuatu'.

Similar themes extended to OIs beyond Tanna: 'Be prepared for a dunking as you explore the beaches and waterholes of Espiritu Santo on the back of a horse' (S-H, 1 October 2017), 'Undiscovered, unspoilt paradise' (S-H, 4 December 2005), 'Swim in a Vanuatu blue hole' (SMH, 25 June 2023), 'Isle the resorts forgot' (SMH, 25 May 2013), 'The island that time forgot' (ST, 18 July 2004), 'Fantasy island' (A, 14 March 2009), 'Immersed in war treasures' (ST, 3 October 2010), 'Meet the unknown tribe' (SMH, 11 July 2004), and 'Secrets of a mass burial' (S-H, 15 November 2009) and the subtitles followed: 'Rather than playdown old cannibalistic customs, Malakula runs tours on them' and 'The thrills of Vanuatu's land jumps'. The tone, selectivity, and effusiveness varied little over a quarter of a century. Eventually, television travel programmes took up similar themes but necessarily less selectively.

Such stories spoke to the need for newspapers and magazines to emphasise diversity, exoticism, and difference, suggesting the range of possibilities available – with graphic stories a means of boosting their own readership – rather than being directly designed to stimulate tourism. Indeed, several such stories lacked any practical directions on how and where to experience such islands and activities (some of which were seasonal or even biennial). Hints of 'jungles' and 'pygmies' might even have deterred some.

Domestically, Air Vanuatu's bimonthly inflight magazine, *Island Spirit*, also took up such themes with similarly alluring titles such as 'Vanuatu's new wild adventure. Rare thermal waterfalls' (January 2004), 'The mystery of the Rom: Dancing on Ambrym Island' (July 2006), and 'Malekula's magic: Of cannibals, missionaries and vanishing people' (January 2007), all profusely illustrated with colourfully decorated and (un)dressed islanders. Yet *Island Spirit* more realistically included stories centred on Efate. A feature story on what were described as Efate's 'outer islands' opened:

> You don't have to travel to the remote islands of Vanuatu to experience the sensation of being there. In fact you only need an hour on the road from Port Vila to feel that you have flown from one island to another.
>
> (*Island Spirit*, July 2023)

At much the same time, the *Sydney Morning Herald* expressed similar sentiments:

> While there are 83 islands in Vanuatu, you actually don't need to drive more than an hour from the airport to discover some of the Pacific nation's best kept secrets.
>
> (8 July 2023)

These two quotations alone suggested that any effective focus on the OIs was over.

Twenty-first-century outer islands

Despite the promise of exotic cultures and distinctive experiences, in practice those tourists who ventured to the OIs invariably only visited Tanna and Santo. Until 1997 data is unavailable; but, in that year and, to a lesser extent in 2004, tourists who had arrived by air were firmly centred on Vila. In 1997, just 4% had gone to Tanna, and by 2007 that had grown to 10%. Fewer than that had gone to Santo in either year. In contrast, 15% had taken a round-Efate-island tour in 1997. Where tourists had travelled outside Vila it was mainly for such physical activities as diving, snorkelling, fishing, and sailing (Tourism Council of the South Pacific, 1998; Vanuatu NSO, 2007). By 2011, 11% of tourists had visited Tanna and 10% had gone to Santo, with just 6% going to any other islands: nonetheless a small increase in interest in the OIs. In the same year, 79% of tourists had only been to Efate, although two-thirds had visited some part of Efate other than Vila (Scheyvens and Russell, 2013). By 2017, 18% of visitors got to Tanna and 15% to Santo, while as many visited offshore Efate islands (6%) as went to all the other islands. Around 40% of tourists were return visitors and these were no more likely to visit OIs than first time visitors, preferring resorts in Vila (Addison, 2019). Give or take shifts related to local transport and natural hazards, proportions remained much as they were before Covid-19.

In 1997, most visitors were on packages which offered minimal flexibility, with accommodation in Vila being pre-paid for the duration the trip. That inevitably reduced the likelihood of venturing elsewhere. In the first decade of the twenty-first century, the principal reasons given by Australian tourists for not visiting any OI were that, deterred by high airfares, the OIs were too expensive to access (32%), they wanted a "stay put" holiday involving relaxed family time (27%) and did not have enough time (27%). Smaller proportions (16%) 'had not heard of any outer islands', were uncertain about possible experiences and services there, or did not want to fly in a domestic aircraft (15%) while some 16% simply had 'no interest' in the OIs (Cassidy and Brown, 2010). These basic structural explanations never substantially changed.

Over time, cruise ships became more frequent arrivals in Vanuatu, and the number of passengers had matched air arrivals by the turn of the century: a twenty-first-century cruise ship boom followed. All cruises included Port Vila (with many only visiting Vila) but in this century 'Mystery Island' – a tiny island off the coast of Anatom – has been a regular stop, along with Luganville, Champagne Bay (Santo), Lamen Bay (Epi), Wala (Malakula), and Pentecost occasionally being on itineraries. Cruises made limited contributions to local development, since time on shore was limited (Cheer, 2016), except in Efate, where available land-based tours were focused on Vila and its immediate environs.

In 2016, the government instituted a rural bungalow programme to encourage more ni-Vanuatu to set up small tourism businesses. By then, occasional Peace Corps or similar visitors had assisted OIs and local ventures with webpage construction; but few were comprehensive or maintained. While government encouragement brought a profusion of bungalows on some islands, on Ambae, the number of bungalow beds went from 9 in 2007 to 2,512 in 2012 (Scheyvens and Russell, 2013); it did not bring a profusion of visitors. Building them did not make tourists come. By then, the backpacker market had faded away without replacement.

Complications had also emerged for small-scale tourism. Since, by the 1980s, at least one village in Tanna had been encouraged to 'revert to the primitive' by no less than a *National Geographic* stringer in order to encourage tourism (Robinson and Connell 2008), it was unsurprising that when tourism developed in one of the poorest islands in the country, tensions emerged. Villages and islands are never homogeneous and 'tourism has rarely been about empowerment or equity, especially at the grassroots' (Connell and Rugendyke, 2008, p. 279). Patriarchal authority could hinder gains being more widely shared and locally invested and create gender tensions, all centred on where and how culture is enacted, especially where it is linked to critical livelihoods, and, conversely, how it is controlled, portrayed, and dictated by outsiders (Robinson and Connell, 2008; Cheer et al., 2013). Further tensions have followed the extent of participation, frustrations over limited returns from intermittent visits (Bennett, 2017), and arguments over land tenure. Even by 1980 disputes had occurred in places over land rights and access of tour groups, with access to Yasur being occasionally closed to tourists (de Burlo, 1989) and the first significant tourist venture in Tanna, the Tanna Lodge, being burned down in 2015 (Tabani, 2017). On Aniwa Island 'The lagoon is so beautiful that cruise ship companies have asked to visit it. However landowner disputes have so far prevented this from happening' (Harcombe and O'Byrne, 1995, p. 173). That dispute was still place at least in 2010 and had intensified divisions between the two villages on the island. The spectacular land diving ritual (*naghol*) on Pentecost heightened local community tensions as traditional culture was commodified (Cheer et al., 2013). At Wala (Malakula), disputes over access to and distribution of benefits resulted in the closure of Wala as a cruise ship destination, alongside resentment of successful small-scale entrepreneurs (Bennett, 2017; Orsua et al., 2023). Similar issues of involvement, employment, and income distribution in cultural tours to Lelepa island brought by now familiar tensions (Trau and Ballard 2023). Likewise, villages in Tanna have competed to be 'custom villages' and disputes have led to the closure of some villages to tourism – through tensions of ownership – circumstances much like those in Fiji (Connell and Rugendyke, 2008; Movono and Dahles, 2017). Such circumstances have discouraged and ended village tours in some contexts and made it difficult for distant agents and entrepreneurs to include them in tours and hence promote the OIs. At the same time, bungalow owners and tour promoters on small islands were often not easily able to contact urban entrepreneurs or the tourists themselves (Bennett, 2017). Bottom-up initiatives thus faced enormous challenges, making itineraries and presences uncertain.

Cultural villages

If tourists were not to seek distinctive environments and cultures on outer islands, or had no time to visit them, then versions of OI culture would come to them. From early in the twenty-first century, mainly migrants in Port Vila established 'custom' or 'cultural villages', where local people provide and play out (and modify) some facets of local culture: such as a welcome dance, a 'greeting' by 'warriors', a taste of kava and local foods, bamboo band music, medical techniques and other possibilities. Such shows emerged in the suburbs of Port Vila and in other parts of Efate (and, to a lesser extent, around Luganville). Consequently, for a time in the 2000s, it was possible to "see people perform from a real kastom village in Tanna", dramatically illustrated and fulsomely described on a local flier (Robinson and Connell, 2008) as cultural villages came and went, some beset by the same problems and tensions that faced their OI counterparts.

Ekasup Cultural Village, about 5 km from the centre of Port Vila, has been in existence since the start of the twenty-first century. Its promotional brochure stresses:

Upon arrival, you are greeted by a native tribesman who takes you on a 5 minute walk through a tropical rainforest. Visitors will be greeted by the chief and his warriors with clubs, spears, bows and arrows. This is a great opportunity to learn and experience their ancestral way of life. This includes experiencing food preservation, weaving mats/ baskets, hunting traps, fishing techniques, preparation of *laplap* (local dish), natural medicine. The trip relates tales of cannibalism in the not so distant past, the primitive but effective spears and other methods for trapping wildlife including pigs, chicken and fish ... The Chief and his professional guides will take you to learn the different village life styles from farming, home economics, architectural, medicament, ways of their cannibalism and many more.

Mele (probably the richest village in Vanuatu), on the fringes of Port Vila, advertised its Outdoor Cultural Centre and Nature Reserve in similar vein:

See a wide range of wildlife including a range of Vanuatu birds, wild ducks, parrots and jungle fowl. See flying foxes, huge coconut crabs, snakes, rare lizards and turtles. See beautiful custom house ... see and learn about Vanuatu magic and sorcery. Go inside our cannibal house and learn about the strange customs of old. Walk around our sacred kava garden and explore a real chief's house.

Other cultural village structures and 'performances' take similar forms (often derived from that of Ekasup), last for no more than a couple of hours, and are as much immersion in local culture as most tourists want (and rather more than do the majority).

Day tours around Efate, or parts of it, are more comprehensive. They may include a cultural village, alongside a 'traditional' lunch, swimming and snorkelling, and may also enable a short visit to a nearby island. Such tours may also include small chocolate and coffee factories. A tour to Lelepa 'your only closest island you can get to' offers 'unspoilt rain forest', bush walks and snorkelling, and was appreciated by tourists for water-centred activities; one tourist enthusiastically stated: 'Boys went fishing while we relaxed on the beach, great fishing and great snorkelling'. In contrast, the significance of Lelepa as a unique national site of cultural heritage (Trau and Ballard, 2023) was almost entirely ignored despite its 'historical caves with drawings and handprints'.

Other islands off Efate, such as Pele and Nguna, are included in day tours, with the latter being marketed as an 'outer island adventure' where

The Volcano Walk is a steep but rewarding hike, lasting 3.5 hours up the side of one of Nguna's extinct volcanoes. Once atop the hulking mass of rock, take in sweeping views across the Shepherd Islands to the north, and south to Efate. If you're in the mood for something less intense, choose the Environment Walk, perfect for families. Explore a community conservation area at Nguna's smallest village, a fascinating mixture of old forest, historic relics, and community gardens.

Tours run by Ecotours (out of Vila) are not greatly different since there is no real tourist interest in biodiversity, other than exotic flora and fauna. Ecotours' most successful tour is of Vila itself, which culminates at the central market, enabling some understanding of the diversity of local agricultural products. What distinguishes Ecotours is a greater walking component. Nature is more likely to be experienced dramatically: 'adventure quad tour in the jungle' or a 'two-hour rainforest walk' where some tropical marine animals may play a role: 'blue lagoon and baby turtle tour'. At Nguna:

> experience one of Vanuatu's outer islands through a half-day trek, followed by an afternoon snorkel. Support local communities who are conserving reef and forest resources.

Every island off Efate, other than Emao (the least accessible), has some interest(s) in promoting tourism.

Adverts for Efate's 'outer islands' take the same form as those for national 'outer islands'. For most tourists, experiencing these nearby OIs and cultural villages is enough to provide a degree of intrigue, exotic scenery (lagoons and 'jungles') and culture (grass skirts, bamboo bands, painted cheeks and bare chests), alongside drinking kava and eating distinctive fruit, and even boutique chocolate and coffee production: an interlude from urban and resort (in) activities. That is as far into the 'outer islands' as was necessary, in comfortable vehicles and at less expense, yet too far for the majority of tourists.

Covid-19 and its aftermath: A slow recovery

Covid-19 brought the temporary but total collapse of tourism, since no significant domestic tourism existed (unlike in larger and wealthier island states and territories, such as Malta and New Caledonia), critical in an economy increasingly and heavily dependent on tourism. The collapse lasted two years. It involved hotel and resort closures, selective losses of employment and incomes, with women (who made up around three-quarters of the workforce in tourism related activities, whether formal or informal) bearing the brunt. Markets, car hire, taxis, and the urban informal sector were particularly impacted. Several thousand workers lost formal sector employment and the informal sector was virtually wiped out. Handicraft makers and hair braiders had no local market. The domestic value chain of resorts such as The Havannah was no longer relevant. That raised complex questions over the duration of the crisis, and over self-reliance, diversification, sustainability, the future of tourism, the potential for a 'new normal', and therefore of tourist development in small island states (Connell and Taulealo, 2021; Naupa et al.,

2021; Connell, 2023). Such questions resurfaced during the post-crisis revival and especially after the twin cyclones of March 2023 that caused considerable damage to tourism infrastructure (and much else).

When the immediate health crisis was over, it was recognised that the main losers were OIs where tourism had just as abruptly ended, despite the greater ability of workers there to return to and revive an agricultural economy. The Ministry and the Vanuatu Tourism Office once again sought to assist the OIs, the main evidence of this being the production in 2021 by the Vanuatu Tourism Office of a *Domestic Tourism Marketing Strategy, 2021–2023*, followed by a glossy quarterly *Nawimba, Your Tourist Guide*, and a monthly *Vanuatu Life and Style*, both further oriented to *Sapotem Lokol Turisim*. The Marketing Strategy focused on Efate, Santo, and Tanna sought to encourage local people to experience other places and activities, especially sporting and cultural activities (Vanuatu National Tourism Office, 2021). However, just as in the early guidebooks, subsequent directions on where to stay could be limited. At Sola, on Vanua Lava, a promising destination for surfing, 'accommodations can fill up with government travellers so it is best to book your accommodation in advance' possibly at Panglap Homestay which 'has two bungalows with two single beds each and can house four people'; but no instructions were provided on how to contact the owner (*Nawimba*, April 2023). Otherwise phone numbers were provided but no information existed on accommodation style, catering, and the specific location of bungalows. On Maewo, 'hospitality is boundless ... Before you go, connect with nearby Santo Travel Centre for accommodation, ideas and phone numbers' (*Vanuatu Life and Style*, March 2023). This reversal to Moon and Lonely Planet days was not obviously successful, especially because of serious transport constraints. More successful were promotions of a Vanuatu Volcano Run in Tanna, a similar run in nearby Nguna, and other physical activities. Otherwise, there was minimal indication that the magazines (only apparently to be found in the Port Vila tourist office, which local residents had no need to visit) had any effect. No tradition existed of ni-Vanuatu people travelling for pleasure within the country – other than some public servants and NGO workers – and no great interest extended to visiting remote places with uncertain facilities.

During the Covid-19 crisis, the national airline, Air Vanuatu, remained in business only though government subsidies but was in a dire financial position when the border finally reopened. Its domestic operations were dysfunctional with, by 2023, its website mentioning only Tanna, Santo, and Pentecost as destinations (and suggesting "click here for our 2018 packages"). The largest travel agent in Australia was not taking bookings on Air Vanuatu, since it had yet to pay earlier bills, leaving only Qantas flying to Vila. That setback rebounded on domestic flights which have rarely if ever been profitable, and flights to and from OIs became increasingly fickle. Only Tanna and Santo retained some flight security. Coincidentally, during this time, Santo was increasingly referred to as an outer island. That was exacerbated in 2023 when Air Vanuatu's single Boeing 737 more than once developed technical problems, the necessary spare parts were unavailable and flights experienced delays of several days, at some cost to the airline. Such combined problems reduced capacity on flights into Vanuatu, disrupted links to OIs and created a lack of confidence in the airline and thus the travel industry. Cruises too had been as slow to restart and once again bypassed the OIs (while their brochures and websites, and media reviews, were as 'placeless' as ever, increasingly focusing on on-board amenities and activities, cabins and cuisine).

As in Fiji, the first returnees went to the resorts, whether boutique resorts such as The Havannah or international chains such as Holiday Inn, which offered the pleasures of enclosure and introspection. Urban resorts led the rebirth, assisted by tourist conservatism, returnees, and potentially accessible healthcare. However, many components of the tourism industry experienced labour shortages and a skill drain as some of their workers had found other jobs, or returned to villages. More significantly, many had left for Australia and New Zealand to become guestworkers in what were now more attractive and better-paid jobs, potentially outside agriculture (Petrou and Connell, 2023). Urban resorts were losing skilled chefs and others (some at the rate, in mid-2023, of one or two resignations per month) and replacing them at additional cost with migrants mainly from south Asia and the Philippines. Cyclone damage had reduced access to fresh foods of various kinds (notably tree crops such as mangoes, pawpaws, and bananas) and hotels were more dependent on imported foods, which were more costly and disrupted by fractured supply chains, and less frequent shipping. Some resorts were damaged by the cyclones and one at least had its capacity halved and, six months afterwards, was still awaiting replacement air-conditioning units. Post-Covid-19 return to normal was thus hampered as much by the cyclones, and by transport and labour supply factors, as by Covid-19 itself.

Vanuatu's priority was rebirth, but initially with some thought for the OIs and possible links to more sustainable structures, so that the local community would benefit not just in economic terms, but as a society. A *Vanuatu Sustainable Tourism Strategy (2021–25)* was formulated to follow on from the *Vanuatu Sustainable Tourism Policy 2019–2030* of two years earlier. In being oriented to 'supporting and regenerating our traditional economy', its objective was to 'lessen our dependence on tourism' and foster a tourist economy that was more diversified and more supportive of local industries and livelihoods. While the Strategy stressed such themes as resilience and well-being, it was unclear how these policies would be translated into practice or by whom. In a relatively poor and small archipelago state, simply getting the tourism economy underway quickly took precedence.

A few past efforts had been made to generate a more 'green tourism' that might support conservation, food security and livelihood outcomes for rural agriculturalists, and promote sustainable farming practices (Addinsall et al., 2017). For a time, Fansa Farm Foodie Tours provided an interpretative tour based on visiting traditional gardening systems, as well as harvesting, preparing, and eating 'traditional' crops. But it was knocked out by Cyclone

Pam in 2015 and was never revived. Nor were such touristic practices linked to the provision of local foods to tourist enterprises. A rhetoric of sustainable and regenerative tourism was divorced from practice and agricultural enterprises were too small even to provide regular supplies of such basic foods as eggs. Achieving any locally more beneficial 'new normal' was in serious doubt.

'For adventurous travellers' (Vanuatu National Tourism Office website)

Tourism in Vanuatu has been affected by political shifts, changes in airline and cruise ship itineraries, Covid-19, cyclones, labour migration, and not least by fractured supply chains (and the very basic supply chain of tourists, with limited airline capacity). It has rarely been affected by marketing strategies that emphasise the diversity of cultures and landscapes, offering anything more than a token engagement with outer islands. Guidebooks, media representations, tourist brochures, and webpages stress diversity and exoticism, seemingly most evident in the OIs, but rarely translated into travel to those islands.

Only around the turn of the century, as national tourism itself grew rapidly, was there real growth in OIs, but in a very few islands (despite local interest). More accessible Tanna cornered the market for the exotic, whether the Yasur volcano or 'traditional' Yakel village (Robinson and Connell, 2008). Many islands have small populations, limited infrastructure, and poor connectivity for them to achieve greater agency. Indeed, the tourist economy has contracted to urban resorts and the token 'outer islands' off Efate. Cultural villages have moved to the market, where they portray the supposed 'exotic' elements of the OIs. Resorts and cruises emphasise urban primacy. The 'famous' 83 islands of Vanuatu are a shrunken tourist archipelago: urbanisation proceeds and the peripheries decline, while tourism is more centralised despite official commitments to the outer islands. Tourism makes minimal contributions there, emphasising inequality. Even when visitors do travel off Efate, much of their expenditure remains in or returns to Vila.

By 2004, the internet had become the most important source of information on Vanuatu after travel agents (Vanuatu NSO, 2007); but by 2018, that had been displaced by advice from friends and family. General websites such as Tripadvisor were almost as important and Vanuatu's own website was of no importance. Magazine and newspaper articles too had trivial significance. The most critical source of information for those who chose to stay in island bungalows was friends and family; least important were travel agents (Addison, 2019). Overseas travel agents are not (and cannot be expected to be) well informed on OIs. In both the cheapest forms of accommodation and in the elite resorts, tourists themselves were making their own decisions based on personal advice and previous experience.

Online marketing provides minimal information on other islands. Tourists want a secure resort package – a familiar certainty rather than insecurity, especially in the wake of Covid-19. Precisely the same is sought by tourism workers, whether chefs, hair braiders, or gardeners, anxious to get their jobs back (or migrate overseas). Indeed, Covid-19 and cyclones have in effect conspired against any new normality – no green sustainable tourism in a dynamic blue economy – and required a range of efforts to get back on track. No evidence exists that *lokol turisim* has occurred, and for similar reasons that deter overseas tourists: a lack of infrastructure (including hotels), shortages of time, uncertainties (such as infrequent and irregular planes), and cultural differences. Likewise a more sustainable tourism remains rhetorical. Tourists have no duty to visit the OIs; no pressure is exerted on them to do so, whether from the government or the predominantly urban private sector, which itself has no interest in decentralisation or regional development.

Optimism about the future is centred on restoring the Australian, New Caledonian, and New Zealand markets, and developing a Chinese market (nascent before Covid-19 struck), thus focused on expenditure and employment in urban resorts, restaurants, and casinos. It is not oriented to bungalow owners in remote islands without webpages (or smartphones). A steady attrition of skilled workers and island leaders from OIs – often out of frustration with 'development' and urban bias, in search of better access to schools, jobs and shops – has left some places bereft of management and leadership skills and added to local conflicts over the few economic inputs of tourism.

Decades later, Tisdell and McKee's (1988) cautionary words over the possibilities of archipelagic tourism have been vindicated. Under pressure the limited gains of the late twentieth and early twenty-first centuries were confounded by urban bias alongside a multitude of factors affecting Vanuatu – whether international investment, cruises, or cyclones – over which Vanuatu and ni-Vanuatu have little, if any, control. Unusually fragmented politics, rapidly changing governments (and thus Ministers of Tourism), limited development finance (to support an ailing airline industry), and slow recovery from hazards add unusual burdens in and to the small state. Vanuatu is a country where three-quarters of the people live outside Efate, and thus where three-quarters of politicians come from other islands; and yet, few are much interested in regional equity, which is true also of access to overseas employment opportunities. Moreover, in tourism, the interests of the urban private sector, some of it multinational, dominate. Covid-19 and cyclones have joined forces with capitalism to create a centralised tourist economy. The climate of the times is against small islands, even for tourism; while, ironically, the images of these islands still pervade brochures and websites. As confided by a Malakula island tourist operator: "We are at the bottom and a long way away" (quoted in Bennett, 2017, p. 46).

It is not Covid-19 but an ever-present, ongoing normality that constantly changes, challenges and undermines the equitable development of tourism in archipelagos.

References

Addinsall, C., Scherrer, P., Weiler, B., and Glencross, K. (2017). An ecologically and socially inclusive model of agritourism to support smallholder livelihoods in the South Pacific. *Asia Pacific Journal of Tourism Research*, 22(3), 301–315. https://doi.org/10.1080/10941665.2016.1250793.

Addison, A. (2019). *Tourist accommodation choice and destination development: The case of Vanuatu.* Master's dissertation Auckland, New Zealand: School of Hospitality. Auckland University of Technology. http://openrepository.aut.ac.nz/items/e0d5d767-69ee-4123-85af-2b0427a892e2.

Bennett, A. (2017). *Tourism bisnis and 'making ples': An ethnography of ni-Vanuatu bungalow and tour owners on Malekula.* Master's dissertation Wellington, New Zealand: Victoria University of Wellington. https://openaccess.wgtn.ac.nz/articles/thesis/Tourism_bisnis_and_making_ples_An_ethnography_of_ni-Vanuatu_bungalow_and_tour_owners_on_Malekula/17064674.

Cassidy, F. and Brown, L (2010). Determinants of small Pacific island tourism: A Vanuatu study. *Asia Pacific Journal of Tourism Research*, 15(2), 143–153. https://doi.org/10.1080/10941661003629953.

Cheer, J. (2016). Cruise tourism in a remote small island: High yield and low impact? In R. Dowling and C. Weeden (Eds), *Handbook of cruise ship tourism* (pp. 408–423). Wallingford, UK: CABI.

Cheer, J., Reeves, K., and Laing, J. (2013). Tourism and traditional culture: Land diving in Vanuatu, *Annals of Tourism Research*, 43(1), 435–455. https://doi.org/10.1016/j.annals.2013.06.005.

Connell, J. (2023). COVID-19 and tourism in the island Pacific: Gender tribulations and transformations in different seas. In A. Datta, J. Momsen, and A. Oberhauser (Eds), *Bridging worlds: Building feminist geographies* (pp. 212–220). New York: Routledge.

Connell, J. and Rugendyke, B. (2008). Tourism and local people in the Asia-Pacific region. In J. Connell and B. Rugendyke (Eds), *Tourism at the grassroots: Villagers and visitors in the Pacific* (pp. 1–40). New York: Routledge.

Connell, J. and Taulealo, T. (2021). Island tourism and Covid-19 in Vanuatu and Samoa: An unfolding crisis. *Small States & Territories*, 4(1), 105–124. www.um.edu.mt/library/oar/handle/123456789/74986.

Cruz, P. and Passmore, K. (2023). Vanuatu: Disasters and resilience during the pandemic. *Pacific Economic Monitor*, August, 29–31.

Dalton, B. and Stanley, D. (1979). *South Pacific handbook*. Rutland, VT: Moon Publications.

de Burlo, C. (1989). Land alienation, land tenure and tourism in Vanuatu: A Melanesian island nation. *GeoJournal*, 19(3), 317–321. www.jstor.org/stable/41144933.

de Kadt, E. B. (1989). *Tourism: Passport to development?* Oxford: Oxford University Press for World Bank and UNESCO.

Harcombe, D. and O'Byrne, D. (1995). *Vanuatu*. Melbourne, Australia: Lonely Planet.

Movono, A. and Dahles, H. (2017). Female empowerment and tourism: A focus on businesses in a Fijian village. *Asia Pacific Journal of Tourism Research*, 22(6), 681–692. https://doi.org/10.1080/10941665.2017.1308397.

Naupa, A., Mecartney, S., Pechan, L., and Howlett, N. (2021). An industry in crisis: How Vanuatu's tourism sector is seeking economic recovery. In J. Connell and Y. Campbell (Eds), *COVID in the islands: A comparative perspective on the Caribbean and the Pacific* (pp. 231–252). New York: Palgrave Macmillan.

Orsua, N., Cheer, J., and Blaer, M. (2023). Gender empowerment in tourism development: Female bungalow hosts in Vanuatu. In M. Stephenson (Ed.), *Routledge handbook on tourism and small island states in the Pacific* (pp. 177–190). New York: Routledge.

Petrou, K. and Connell, J. (2023). Our 'Pacific family': Heroes, guests workers or a precariat? *Australian Geographer*, 54(2), 125–136.

Robinson, P. and Connell, J. (2008). 'Everything is truthful here'. Custom village tourism in Tanna, Vanuatu. In J. Connell and B. Rugendyke (Eds), *Tourism at the grassroots: Villagers and visitors in the Pacific* (pp. 77–97). New York: Routledge.

Scheyvens, R. and Russell, M. (2013). *Sharing the riches of tourism*. Palmerston North, New Zealand: Massey University School of People, Environment and Planning.

Tabani, M. (2017). Development, tourism and commodification of cultures in Vanuatu. In E. Gnecchi-Ruscone and A, Paini (Eds), *Tides of innovation in Oceania: Value, materiality and place* (pp. 225–260). Canberra: ANU Press.

Tisdell, C. and McKee, D. (1988). Tourism as an industry for the economic expansion of archipelagos and small island states. In C. Tisdell, C. Aislabie and P. Stanton (Eds), *Economics of tourism: Case study and analysis* (pp. 181–204). Newcastle: Institute of Industrial Economics, Australia.

Tourism Council of the South Pacific (1998). *1997 Vanuatu visitor survey*. Suva, Fiji Islands: Tourism Council of the South Pacific.

Trau, A. and Ballard, C. (2023). Community management of cultural tourism at a World Heritage Site: Intersections of the 'local' and 'global' at Chief Roi Mata's Domain, Vanuatu. In M. Stephenson (Ed.), *Routledge handbook on tourism and small island states in the Pacific* (pp. 205–218). New York: Routledge.

Vanuatu (2021). *Vanuatu sustainable tourism strategy, 2021–2025*. Port Vila: National Tourism Office.

Vanuatu NSO (2007). *2004 Visitor Survey Report*. Port Vila: National Statistics Office.

Vanuatu National Tourism Office (2021). *Domestic tourism marketing strategy, 2021–2023*. Port Vila: National Tourism Office.

Part III

Trans-national, regional archipelagos

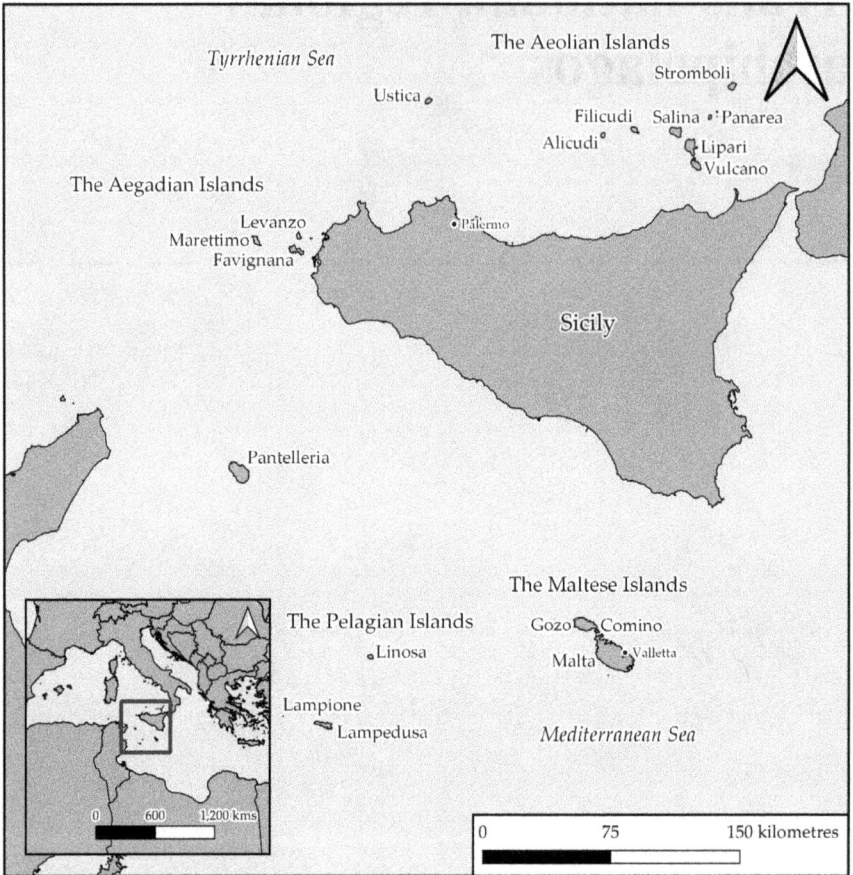

Figure 10.1 The Central Mediterranean

10 Challenging the 'status quo'

Archipelago tourism in the central Mediterranean after the pandemic

Karl Agius

Introduction

Many small islands are highly dependent on the revenue and employment generated by tourism (Omarjee, 2022); McElroy and Parry (2010) proposed the category of 'small island tourist economies' to illustrate this dependence. Most of the tourism activity on small islands takes place in coastal areas with 'sun, sand and sea' (3S) tourism being a leading attribute of such destinations (Alipour et al., 2020). Beyond economic gains, tourism development on Mediterranean islands is impacting negatively on their fragile ecosystems, increasing pressure on local biodiversity (and endemic species), especially along the coastal fringe (Omarjee, 2022; Boissevain and Selwyn, 2004). Furthermore, small islands and their tourism sectors are more susceptible to climate change which presents multiple risks for coastal-based tourism activities and related infrastructure (Wolf et al., 2021; Agius, 2022a).

The reliance of small islands on tourism and the travel sector meant that they suffered strongly because of the Covid-19 pandemic (Gu et al., 2022). The pandemic brought tourism to a complete halt, even if several small islands embraced their insularity to better manage the pandemic and then took steps to restart their tourism activities quicker than their corresponding mainlands (Agius et al., 2022). Although the pandemic has largely had a negative impact on the tourism sector, it has been considered a unique opportunity for many islands to review existing models of tourism development, in particular for them to be more resilient to future economic, health, and environment-related shocks. Responsible governance and management of islands' natural resources and their tourism activities, addressing climate change impacts, the diversification of islands' economies, and the promotion of innovative and personalised tourist experiences have all been considered as necessary measures towards increasing small island resilience (Figueroa and Rotarou, 2021).

In the immediate aftermath of the pandemic, some islands responded by means of innovative approaches, such as tourist schemes that capitalise on the rise of remote working, as in the case of Barbados and Mauritius (Foley et al., 2022). Other small islands fell back on tried and tested approaches: for example, Akaroa (South Island, New Zealand) shifted from 'cruise tourism'

DOI: 10.4324/9781003451037-14

to 'domestic tourism' (Hussain and Fusté-Forné, 2021). In the case of archipelagos, the latter took the form of island-to-island travel (Agius, 2022b).

Given their economic challenges and heavy dependence on tourism for their prosperity, there is already clear evidence that – now that the pandemic is considered as a thing of the past – small islands are rushing to reach pre-pandemic figures in terms of visitors, nights and expenditures; or even seek to build on the upward trend experienced just before the pandemic. There is also the risk that, rather than adopting new approaches, islands will follow previous models which in some cases have proved to be problematic. In addition, there is the possibility that new approaches encounter 'old' challenges associated with archipelago tourism, including limited marketing visibility for smaller islands/entire archipelagos, domination by the mainland or the largest island in the archipelago for tourism flows, as well as dependence on mainland or the largest island in the archipelago for essential services (Baldacchino, 2015a). This chapter presents the case of islands in the central Mediterranean region to assess what changes (if any) have taken place and to what extent they have managed to address, or even overcome, pre-pandemic tourism (and sustainability) related challenges.

Area of study

The area of study consists of archipelagos and islands, all situated in the central Mediterranean region between Southern Europe and North Africa, and which straddle the territorial and maritime boundaries of two European sovereign states: Italy and Malta. These are: the Pelagian Islands (comprising Lampedusa and Linosa); the Aegadian islands (comprising Favignana, Levanzo, and Marettimo); the Aeolian Islands (comprising Lipari, Vulcano, Salina, Filicudi, Alicudi, Panarea, and Stromboli); the island of Pantelleria; and the island of Ustica (all in Italy); as well as the islands of Gozo and Comino (forming part of the Maltese archipelago). The islands in the central Mediterranean region have been considered as a dynamic network, boasting a long-shared history. They are known collectively as the Sicilian archipelago (Baldacchino, 2015b); or the Sicilian-Maltese archipelago (Camonita, 2019).

Lampedusa and Pantelleria have their own airport; but all the other islands under scrutiny are reachable only by sea. Sicily serves as the sole gateway island for the Aegadian islands, Ustica and the Aeolian Islands. Malta is the main access point to the Maltese archipelago, since the only international airport and cruise liner terminal are both found here (Agius, Theuma, and Deidun, 2021). While each standalone island and archipelago under Italian jurisdiction has their own municipality (with the exception of the Aeolian islands that have a municipality on Lipari and three municipalities on Salina) (Andaloro et al., 2012), they administratively belong to the Region of Sicily dominated by the biggest island – Sicily. In the case of the Maltese Islands, a small island state and member of the European Union, the central government is found on Malta but Gozo has a Ministry dedicated to Gozo affairs and thus a strong voice in the national government. The island of Comino, with just two residents, administratively falls under the local government of the village of Għajnsielem, Gozo. The islands are a hotspot for biodiversity and their terrestrial and marine environment are protected through multiple designations (Agius, 2022). Table 10.1 sums up the different demographic characteristics of the islands in the area of study.

Methods

During June and July 2023, 11 interviews were held with stakeholders across the entire area of study in order to obtain their views on various aspects related to archipelago tourism. Stakeholders interviewed included locals, tourism industry representatives (including tour operators, guides, private marketing agencies, service providers [such as transport services], and the hospitality industry), government officials (tourism policy makers and politicians), leaders from civil society/non-governmental organisations (NGOs), and academics.

A strategic informant sampling technique was adopted to recruit interviewees. This is known as expert sampling and involves the selection of 'typical' and 'representative' individuals (Finn, Walton, and Elliott-White, 2000). Interviews were held over the phone or via videoconference to obtain information on developments following the Covid-19 pandemic. The use of online interviews has been used in tourism research (Power, Di Domenico, and Miller, 2017) since the use of virtual platforms also permits valid and high-quality interviews (Suryani, 2013). Interviews lasted between 30 to 45 minutes each: they were kept semi-structured and informal, and conducted in the respondent's native language. This exploratory approach allowed the researcher to obtain as much information as possible on a topic which has not received much attention in both the academic and policy literature.

No formal questions were prepared; but a checklist of topics derived from the literature review and the research plan was kept in hand to guide the researcher

Table 10.1 Characteristics of the Maltese-Sicilian archipelago

Factor	*Island/archipelago*					
	Sicily					Malta
Peripheral island(s) / archipelago	Aegadian islands	Pelagian Islands	Aeolian Islands	Pantelleria	Ustica	Gozo Comino
Tourist arrivals in 2022	207,843	253,710	143,452	151,917	32,784	460,514
Resident Population	4,468	6,462	15,021	7,407	1,307	39,287
Land area (km²)	37.7	25.5	115.2	84.5	8.7	69.9

Source: Agius, 2022c; Battistelli et al., 2023; NSO, 2023; Palermo Today, 2021.

throughout the interview. There is a gap in the literature on changes experienced by the tourism sector on archipelagos following the pandemic. Therefore, a series of topics were identified so as to respond to the research question. Issues tackled during the interviews included: (1) how and to what extent did tourism take off again following the pandemic; (2) what novel approaches were developed, if any, to address past mistakes, and to embrace sustainable tourism; (3) what changes occurred, if any, in the political-economic relationship between main and outer islands during the pandemic; and, if so, what was the impact that this left on tourism policy; (4) what type of tourist visited the islands and has this been transformed post-pandemic; (5) have transport connections been lost and gained; and (6) what new approaches, if any, have been piloted and implemented in the branding and marketing of the islands/archipelagos in the area of study.

The use of a checklist ensured that a consistent range of topics was covered in each interview (Wearing et al., 2002). This also allowed the researcher to ask supplementary questions or to ask the interviewee to explain the answer provided (Veal, 2006). Data collection through interviews was considered to be completed when experiencing exhaustion of sources, saturation of categories and the emergence of regularities or repetitions (Dooley, 2002).

Secondary data from online news portals and official social media accounts of local government/authorities was used to complement the research findings gleaned from the primary data collected from the elite interviews.

Once data collection was completed, the collected information was analysed manually using the coding scheme adopted by Stoffelen (2019) for qualitative studies in the tourism sector. This consisted in selecting descriptive codes and summarising data under respective codes, adding content from secondary data and field notes as well as selecting key quotes from interviews held. Research ethics considerations were fulfilled through the University Research Ethics Committee (UREC) of the University of Malta. Necessary precautions were taken to respond to challenges encountered when conducting research on small islands and within archipelagos where rivalry can play a dominant role (Baldacchino, 2015a). These included engaging stakeholders from different islands within archipelagos, ensuring anonymity, and not recording the interviews (Agius, 2023).

The researcher is himself island born and bred, has intermittently lived in the Maltese Islands, and has conducted research in the area of study for over a decade (Agius, 2018). This has led to the building and cultivation of an extensive network of contacts and informants, local knowledge, as well as access to secondary data. It has also placed the researcher in a 'privileged' position of having insider information and understanding, as well as being able to look at island-related issues "from the inside out", with an 'island imagination' (Baldacchino, 2008, p. 49).

Results and discussion

The numbers game: islands succeed in reaching/exceeding pre-pandemic figures

The pandemic had different effects on small islands. However tourism across the area of study did kick off immediately after most travel restrictions were lifted (June 2020). Those islands that normally had visitors in the shoulder seasons (such as March–May) lost several visitors in the first summer (2020) but made up for the losses due to a surge in tourism experienced in the peak season (Agius et al., 2022).

The situation was even more favourable on the most peripheral islands forming part of archipelagos. These tend to experience extreme seasonality, with the tourism season lasting two or even one month. Hence the delay in restarting tourism did not negatively impact such islands. On the contrary, such islands saw an immediate surge in tourists during their short tourism season when travel was permitted: a period which coincided with a slump in Covid-19 cases. Interviewees from Marettimo and Linosa confirmed that, right after the lifting of restrictions, these islands experienced even more tourists than in previous years.

Stakeholders interviewed across the area of study confirmed that the positive trend continued in 2021 and even more so in 2022. The Aegadian islands were described as 'sold out' during the summer of 2021 (Spanò, 2021). Between 2020 and 2021, tourist arrivals (June till August) on Ustica went up from 18,671 to 21,780 (Palermo Today, 2021). The positive trend continued, with operators from Pantelleria reporting that packages for the first two weeks of October 2022 were 'sold out' (Guida Viaggi, 2022). An interview from Lampedusa said that 2022 was a positive year for tourism, with tourists arriving between May and October. The situation has also improved in Gozo. In 2021, Gozo exceeded 2019 figures (NSO, 2022). According to a survey conducted among Gozitan tourism operators, 62% declared that the performance of their respective businesses during the summer months of 2022 was better than that of the same period of 2021, and with the Gozo Tourism Association (GTA) describing 2022 as their 'year of recovery' (Gozo News, 2023). Yet, this recovery is mostly due to the rise of domestic tourism (see Table 10.2).

Table 10.2 Overview of number of tourists visiting Gozo (2019–2022)

	2019	*2020*	*2021*	*2022*
Visitors to Gozo	396,251	388,213	435,907	460,514
Domestic tourism (Number and %)	215,272 54.3%	348,489 89.8%	365,252 83.8%	331,455 72%
International tourism (Number and %)	180,979 45.7%	39,724 10.2%	70,655 16.2%	129,059 28%
Visitors to Gozo only	92,715	20,659	29,889	64,687

Source: NSO (2022; 2023).

The outlook for the coming years varies. In Gozo, 69% of tourism stakeholders envisaged the same level of business for 2023 (Gozo News, 2023). On the other hand, local operators from some Sicilian islands such as Pantelleria said that they are expecting less tourists to visit the islands following the surge experienced by domestic/proximity tourism. An interviewee from the Aeolian islands said that, following the boom experienced during the first year of the pandemic, trends might revert to the levels of 2019 and therefore the islands might experience a slight drop in numbers.

An interviewee from Favignana said that, while mass tourism had returned to the island just after the pandemic, inflation experienced on the islands might make them less competitive, with tourists opting to 'staycate' on the mainland. Climate change is also a concern. According to an interviewee from Favignana, while in summer of 2020 pandemic-related restrictions contributed to the late kick off of the tourism season, in 2023 climate change and the presence of storms/bad weather in May and June 2023 led to a similar scenario in terms of tourism activity. This is not an isolated case: in August 2022, a storm caused substantial damage on the island of Stromboli (Amato, 2022). An interviewee from Lampedusa stated that 2023 looked promising but, after undocumented migration-related events started to gain attention on the news and other media, bookings started to drop. The interviewee said "Immigration is like a pandemic for the tourism sector on Lampedusa." The impact of immigration on tourism in the Pelagian islands has already been reported before the onset of the coronavirus pandemic (Agius, 2021). The pressure faced by the local community has also been acknowledged by the President of the European Commission Ursula von der Leyen during her visit on the island in September 2023, following the umpteenth immigration crisis (European Commission, 2023).

Tourism on islands experiencing inertia: overtourism and seasonality regain centre stage

During the pandemic, there has been much talk that 'business as usual' tourism may not be the best way forward. Rhetoric focused on the opportunity to change tourism models with novel approaches, address past mistakes, and instead embrace sustainable tourism. Interviews conducted have shown that, while there is willingness by some stakeholders to bring about change, tourism continues to experience island-related challenges which hamper the development of sustainable tourism.

In the case of the Aegadian islands, during lockdown, much was said about the need to re-dimension tourism as a more sustainable model based on local knowledge, skills, and quality rather than on numbers. However, the wave of tourism experienced once restrictions were lifted (especially on Favignana) was welcomed with open arms (Sajeva, 2020). Although new initiatives have promoted ecotourism activities such as snorkelling (Balistreri, 2022), mass tourism continues to dominate on Favignana. An interviewee from Levanzo said that the island still faced the challenges of the past, especially in the summer months. This included failure to treat sewage water properly and difficulties with waste management. Due to a surge in tourism, 146 tons of additional waste were collected in July 2021, even if facilities for solid municipal waste collection left much to be desired (Spano, 2021). An interviewee from Lampedusa said that there was no change or novelty because investment on the island is limited and thus it is difficult to attract quality tourism. There is also an alleged lack of planning and a dependence on airlines operating to the islands; so, these issues cannot really be addressed.

According to an interviewee from Linosa, there has been no change or novel approaches as tourism on the island has always revolved around the sea, nature, and the tranquil environment. Similarly, an interviewee from Pantelleria said that the island was trying to target more international tourists; but, in terms of product, nothing has changed since the market of the island was already different from that of other destinations. The island had well identified its "nature of differentiation" (Baldacchino, 2015a) in order to attract medium to high-end travellers. On the other hand, the island has failed to learn from past mistakes and there was still the need to give greater attention and value to the natural assets that the island has to offer, such as thermal attraction and hiking and adventure trekking routes.

Another interviewee from Pantelleria stated that operators had learned more about the fragility of the tourism sector and the need to diversify. This, however, is not always possible due to the skills of the person; but the pandemic served as an eyeopener to ensure some liquidity to serve as a safeguard in case of crises.

An interviewee from Ustica said that several events are being organised, mostly by the private sector, to attract tourists as well as address seasonality. These included events related to yoga, running tours, and water sports such as apnoea (diving without scuba gear). Other events being organised are taking sustainability into account. Pantelleria organised a festival, branding it as a 'zero emissions' event (Vivere Pantelleria, 2023). The Marine Protected Area (MPA) on the Aegadian islands has organised an 'Ecological Beaches' initiative to eliminate plastic on the beach (ISPRA, 2021).

An interviewee from the Aeolian islands emphasised the need to make changes to current tourism trends and the difficulty policy makers find is to shift from a system based on numbers to one based on quality. Most tourists attracted to the archipelago, especially Lipari, in August are young people who cause havoc in the centre late at night. There is also a challenge with day visitors and mini-cruises. Between August and September, the local government of the Aeolian Islands set a limit on the number of excursionists that can disembark on the islands of Panarea and Stromboli due to overtourism (Sarpi, 2022). The archipelago is more interested in quality tourism based on nature, volcanic activity, culture, and the different attractions offered by the entire archipelago. The municipality of Lipari has entrusted a foundation to design a strategic plan for sustainable tourism in the Aeolian Islands

(Fondazione Santagata per l'Economia della Cultura, 2023). One proposal has been to institute a national park, similar to Pantelleria. The proposal has been on the table for several years but blocked out of fear of restrictions (Di Pazza, 2020). The Aeolian islands are not alone in experiencing overtourism. Those visiting Comino, which is itself a Natura 2000 site, especially the Blue Lagoon, experience crowdedness, noise, and the smell of fried food from nearby kiosks. In a clear example of private interests encroaching on a public good, sunlounges, and umbrellas hog the very small beach of the Blue Lagoon, which can receive thousands of visitors on a single summer's day (Moviment Graffitti, 2022). To date, there is no plan to identify or set a carrying capacity here.

The small islands in the study area continue to face seasonality issues. Italy's National Recovery and Resilience Plan identified the "seasonal adjustment" of local economies as a key objective (Gallia and Malatesta, 2022). Yet, these islands – and the peripheral ones forming part of archipelagos, such as Linosa in particular – continue to experience a seasonality pattern similar to that reported before the onset of Covid-19 (Agius and Briguglio, 2021). This is because domestic tourism, associated with the summer months, continues to play a leading – perhaps even a bigger – role than before the pandemic (see Table 10.3). An interviewee from the Aeolian Islands said that the islands of Vulcano and Stromboli are a major attraction for those interested in volcanoes and that the islands attract tourists even until October; but the other islands had a shorter tourism season, and hence more efforts were needed to address seasonality.

Other problems experienced in the past have come back to haunt these islands. Since outer islands are dependent on the short tourism season, jobs that last all year round are limited. As a result, the outermost islands experience changes in population. This is also confirmed by the lower number of students in schools. An interviewee from the Aeolian islands said that, since tourism mostly takes place between March and October, it was not worth keeping hotels open to cater for just a few visitors beyond that window, especially in the context of the energy crisis: a clear case of a 'chicken and egg' dilemma. The result is less services for tourists off-season. However, according to an interviewee from Marettimo, there is a drive to increase more

Table 10.3 Core-periphery relationships in the Maltese-Sicilian archipelago

Inner core			Sicily			Malta
Outer core	Favignana	Lampedusa	Lipari			
	Levanzo	Linosa	Vulcano	Pantelleria	Ustica	Gozo
	Marettimo		Salina			Comino
			Stromboli			
			Filicudi			
Periphery			Alicudi			
			Panarea			
			Stromboli			

services; and there is an initiative to upgrade existing hotels on various islands, including Ustica. An interviewee from Levanzo said that new places are opening offering further accommodation and facilities to visitors. However, works to reopen a disused hotel were also leading to controversies. Authorities have been criticised for allowing construction of a therapeutic solarium due to an alleged negative impact on the coast (Lo Porto, 2023). In the case of Comino, the redevelopment of the existing bungalows has been slammed by NGOs due to the environmental impact this will leave (Croce, 2023). This shows that sustainable tourism is still not at the top of the development agenda.

Interviewees from Levanzo and the Aeolian islands outlined new challenges encountered, such as difficulty to find staff, especially for short contracts. It is also hard to find personnel capable of communicating in various languages including English. This could be addressed by stretching the tourism season into the shoulder months and offering better job contracts. Similarly, in a survey conducted among Gozitan tourism operators, 60% claimed they were struggling to recruit new employees; while 77% declared that they had to employ foreigners to provide services (Times of Malta, 2023).

Core-periphery relationships continue to characterise archipelago tourism

Local communities on the Sicilian islands feel a sense of abandonment by national and regional governments (Sajeva, 2020). Interviews conducted suggest that core-periphery dynamics continue to play a significant role in governance and to condition decisions that somehow impact tourism. An interviewee from Pantelleria said that the regional government was focused on regional policy with limited attention to local issues. Another interviewee said that Pantelleria was always "the third world of Sicily" and that the island never benefited from regional policy. An interviewee from Ustica proposed that the regional government should have a department directly responsible for the islands forming part of the Region of Sicily rather than adopting a general approach by each department since the realities on the small islands were different from those of Sicily. An interviewee from Favignana said that there is no dialogue with the regional government. For example, because of decisions taken by central and national authorities, water consumption on islands has had to be reduced. This did not make sense and instead authorities should have planned for and invested in a reverse osmosis plant to ensure water supply. The interviewee added that, while a Marine Protected Area (MPA) was instituted, waste water was not being treated; this was not beneficial to tourism (while keeping in mind that water consumption by tourists is typically much higher per capita than by locals).

Political relationships within archipelagos have not changed much either, with domination/subordination still evident. In the Aegadian islands, the vice-mayor, who hails from the second largest and most remote island of the archipelago (Marettimo), resigned citing lack of attention and failure to address pertinent matters such as waste management. An interviewee from Levanzo also criticised the local government for lack of attention given to the needs of the island. Similarly, an operator from Linosa said that there is no attention given to the peripheral islands of the Pelagian archipelago, to the extent that while there are investments in infrastructure (such as roads) on Lampedusa, nothing has changed on the smaller island of Linosa. The nested core-periphery relationships are summarised in Table 10.3.

Domestic/seaside tourism continues to prevail

Tourism on islands has restarted thanks to visiting friends and family and domestic/proximity tourism. In fact, Covid-19 has helped to strengthen pre-pandemic trends in terms of the range of tourist profiles. In the case of the Aeolian islands, Aegadian islands, Pelagian islands, Ustica, and Pantelleria, before the pandemic most tourists were domestic, hailing predominantly from either the main island of Sicily or from mainland Italy. The pandemic helped to boost such trends. Studies have shown that tourists in Italy tend to travel close to home in times of economic crisis (Cafiso, Cellini, and Cuccia, 2018). The pandemic and war in Ukraine have all contributed to such a scenario favouring travel from the closest region, especially in the peak season. Such tourists are mostly interested in seaside tourism. Interviewees from the Aegadian archipelago and Pantelleria confirmed that international tourism has now started to gain pace but remains quite limited.

Some of the islands have reinforced their policy to push forward nature-based tourism and to attract tourists interested in nature, especially in the immediate aftermath of the pandemic. In the case of Pantelleria, the National Park, whose institution was contested by some locals, has now become considered as a catalyst to restart tourism after the pandemic and as a driver for the sustainable development of tourism on the island. Proposals have been made for islands such as Pantelleria to capitalise on their cultural heritage and to go beyond the attractions related to the sea and coast (Culoma, Baráth, and Morini, 2022). Other suggestions have been to embrace the geo-volcanological heritage through the promotion of thermal tourism, given the presence of thermal springs, natural saunas, mud treatments, and caves with therapeutic effects located across the island (Beggio and Ongaro, 2020).

The MPA surrounding the Aegadian archipelago is also considered as a major attraction for tourists. Furthermore, given its richness and diversity in terms of abundance and peculiarity of landforms, Favignana is considered as an excellent candidate for the exploitation of its geoheritage for touristic purposes (Pappalardo et al., 2021). An interviewee from Levanzo said that tourists arriving in April and May were mostly interested in sustainable tourism and had an interest in cultural attractions; whereas those visiting the island in the peak season were mostly interested in the sea. An interviewee from Levanzo said that, while local communities have long been discussing the need to attract more international tourists off season, nothing has changed after the pandemic.

An interviewee from Linosa said the MPA office on Linosa had closed and likewise the turtle rescue centres on Linosa and Lampedusa which used to attract researchers to the islands. The interviewee added that this was not the right approach to attract tourists. As a result, most tourists were visiting the island between July and August and several (around 100) visited the island for one day, bringing packed lunches with them, spending little on the island and leaving mostly trash.

An interviewee from Lampedusa said that the profile of the tourists did not change as there are no five-star hotels on the island and so it is difficult to attract high quality tourism. On the other hand, the relatively expensive flight ticket implicitly serves as a visitor selector. Supporting this argument, an interviewee from Pantelleria confirmed that most visitors are medium to high-end tourists from the north of Italy, and that this profile is self-determined by the relatively expensive costs of the flights.

An interviewee from the Aeolian Islands said that in August those who visit the main islands such as Lipari are mostly domestic tourists and do so mostly for nightlife and for seaside tourism. However, the archipelago is keen on attracting more tourists who are interested in nature (including volcanoes) as well as in cultural and historical attractions, since Lipari has an interesting museum spanning over a number of buildings and showcasing artefacts from the Neolithic age to the Roman period. It also includes a section dedicated to underwater archaeology and volcanology (Italy this Way, 2023).

Connectivity remains the Achilles heel of archipelago tourism

Small islands have been rather swift to reinstate connectivity following the pandemic. This is due to the vital role that connectivity plays in their local economy. Some islands, such as Gozo, have gained connections which had been suspended long before the pandemic, such as the use of a fast ferry. However, the services offered by the two companies competing on this route had been described as a "quasi-monopoly" as the two operators agreed on a shared time schedule. Just after a year from the launch of the service, the two fast ferry companies admitted that the service was not financially feasible (Cordina, 2022). As a result, the companies had to reduce the frequency of crossings, causing discontent among users and triggering the Maltese government to provide them with temporary financial assistance and to seek approval by the European Commission to allow a Public Service Obligation (PSO) to be put in place. Since then, the two companies have merged the service under the name Gozo Highspeed, offering customers one consolidated ferry schedule, ticketing system, and brand identity (Balzan, 2023). Gozo is also vying for other modality options such as the introduction of an air-link. On the other hand, the pandemic and successive crises have served as an excuse to once again suspend discussion on a fixed-link between the two main islands (Malta and Gozo). The Gozo Tourism Association has shifted its position about the proposed tunnel from positive to cautiously neutral, in order to better safeguard Gozo's island character (GTA, 2022).

The situation in the Sicilian Islands seems to be a *deja vu* and challenges with connectivity persist. An interviewee from Linosa said "Tourism depends on connectivity and, with no secure connections, it is difficult to bring tourists to the island." An interviewee from the Aeolian islands said that the absence of an airport means there is no alternative to maritime transport which is thus fundamental to that archipelago. This explains why some 40 associations from the Sicilian islands have together repeatedly written to the Region of Sicily copying the Ministry for Infrastructure, mayors of the islands/archipelagos, and shipping companies to collectively denounce the lack of attention given by the Italian political authorities to the connectivity issues of the Sicilian islands. They have claimed that, rather than adapting to new needs, services were being reduced (QDS, 2022). Moreover, they have called for the restoration of original tariffs on all state-subsidised vessels which are more expensive to passengers than the regionally subsidised trips on roll-on, roll off (RO-RO) ferries and hydrofoils (QDS, 2022; Leone, 2023). The reduction in services off-season causes serious inconvenience to local communities, as well as discourages the arrival of tourists off-season, further locking in the industry's seasonality.

An interviewee from the Aeolian Islands said that the islands are well connected with the region and the national government, spending €130M on an annual basis to secure services. This archipelago is connected to Sicily via Milazzo, Messina, and Reggio Calabria; and throughout summer there are trips to and from Palermo. The archipelago is also connected to mainland Italy via Naples. The challenge with connectivity is thus a matter of finding the right equilibrium of having a fleet of vessels which can respond to the huge demand in summer and the limited demand off season. For example, demand for Alicudi and Filicudi is very low; but the hydrofoil and RO-RO ferries still need to visit such islands. This equilibrium is difficult to reach; but by addressing seasonality and increasing demand in the winter period, connectivity services can become more sustainable and reduce inconvenience for locals.

Residents face more limited crossing options throughout winter because of a drop and then risk not finding a seat once tourist numbers start to increase in summer. As of May, and until October, Aegadian island residents may find it difficult to buy a ticket to cross from Sicily to their respective homes (Cozzo, 2023).

RO-RO ferry boats connecting the Aeolians islands and Ustica to mainland Sicily have been impounded by the authorities due to their inadequacy to carry passengers with reduced mobility including people with a disability, the elderly, as well as women and men with strollers (ANSA, 2023). Similarly, interviewees from Linosa and Pantelleria said that the boats used are old, slow and not ideal for overnight trips. Apart from the lack of local comforts, trips were taking longer due to the dismal state of the ships. Furthermore, an interviewee from Pantelleria said that the port is inadequate and the regional government is not interested in investing. Hence, even in moderate weather conditions throughout the summer period, the service may be suspended. They called for new and fast vessels and for better ports. To respond to such challenges, the regional government has committed to commission a new RO-RO vessel to serve Pantelleria, Lampedusa, and Linosa. The vessel would use cleaner energy, be able to work in challenging weather conditions, as well as have increased vehicle and passenger capacity (Ditta, 2023). However, there is little hope this will really materialise. An interview from Linosa recalled that in 2019 the President of the region visited the island and promised that the port will be upgraded to facilitate mooring of ships and island-mainland/ island-island connectivity (Regione Sicilia, 2019); but, so far, no progress has been made on these pledges.

In the case of Lampedusa and Pantelleria, while air travel is considered as an asset by both islands to link such remote destinations to the mainland, there is disagreement on the quality of the services. An interviewee from Lampedusa said that, while the islands are connected via an airport, there are very few flights. An operator from Pantelleria said that some airlines had increased the frequency of their flights and this has improved the connectivity of the island. Another interviewee added that, in Pantelleria, air travel depends heavily on private companies. Supporting this argument, an interviewee from Lampedusa explained that regular news about the arrival of undocumented foreign migrants by boat from North Africa have led to less tourist bookings and thus to less frequent flights. Interviewees added that most flights connect the islands to mainland Italy. The flights are available in summer but cease to operate off season. More connections are needed especially off season to stretch the tourism season, at least to one city in the north of Italy which is a major market for the island. Flights to international destinations are limited or have been lost: a case in point is the flight service linking Lampedusa to Malta, which is no longer operative (Times of Malta, 2017). This indeed has been the only, short-lived example of a route that connected the Maltese and Italian archipelagos without involving Sicily.

Flights secured via a PSO (funded by the national and regional governments) that enable an airlink between Pantelleria as well as Lampedusa with Sicily (Trapani, Palermo, Catania) have also proved problematic. Such flights are crucial as they ensure 'territorial continuity' throughout the entire year, unlike flights offered by the private sector. In December 2022, the PSO was only extended for seven months. This caused uncertainty, preventing tour operators from planning their work longer term and for travellers from organising their holidays. It was only in April 2023 that flights from July were confirmed through a new tender that will ensure flights until 2025.

Promoting archipelagos is like 'a drop in the ocean'

One can say that single island, archipelago, and regional promotion of islands are all taking place in the central Mediterranean and in some cases concurrently. This is due to the different interests and initiatives of both the public authorities and the private sector.

In the case of Gozo, in addition to coverage via the portal *Visit Malta*, the island is promoted through the portal *Visit Gozo* managed through the Ministry for Gozo, with emphasis placed on the more relaxed, natural, and rural character of the island. Whether the segregation of marketing efforts for Gozo to position itself as a distinct destination is bearing fruit remains to be seen since domestic tourism remains the prevalent form of tourism while numbers of those visiting solely Gozo (single centre) have remained quite low (Table 10.2). Moreover, in 2022, same-day visitors to Gozo and Comino accounted for 90% of the total inbound visitors to Gozo and Comino, totalling 1,203,890 visitors (NSO, 2023).

Tourism entities in the Sicilian region are opting for more strategies that encompass the entire archipelago in their marketing pitches. The Aeolian islands work closely together when it comes to promotion, even if there are four different local governments involved in the administration of the archipelago. One example is events organised together, such as one targeting those who emigrated from the archipelago (Loveolie, 2023).

In some cases, a regional dimension approach is adopted to market a group of islands/ archipelagos together. This is the case of the joint marketing strategy adopted by the 'Islands of Sicily' which has promoted the islands in fairs held in Italy (Milan and Bergamo) as well as overseas in such cities as Berlin and London (Islands of Sicily, 2023).

There is palpable discontent on the efforts of local and regional governments in promoting the islands. This discontent was more vociferous on smaller islands forming part of an archipelago. An interviewee from Lampedusa said that, whereas a beach in Lampedusa had been voted as one of the best beaches in the world (Hughes, 2023), the region did not give this accolade much attention. The operator said that authorities must promote the island and showcase what it really is, since Lampedusa was being too closely associated in the press with being a detention site for undocumented migrants (Agius, 2021). In addition, national authorities should compensate the island through marketing support. According to an interviewee, the Local Government on Pantelleria was not promoting the territory adequately both qualitatively and quantitatively. The right markets, including those in northern Europe, were being overlooked. Criticising the local government for its role in marketing efforts, an interviewee from Favignana complained that there were no strategies in place or a programme of events published well ahead of summer to attract visitors to the archipelago. This is vital to encourage the arrival of visitors to the archipelago, especially off season (spring and autumn) (Sammartano, 2023). An interviewee from Pantelleria said that there is the need for a strategy and a multi-annual plan.

Some positive elements were attributed to non-local/regional efforts. Whereas there was a rise in French tourists on Pantelleria, this was not due to the initiatives of local operators or authorities but due to the work of French actress Carole Bouquet who produces wine on the island (TF1, 2020). An interviewee from Pantelleria said that, in terms of marketing, the pandemic had left an overall positive impact. Those who could not travel abroad visited either another region or travelled to the islands. Several of these visitors shared a positive word of mouth on the island and hence the impact was still being felt. An interviewee from Lampedusa said that marketing was mostly done by the private sector and that most promotion is done via word of mouth by past visitors. On the other hand, according to some stakeholders interviewed, there is a better understanding of the importance of the natural environment to promote the destinations. The board in charge of managing the MPA around Ustica, together with the municipality, promoted the island and its diving niche in various fairs, including Paris (Resto al Sud, 2023).

Whatever the efforts, budgets remain limited and hence marketing mostly takes place in national markets through tourism fairs, news articles/blogs and TV documentaries. Operators from Ustica and Linosa said that promotion, including that spearheaded by the private sector, such as initiatives via social media and joint marketing strategies, was not necessarily delivering results.

Conclusion: Plus ça change ...

While there is an evident attempt to promote archipelago tourism (including selling islands in different ways) both in the region of Sicily as well as in the Maltese Islands, data (where available) and insights provided by interviewees show that archipelagos in the area of study are yet to fully benefit from this opportunity. The 'Islands of Sicily' project (Agius, 2022c) remains to date the best attempt by the private sector to promote archipelago tourism in the area of study, to complement existing tourism offers and to put the smaller and more remote islands under the spotlight. This may serve as an inspiration to policy makers and advocate against single island promotion which has been described as a 'zero-sum game' (Baldacchino, 2015a).

The World Health Organization (WHO) has declared the end of the global emergency status for Covid-19. However, new crises – such as the conflicts in Ukraine and Palestine, increases in the cost of living, as well as new regulations on the use of sustainable aviation fuels – are leaving an impact on the disposable income of consumers and the cost of flight and ferry services. Hence, as with other crises, 'staycations', travelling closer to home or domestic tourism are likely to continue to comprise a substantial component of tourism in the area of study in the years to come. This scenario may favour the development of archipelago tourism and make up for a reduction in international tourism on small islands. Taking into account current travel patterns of domestic tourists (mostly in summer) as well as limited availability of services off season, seasonality is expected to become more evident.

Small islands remain heavily dependent on higher levels of governance (regional/national) for key aspects of their tourism development, and transport in particular. Little has changed since before the pandemic, other than the fact that challenges have increased due to a more turbulent international context and more onerous EU environmental obligations and targets.

The interviews held suggest that, while there is willingness by some stakeholders, including local authorities, to transition to a sustainable form of tourism, the economic dependency on tourism has triggered a 'business-as-usual' approach. Post pandemic, not much has changed. Hence, the push for (more) sustainable tourism practices on the islands needs to take off in parallel with a welcome diversification of the local island economies, preventing them from putting all eggs in the tourism basket. This is vital to address challenges such as overtourism, fluctuations in population due to lack of secure jobs, and the (over) development of their very finite coastal areas.

References

Agius, K. (2018). *Assessing the ecotourism potential of central Mediterranean islands with a case study on marine ecotourism.* Unpublished PhD thesis. University of Malta. www.um.edu.mt/library/oar/handle/123456789/82874.

Agius, K. (2021). Doorway to Europe: Migration and its impact on island tourism. *Journal of Marine and Island Cultures*, 10(1), 21–33.

Agius, K. (2022a). Climate change adaptation and mitigation in a small archipelago: The role of ecotourism in Malta. *International Journal of Islands Research*, 3(1), 6. https://arrow.tudublin.ie/ijir/vol3/iss1/6.

Agius, K. (2022b). Island to island travel: The role of domestic tourism for the swift recovery of island tourism. In: P. Mohanty, A. Sharma, J. Kennell, andA. Hassan (Eds), *The Emerald handbook of destination recovery in tourism and hospitality* (pp. 397–415). Leeds, UK: Emerald Publishing. doi:10.1108/978-1-80262-073-320221023.

Agius, K. (2022c). The ecotourism hub: A joint cross-border marketing strategy for peripheral islands. *Shima,* 16(1), 304–324. doi:10.21463/shima.124.

Agius, K. (2023). Island settings and their influence on geographical research methods. *Geographical Research*. doi:10.1111/1745-5871.12571.

Agius, K. and Briguglio, M. (2021). Mitigating seasonality patterns in an archipelago: The role of ecotourism. *Maritime Studies*, 20(4), 409–421. doi:10.1007/s40152-02100238-x.

Agius, K., Sindico, F., Sajeva, G., and Baldacchino, G. (2022). 'Splendid isolation': Embracing islandness in a global pandemic. *Island Studies Journal*, 17(1), 44–65. doi:10.24043/isj.163.

Agius, K., Theuma, N., and Deidun, A., (2021). So close yet so far: Island connectivity and ecotourism development in central Mediterranean islands. *Case Studies on Transport Policy*, 9(1), 149–160. https://doi.org/10.1016/j.cstp.2020.11.006.

Alipour, H., Olya, H. G., Maleki, P., and Dalir, S. (2020). Behavioural responses of '3S' tourism visitors: Evidence from a Mediterranean island destination. *Tourism Management Perspectives*, 33, 100624. https://doi.org/10.1016/j.tmp.2019.100624

Andaloro, A. P. F., Salomone, R., Andaloro, L., Briguglio, N., and Sparacia, S. (2012). Alternative energy scenarios for small islands: A case study from Salina

Island (Aeolian Islands, Southern Italy). *Renewable Energy*, 47, 135–146. doi: 10.1016/j.renene.2012.04.021.

Amato, C. (2022). Ondata di maltempo a Stromboli, l'acqua travolge tutto. [Spate of bad weather in Stromboli: The waters overwhelm everything.] *LiveSicilia*. https://livesicilia.it/maltempo-stromboli-video/.

ANSA. (2023). Traghetti non a norma, tre sequestrati a Caronte & Tourist. [Ferries not in order: Three detained at Carone and Tourist companies.] *Ansa*. www.ansa.it/sicilia/notizie/2023/06/06/traghetti-non-a-norma-tre-sequestrati-a-caronte-tourist_2e92f514-8bdb-4fb4-83d7-5e6b04623a0e.html.

Baldacchino, G. (2008). Studying islands: On whose terms?: Some epistemological and methodological challenges to the pursuit of island studies. *Island Studies Journal*, 3(1), 37–56. https://doi.org/10.24043/001c.81189.

Baldacchino, G. (2015a). More than island tourism: Branding, marketing and logistics. In G. Baldacchino (Ed.), *Archipelago tourism: Policies and practices* (pp. 1–18). Farnham: Ashgate. https://doi.org/10.4324/9781315567570-7.

Baldacchino, G. (2015b). Lingering colonial outlier yet miniature continent: Notes from the Sicilian archipelago. *Shima: The International Journal of Research into Island Cultures*, 9(2), 89–102.

Balistreri, P. (2022). *Favignana: Sentieri d'acqua. Guida snorkeling*. [Favignana: Paths of water, Snorkeling guide.] Palermo, Italy: Edizioni Danaus.

Balzan, J. (2023). *Rival fast ferry companies to merge into 'Gozo highspeed'*. https://newsbook.com.mt/en/rival-fast-ferry-companies-to-merge-into-gozo-highspeed/.

Battistelli, F., Minutolo, A., Nanni, G., Laurenti, M., Montiroli, C., Tomassetti, L., and Petracchini, F. (2023). Suolo, rifiuti, acqua, energia, mobilità, depurazione: Le sfide della sostenibilità nelle isole minori. [Soil, waste, water, mobility, pollution: The challenges of sustainability in the smaller islands]. *Legambiente*. www.legambiente.it/wp-content/uploads/2023/06/Report-Isole-Sostenibili-2023.pdf.

Beggio, F. and Ongaro, V. (2020). *Cossyra, figlia del vento: Riscoprire l'isola di Pantelleria attraverso il suo patrimonio termale*. [Cossyra, daughter of the wind: Rediscovering the island of Pantelleria through its thermal heritage.] Master's dissertation, Politecnico di Milano.

Boissevain, J. and Selwyn, T. (Eds). (2004). *Contesting the foreshore: Tourism, society and politics on the coast*. Amsterdam: Amsterdam University Press.

Cafiso, G., Cellini, R., and Cuccia, T. (2018). Do economic crises lead tourists to closer destinations? Italy at the time of the Great Recession. *Papers in Regional Science*, 97(2), 369–386.

Camonita, F. M. (2019). Envisioning the Sicilian-Maltese archipelago: A Braudelian inspired triple-level analysis of a European cross-border region. *Island Studies Journal*, 14(1), 125–146. https://doi.org/10.24043/isj.82.

Croce, S. (2023, 7 August). The revised plan for Comino is still bad news. *Times of Malta*. https://timesofmalta.com/articles/view/the-revised-plan-comino-still-bad-news.1047655.

Cordina, J. P. (2022). Fast ferry operators insist service is not financially viable. *Newsbook*. https://newsbook.com.mt/en/fast-ferry-operators-insist-service-is-not-financially-viable/.

Cozzo, M. (2023). Egadi, residenti non trovano posto su aliscafi: chiesto incontro urgente con Liberty Lines. [Egadi, residents find no room on the hovercraft: Urgent meeting with Liberty Lines called.] *Il Giornale di Pantelleria*. www.ilgiornaledipantelleria.it/egadi-residenti-non-trovano-posto-su-aliscafi-chiesto-incontro-urgente-con-liberty-lines/.

Culoma, A., Baráth, S. S., and Morini, G. (2022). Rediscovering Pantelleria beyond the sea. *Journal of Art Historiography*, (27s). https://arthistoriography.files.wordpress.com/2022/11/13_barath.pdf.

Di Pazza, A. (2020). È ora di completare l'istituzione dei tre parchi nazionali siciliani: il caso delle isole Eolie. [It's time to complete the institution of three national parks in Sicily: the case of the Aeolian Islands.] *RivistaNatura*. https://rivistanatura.com/e-ora-di-completare-listituzione-dei-tre-parchi-nazionali-siciliani-il-caso-delle-isole-eolie/.

Ditta, D. (2023). Trasporti, Fincantieri costruirà a Palermo una nuova nave per i collegamenti con le isole minori. [Transport: Fincantieri will build a new ship in Palermo to provide ferry services to the smaller islands.] *Palermotoday.* www.palermotoday.it/cronaca/isole-minori-costruzione-nuovo-traghetto-cantieri-navali-palermo.html.

Dooley, L. M. (2002) Case study research and theory building. *Advances in Developing Human Resources*, 4(3), 335–354. doi:10.1177/1523422302043007.

European Commission (2023, 17 September). Press statement by President von der Leyen with Italian PrimeMinister Meloni in Lampedusa. https://ec.europa.eu/commission/presscorner/detail/en/statement_23_4502.

Figueroa B. E. and Rotarou, E. S. (2021). Island tourism-based sustainable development at a crossroads: Facing the challenges of the Covid-19 pandemic. *Sustainability*, 13, 10081. https://doi.org/10.3390/su131810081.

Finn, M., Walton, M., and Elliott-White, M. (2000). *Tourism and leisure research methods: Data collection, analysis, and interpretation.* Harlow: Pearson Education.

Foley, A. M., Moncada, S., Mycoo, M., Nunn, P., Tandrayen-Ragoobur, V., and Evans, C. (2022). Small island developing states in a post-pandemic world: Challenges and opportunities for climate action. *Wiley Interdisciplinary Reviews: Climate Change*, 13 (3), e769. http://dx.doi.org/10.1002/wcc.769.

Fondazione Santagata per l'Economia della Cultura. (2023). *Piano strategico per il turismo sostenibile delle Isole Eolie.* [Strategic plan for sustainable tourism in the Aeolian Islands.] Patrimonio UNESCO. www.fondazionesantagata.it/in-evidenza/piano-strategico-turismo-isole-eolie-unesco/.

Gallia, A. and Malatesta, S. (2022). Le isole minori italiane nelle Missioni del PNRR: Una visione sul futuro. *Documenti Geografici*, 1, 161–174. www.documentigeografici.it/index.php/docugeo/article/view/349.

Gozo News (2022). 2022 is the recovery year for tourism in Gozo: GTA survey. https://gozo.news/100838/2022-is-the-recovery-year-for-tourism-in-gozo-gta-survey/.

GTA. (2022). Recover and consolidate growth: Budget 2023. *Times of Malta.* https://cdn-others.timesofmalta.com/2abe038b8d1d06c40106152356a5d5bc4b73a922.pdf.

Gu, Y., Onggo, B. S., Kunc, M. H., and Bayer, S. (2022). Small Island Developing States (SIDS) COVID-19 post-pandemic tourism recovery: A system dynamics approach. *Current Issues in Tourism*, 25(9), 1481–1508. http://dx.doi.org/10.1080/13683500.2021.1924636.

Guida Viaggi. (2022). Casano, Pantelleria island: Bene l'estate. Forte interesse dei gruppi per i periodi di spalla. [Casano, Pantelleria Island: A good summer. Strong interest by groups in the shoulder season.] *Guidaviaggi.* www.guidaviaggi.it/2022/06/13/casano-pantelleria-island-gruppi-forte-interesse-per-i-periodi-di-spalla/.

Hughes, R. (2023, March 1). One of the world's best beaches in Italy. Here's why you should visit in 2023. *Forbes.* www.forbes.com/sites/rebeccahughes/2023/03/01/one-of-the-worlds-best-beaches-in-italy-heres-why-you-should-visit-in-2023/?sh=2003352d59b5.

Hussain, A. and Fusté-Forné, F. (2021). Post-pandemic recovery: A case of domestic tourism in Akaroa (South Island, New Zealand). *World*, 2, 127–138. https://doi.org/10.3390/ world2010009.

Islands of Sicily. (2023). Una geografia insulare [An island geography]. www.islandsofsicily.com. Video at: https://youtu.be/BB5unEE3g3M.

ISPRA (2021). La spiaggia ecologica di Favignana per la protezione ambientale ed il turismo consapevole. [The ecological beach at Favignana for environmental protection and smart tourism.] *Isprambiente*. www.isprambiente.gov.it/it/archivio/eventi/2021/09/la-spiaggia-ecologica-di-favignana-per-la-protezione-ambientale-ed-il-turismo-consapevole.

Italy this way. (2023). Lipari Aeolian museum, an important archaeological museum in Lipari, Sicily. *Italy this Way*. www.italythisway.com/places/lipari-museum.php.

Leone, B. (2023). Isole della Sicilia, tagli alle corse delle navi: 40 associazioni scrivono ai governi. [Island of Sicily cuts to ferry schedules: 40 associations write to the governments.] *Giornale di Sicilia*. https://messina.gds.it/articoli/economia/2023/02/28/isole-della-sicilia-tagli-alle-corse-delle-navi-40-associazioni-scrivono-ai-governi-a80f8789-d412-4fff-94c2-01ebd44e77b2/.

Lo Porto, G. (2023). Solarium di Levanzo, dal comune via libera ma con il nulla osta scaduto. [Levanzo solarium: All clear from the council but with expired authorization.] *La Repubblica*. https://palermo.repubblica.it/cronaca/2023/06/09/news/solarium_levanzo_concessione_scaduta_ambientalisti_sindaco_forgione-403810526/.

Loveolie. (2023). *Isole festival: Fifth edition*. www.loveolie.com/eventi/eventi-culturali-eolie/isole-festival-v-edizione.

McElroy, J. L. and Parry, C. E. (2010). The characteristics of small island tourist economies. *Tourism and Hospitality Research*, 10(4), 315–328. www.jstor.org/stable/23745403.

Militello, S. (2021, 14 September). Ustica conquista i turisti: da giugno ad agosto 21,780 arrivi. [Ustica conquers the tourists: 21,780 arrivals between June and August.] *Palermotoday*. www.palermotoday.it/economia/turismo-estate-2021-ustica-bilancio-presenze.html.

Moviment Graffitti (2022) Registration Reclaim Blue Lagoon 13.08.22. www.comino.movimentgraffitti.org/.

NSO. (2022). *Regional Tourism: 2021*. Valletta, Malta: National Statistics Office. https://nso.gov.mt/regional-tourism-2021/.

NSO. (2023). *Tourism*. Valletta, Malta: National Statistics Office. https://nso.gov.mt/regional-tourism-2022/.

Omarjee, Y. (2022). *Report on EU islands and cohesion policy: Current situation and future challenges*. European Parliament: Committee on Regional Development. www.europarl.europa.eu/doceo/document/A-9-2022-0144_EN.html.

Pappalardo, M., Bevilacqua, A., Luppichini, M., and Bini, M. (2021). Geomorphological features of Favignana Island (SW Italy). *Journal of Maps*, 17(2), 30–38. https://doi.org/10.1080/17445647.2020.1866699.

Power, S., Di Domenico, M., and Miller, G. (2017). The nature of ethical entrepreneurship in tourism. *Annals of Tourism Research*, 65, 36–48. doi:10.1016/j.annals.2017.05.001.

QDS. (2022). Isole minori, 31 associazioni scrivono a Schifani e Salvini: 'Tutelare collegamenti'. [The smaller islands: 31 associations write to Schifani and Salvini 'Build better connections'.] *QDS*. https://qds.it/isole-minori-lettera-schifani-salvini-tutelare-collegamenti/.

Regione Sicilia. (2019, 9 August). Musumeci a Linosa, sei milioni per il porto. [Musumeci in Linosa: 6 million for the port.] *Regione Sicilia*. www.regione.sicilia.it/la-regione-informa/musumeci-linosa-sei-milioni-porto.

Resto al Sud. (2023, 10 January). L'area marina di Ustica conquista Parigi. [Ustica's marine area conquers Paris.] *Restoalsud.* www.restoalsud.it/viaggi-e-turismo/larea -marina-di-ustica-conquista-parigi/.

Sajeva, G. (2020). Egadi Islands, *COVID-19 Island Insight Series, no 2,* November 2020, University of Strathclyde Centre for Environmental Law and Governance, University of Prince Edward Island Institute of Island Studies and Island Innovation. https:// islandstudies.com/files/2021/04/COVID19-Island-Insights-Series-02.-Egadi.-November-2020.pdf.

Sammartano, F. (2023, 14 April). Favignana, Sammartano: 'Nessun evento nelle isole. Un fallimento totale'. [Favignana, Sammartano: 'No event on the islands: A complete disaster'.] *Trapani Prima Pagina.* www.primapaginatrapani.it/favignana-sammartano-nessun-evento-nelle-isole-un-fallimento-totale.

Sarpi, S. (2022, 19 August). *Panarea presa d'assalto dai turisti: Ordinanza restrittiva sugli arrivi.* [Panarea overwhelmed by tourists: A regulation to restrict arrivals.] *Gazzetta del Sud.* https://messina.gazzettadelsud.it/articoli/cronaca/2022/08/19/ panarea-presa-dassalto-dai-turisti-ordinanza-restrittiva-sugli-arrivi-dda1f298-d58d-4478-8e08-c1cd81d27a6d/.

Spanò, L. (2021). Assalto di visitatori alle Isole Egadi: Il sindaco Forgione invoca un turismo di qualità. [Tourist assault in the Aegadian Islands: Mayor Forgione appeals for quality tourism.] *Giornale di Sicilia.* https://trapani.gds.it/articoli/econom ia/2021/09/07/assalto-di-visitatori-alle-isole-egadi-il-sindaco-forgione-invoca-un-tur ismo-di-qualita-3f546b3f-a429-46c5-891f-06d1e4abe16d/.

Stoffelen, A. (2019). Disentangling the tourism sector's fragmentation: A hands-on coding/ post-coding guide for interview and policy document analysis in tourism. *Current Issues in Tourism,* 22(18), 2197–2210. https://doi.org/10.1080/13683500.2018.1441268.

Suryani, A. (2013). Comparing case study and ethnography as qualitative research approaches. *Jurnal Ilmu Komunikasi,* 5(1), 117–127. https://doi.org/10.24002/jik.v5i1.221.

TF1 (2020). À la découverte de l'île de Pantelleria, le jardin secret de Carole Bouquet. [Discovering the island of Pantelleria: The secret garden of Carole Bouquet.] *TFL*www.tf1info.fr/voyages/video-a-la-decouverte-de-l-ile-de-pantelleria-le-jardin-se cret-de-carole-bouquet-2169305.html.

Times of Malta (2017, 19 February). The tropics of the Med. https://timesofmalta. com/articles/view/the-tropics-of-the-med.640138.

Times of Malta (2023, 15 June). Gozo tourism operators struggle to find workers, worry about poor skills. https://timesofmalta.com/articles/view/gozo-tourism-opera tors-struggle-find-workers-worry-poor-skills.1037739.

Veal, A. J. (2006). *Research methods for leisure and tourism: A practical guide.* Harlow: Pearson Education.

Vivere Pantelleria. (2023). *The Island Festival Pantelleria.* www.viverepantelleria.it/ blog/l-isola/the-island-festival-pantelleria.html.

Wearing, S., Cynn, S., Ponting, J., and McDonald, M. (2002). Converting environmental concern into ecotourism purchases: A qualitative evaluation of international backpackers in Australia. *Journal of Ecotourism,* 1(2–3), 133–148. https://doi.org/10. 1080/14724040208668120.

Wolf, F., Filho, W. L., Singh, P., Scherle, N., Reiser, D., Telesford, J., Božić Miljković, A., Hausia Havea, P., Li, C., Surroop, D., and Kovaleva, M. (2021). Influences of climate change on tourism development in small Pacific island states. *Sustainability,* 13(8), 4223. https://doi.org/10.3390/su13084223.

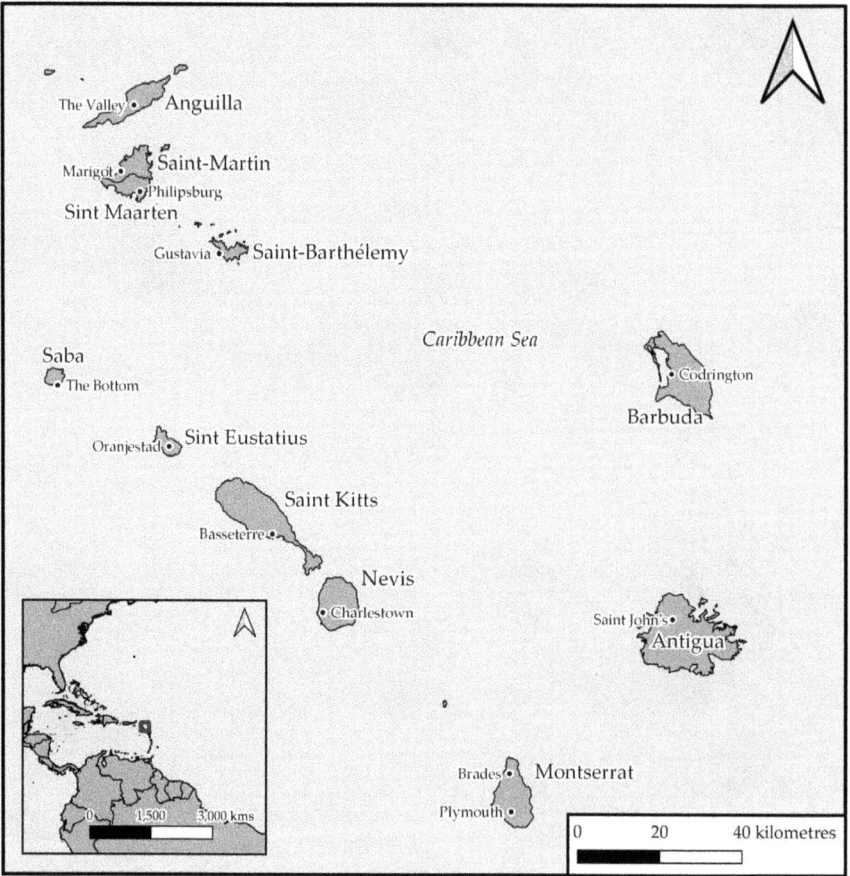

Figure 11.1 The North East Caribbean

11 The impromptu archipelago

Sint Maarten as the hub of the Northeastern Caribbean

Arend Jan (Arjen) Alberts

Introduction

This chapter reflects on the islands of the Northeastern Caribbean as 'Small Island Tourism Economies' (SITEs), following the contribution of Jerome McElroy and others to the taxonomy of small island states and territories. Measured by McElroy and de Albuquerque's "Tourism Penetration Index" (TPI) Sint Maarten was the most intensely developed SITE in the world in the 1990s and again after 2006, challenged intermittently by the UK Virgin Islands (McElroy and de Albuquerque, 1998; McElroy, 2003, 2006; McElroy and Hamma, 2010). The TPI averages three economic and spatial intensity indicators, measuring tourism spending per resident, average number of tourists present per resident, and rooms per km^2.

Sint Maarten is an autonomous country within the Kingdom of the Netherlands. It occupies the southern half of an island in the Northeastern Caribbean, the northern half of which is an overseas collectivity of France (Saint-Martin). Through its geographical endowments and its rapid growth from the 1970s onwards, Sint Maarten developed into the region's most highly concentrated tourism economy. Concurrently, the subnational jurisdiction established itself as a logistical and tourism hub for the surrounding islands, in particular Anguilla (now a British Overseas Territory), the Dutch islands of Saba and Saint Eustatius (presently Dutch overseas special municipalities), Saint-Barthélemy (a French overseas collectivity), and of course for the French side of the same island.

Tourism Area Life Cycle

McElroy and others find that a high score on the TPI mostly coincides with an advanced stage in the Tourism Area Life Cycle (TALC), a framework developed by Richard Butler (Butler, 1980, 1996, 2006a, 2006b). In a model inspired by the Product Life Cycle concept, a cornerstone of product marketing theory, Butler identifies a number of stages tourism destinations typically traverse. Through the initial *exploration, involvement*, and *development* stages, destinations progressively adapt their economy, society, and infrastructure to the

DOI: 10.4324/9781003451037-15

demands of their visitors, fuelled by ever increasing income from and investments into the tourism sector. This eventually challenges the limit of the destination's "carrying capacity". Negative feedback mechanisms slow down growth and jeopardise the tourism product itself. This phenomenon is nowadays widely acknowledged as "overtourism" in most leading destinations around the world, as well as in Caribbean SITEs (Peterson, 2020). In the critical range of development, a destination will enter into the stages of *consolidation* and eventually *stagnation* (Butler, 1980, p. 7). In parallel to the original product life cycle model, a destination will then enter into *decline* and perhaps continue on a subsistence level; or else reinvent itself and start an entirely new cycle in an act of *rejuvenation*, attracting a new audience with a renewed product. A high intensity case like Sint Maarten appears to linger in the stagnation phase, on a slow growth path with increasing carrying capacity pressure, with neither rejuvenation nor outright decline (Alberts, 2016; Pereira and Croes, 2018).

Carrying capacity

Carrying capacity is a widely used, though not narrowly defined, concept. There is consensus however that in the context of tourism development it – at least – comprises dimensions like: the impact on the natural environment, visible through damage to the natural surroundings; the impact of tourism on the host population and its culture, manifesting itself through for instance loss of cultural heritage or a negative attitude towards tourists; and the impact of tourism on the visitor experience itself, most clearly visible in the case of overcrowding of facilities and places of interest. These however comprise mainly the direct impacts: indirectly, tourism development in many SITEs has given rise to massive attraction of foreign labour, causing rapid population growth and a concurrent rise in demand for physical infrastructure like energy, water and roads, but also housing, health care, and education. Generally, top-tier SITEs have not been able to keep up with these demands.

How soon carrying capacity is reached or exceeded depends on a multitude of factors. Of most interest however are: first, the nature of the tourism product, accommodations, and attractions themselves, including the industry and its standards; second, the characteristics of the target market, such as the tourists' culture and behaviour; and third, the government policies and management of tourism development, ideally with mitigating measures in the short run and strategic management choices in the long run. Well aware of these issues, the Government of Sint Maarten commissioned an extensive carrying capacity study (TTCI, 2004) as early as 2004, in tandem with its strategic Tourism Master Plan (TTCI, 2005). The findings and recommendations of these reports, however, have not led to visible changes in strategic direction by successive governments over the last two decades.

Vulnerability reconsidered

In the vulnerability and resilience debate related to islands, tourism has generally been considered a sector with a low vulnerability to external economic shocks compared to other branches of industry. Even international crises like 9/11, economic recessions or a major PR crisis like the 2006 disappearance of a vacationing US teenager in Aruba had mild effects on the receiving islands (Alberts and Baldacchino, 2017; Croes, 2006; Ridderstaat, 2015). A key assumption here was the continuous and reliable availability of international travel. That is, until the Covid-19 pandemic struck. With mobility, its cornerstone, challenged, the relative vulnerability and resilience of island tourism needs a reassessment. It needs to be said that the cruise industry did foreshadow some hints of this travel-related vulnerability in the pre-Covid-19 era though, with the highly contagious Norovirus even acquiring the moniker of "cruise ship virus" (Viswanathan, 2023), a label understandably contested by the cruise industry (CLIA, 2023).

Dual disasters in the North-Eastern Caribbean

The archipelago under review was struck by two consecutive disasters, each of an unprecedented scale and depth. September 2017 brought the category 5+ hurricane Irma, the largest cyclone to strike this corner of the Caribbean since scientific measurements began. Irma, closely followed by category 5 Maria, caused considerable direct human suffering and material damage as well as severely interrupting economic life for more than one year, with noticeable effects for several more. Generally, 2019 saw a rebound in the archipelago that turned out to be an intermission before the Covid-19 pandemic brought international travel – and thereby tourism – to a screeching halt at the beginning of 2020. In line with global developments, 2020 and 2021 were years mostly lost to the tourism industry and therefore led to a unique economic interruption to all islands studied here. This chapter thus often refers to 2016 as the most recent 'normal' year, especially for use as a benchmark for comparison between the different territories.

The notion of a northeastern Caribbean tourism archipelago

Island tourism in a tourism archipelago?

All islands under study stand the test of Butler's definition of *island tourism*; they are *physically separated* from any other landmass, they have notable *cultural differences*, showcase an *attractive climate and environment* and have a measure of *political autonomy* (Baldacchino, 2015, pp. 2, 3). All are islands at a limited distance from each other; the obvious semi-exception is the island at the centre of this chapter, occupied by two separate jurisdictions, each complying with three out of four criteria, yet 'joined at the hip' in a geographical sense. The islands under review are all small and comparatively close to each other, but count at least three different official languages plus a myriad of unofficial ones, while each has a population with roots in the colonial past and the transatlantic slave trade, superimposed by layers of nineteenth- and twentieth-century migration, giving each its own cultural personality. Situated in the Caribbean, the attractive climate is a given, as well as beautiful natural surroundings. Not even the serious hurricanes that strike the region every few decades deter visitors in the long run, although they do give rise to a distinct seasonality in the tourism flow. None of the six territories under review are sovereign states. They are rather subnational island jurisdictions (Baldacchino, 2010): they all run their own internal affairs and are in charge of shaping their own tourism futures. They share three former colonial powers to which they have a total of five distinct constitutional ties: French Saint-Martin's status is similar to Saint-Barths, but with different ties to the EU; Sint Maarten's is that of an autonomous country within the Dutch Kingdom; Saba and Statia are Dutch special municipalities with far less authority; and Anguilla has the status of a United Kingdom Overseas Territory (UKOT) (see Table 11.1).

Given the fact that we have a collection of six tourism islands, this chapter tests the notion that a North-Eastern Caribbean tourism archipelago exists, with Sint Maarten as its hub. As all are separate jurisdictions, this group obviously is not going to be an *archipelago state* in the sense of La Flamme (1983, p. 361). However, we acknowledge his fourth archipelago state attribute, the *centrifugal tendency*, since peripheral islands have typically shown the tendency to separate from their centre. In fact, each of the islands under review has a history of secession from a different (political) archipelago centre outside the scope of this chapter; Saint-Martin and Saint-Barths from Guadeloupe; Sint Maarten, Saba and Statia from Curaçao; and Anguilla from St. Kitts. All these disconnections took place successfully. We will now apply this centrifugal notion to the Sint Maarten centred tourism archipelago as well.

The hypothesis to be tested in this chapter is that, presently, the six entities under review form an *archipelago tourist destination* (Baldacchino, 2015, pp. 8–11). This concept, to an extent, presupposes the archipelago constituting one jurisdiction, which in our case is not applicable. Some attributes are therefore in this case less relevant in testing the tourism archipelago conjecture, like *visibility* and *tweaked representation*. This is because they all have a clear separate existence and tourism marketing profile and are unable to control each other's public image to the outside world. *Domination and subordination* do not exist within the archipelago in a political sense, but it is very relevant economically and logistically, which is central to the questions this chapter explores. The fourth attribute of *liminality or layering* of an archipelago is relevant in the historical sense described above, as layers of previous political archipelagos have peeled off to congeal and form a new, economically, and logistically defined tourism archipelago. Finally, the *differentiation between islands* is again quite self-evident, as each has its positioning in a marketing sense and expresses its own product to its own niche in a global tourism market.

Table 11.1 North-Eastern Caribbean: Key facts and figures

	Constitutional ties with	Surface area (km²)	Population	Population density (inh/km²)	Year of pop. density	2016 GDP mlnUS$	2016 per capita GDP	Recent GDP mlnUS$	Year recent GDP
Sint Maarten	Netherl.	34	42,759	1258	2022	$ 1,072	$ 27,200	$ 1,145	2021
Saint-Martin	France	53	34,500	610	2021	$ 715	$ 19,264	$ 632	2021
Saint-Barths	France	21	10,400	496	2021	$ 406 *)	$ 43,135 *)	n/a	
Anguilla	UK	91	15,700	173	2023	$ 320	$ 21.020	$ 260	2020
Saba	Netherl.	13	2,035	157	2023	$ 48	$ 24,500	$ 44	2020
Statia	Netherl.	21	3,293	157	2023	$ 131	$ 40,600	$ 89	2020

*) 2014

Source: Sint Maarten: Ministry of General Affairs (2017); Department of Statistics Sint Maarten (2023b, 2017; IMF, 2022); Saint-Martin: IEDOM (2022b); Saint-Barths: IEDOM (2022a); Anguilla: Anguilla Statistics Department (2023a, 2023b); Saba and Statia: CBS (2023a, 2023b, 2023c); Ministry of General Affairs (2017). Exchange rates used: 2016 US$1=€0.904, 2021 US$1=€0.8458, all years US$1=EC$2.6882.

Table 11.1 shows the relative population and economy sizes within the proposed archipelago, with Sint Maarten emerging as the dominant entity. Its position as a logistic 'hub' of and to the island group is examined below. A preliminary test of the existence of a North-Eastern Caribbean tourism archipelago is of course, whether it is recognised as such in the public eye. Sint Maarten is presently indeed advertised as a hub for visitors to the North-Eastern Caribbean. Not just by Sint Maarten itself (St. Maarten Tourist Bureau, 2023; SHTA, 2023) but also internationally (Creedy, 2015; Premium Caribbean, 2023; Reid, 2012). Conversely, most of the peripheral islands' tourism authorities make no secret of their logistical position; Saba and Statia plainly mention their 'international hub' (Getting Here, Saba Tourism, 2021; Statia Tourism Bureau, 2023), while Saint-Barths also mentions Sint Maarten as the primary waypoint to reach the island (St Barts Tourism Committee, 2023). Saint-Martin however positions itself boldly at the centre of the *French and Dutch* Antilles and mentions Anguilla, Saba, and Saint-Barths as secondary destinations, while giving the diminutive Grand Case airport an equal status as 'international airport' to the Princess Juliana airport on the Dutch side where more than 80% of its own visitors land (Office de Tourisme, 2023). Finally, Anguilla stresses its uniqueness and mentions Puerto Rico, St. Maarten, and Antigua – in that order – as gateways for reaching the island and refers to ferries 'from one of the neighbouring islands', carefully downplaying its one-sided logistical dependence on Sint Maarten/Saint-Martin (Anguilla Tourism Bureau, 2018).

Sint Maarten

History and governance

A leading factor in Sint Maarten's development is its remoteness to Curaçao, the administrative centre of the Netherlands Antilles until its dissolution in 2010. The centrifugal tendencies in this archipelago, further fuelled by differences in language and culture, were prominent since the Dutch Kingdom Charter came into force in 1954 (Alberts, 2020, p. 67). In practice, the central government, situated some 1,000 km from Sint Maarten, granted the local administration a great deal of autonomy, in return for Sint Maarten's support to the political coalition of the day. This symbiotic relationship spanned the period until the mid-1990s and helps to explain the vigorous yet in retrospect poorly managed and unsustainable heydays of Sint Maarten's tourism development (Alberts, 2020, p. 54). Since 2010, Sint Maarten is a constituent country within the Kingdom of the Netherlands with a high degree of autonomy.

Economics and tourism

The glory days of Sint Maarten (tourism) development ended in the mid-1990s when, due to a breakdown in local governance, the Netherlands instituted 'higher supervision' on the island in 1993, which was silently ended after the disastrous impact of hurricane Luis in 1995. The quality of the tourism product was already in decline, and quantity – especially in the newly arrived timeshare product – had taken over. After hurricane Luis, there was a radical shift towards cruise tourism, with a new harbour funded by Dutch reconstruction funds. The days of Sint Maarten as a high-end destination were over: tourism development entered a *stagnation* phase with some *decline* elements. An important indicator is real per capita GDP, that sharply increased up to the 1980s, but reached a plateau from 1995 onwards (Alberts, 2016, p. 88).

Pre-Irma and Covid-19, Sint Maarten captured a market share in Caribbean stayover tourism that declined gradually from 3% around the turn of the century to 2.4% in 2016 (IMF, 2023, p. 61), illustrating the latter-stage TALC situation outlined above. Similarly, tourism (export) receipts as a percentage of GDP were in sharp decline from around 80% in the beginning of the 2010s, to around 60% in 2016, the last 'normal' year, although recovering to above 60% in 2022.

As one of the busiest cruise ports, Sint Maarten is a mainstay of the cruise itineraries in the region. Peaking at almost 2 million passengers annually, among the Caribbean islands it is only exceeded by the Bahamas and the US Virgin Islands. Although worldwide a modest part of tourism with 2% of travellers, in the region this group accounts for two thirds of all visitors. Sint Maarten also boasted the highest amount of onshore spending per passenger, in contrast to the overall tendency of this number to fall (Fernandez-Stark and Daly, 2017). However, even before the dual disasters, total cruise numbers were on a gradual decline to about 1.7 million in 2016.

Logistics

Sint Maarten has one of the largest cruise harbours in the region; the outcome of a deliberate strategy starting in the 1990s. Sint Maarten has a large freight port as well, that receives most of the goods for the French side of the island, as well as transit goods for the surrounding islands. The Princess Juliana International Airport (PJIA) however, is even more central to Sint Maarten's position as a tourism hub. The original landing strip was created by the US military in the Second World War to aid the battle against German U-boats that inflicted great damage to merchant shipping in the region. The relocation and upgrade in 1964 were instrumental to Sint Maarten's budding tourism industry. Further extensions in 1985 and 2001 lead to the present 2,300-metre landing strip, the only one in the region to accommodate the largest international planes. In an illustration of its regional hub function, PJIA air traffic control provides approach control services for the airports of Saint-Martin, Anguilla, Saint-Barths, Saba and Statia as well ("Princess Juliana International Airport", 2023).

Saint-Martin

History and governance

Saint-Martin, a French Collectivité, occupies the northern half of the island it shares with Sint Maarten. This circumstance makes for the smallest inhabited island shared by two jurisdictions in the world. Saint-Martin, at the same time as Saint-Barths, left the Département d'Outre-Mer of Guadeloupe in 2008 (IEDOM, 2022a, p. 20) in a move consistent with the centrifugal forces we find elsewhere in (Caribbean) archipelagos.

Economics and tourism

The economy of the island is highly dependent on tourism, but this has led to less favourable outcomes compared to other French territories. Strongly tied economically to the Dutch side of the island and a supplier of cross-border labour, the population initially grew concurrently with Sint Maarten's economy: Sint Maarten's top year in terms of total visitors, 2014, was also the year of its highest recorded GDP (CEROM, 2023, p. 3). Conversely, in times of economic downturn after 2008 and even more after 2017, population shrank. In Sint Maarten on the other hand, neither the 2008 economic crisis, nor Irma or Covid-19 caused a population decrease. Economically, Saint-Martin is among the "poorest of the rich", with a per capita GDP much lower than Sint Maarten or the nearest French territories of Saint-Barths, Guadeloupe, and Martinique, but still in the Caribbean top tier (CEROM, 2023, p. 1). Outmigration apparently kept pace with Saint-Martin's economic downturn, causing the unemployment level to remain high and stable at around 30% before and after the Irma/Covid-19 crises (CEROM, 2023, p. 2).

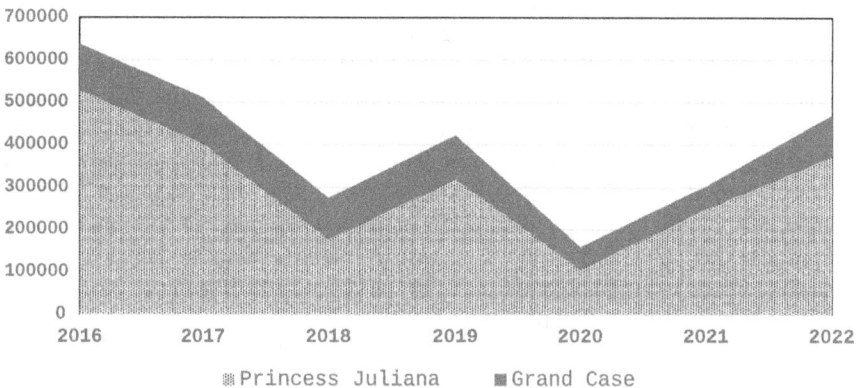

Figure 11.2 Sint Maarten/Saint-Martin airport stayover arrival numbers (cumulative)
Source: Department of Statistics Sint Maarten (2023c; IEDOM (2023a).

Logistics

Logistically, the Saint-Martin tourism industry is highly dependent on the Dutch side of the island, more specifically on its international airport. Of total visitors to Saint-Martin, only 11% arrive directly on the French side, the remainder via the Dutch part (IEDOM, 2022b, p. 79). The Grand Case airport of Saint-Martin, with a runway 1,200 metres long, accommodates regional flights only.

Cruise tourism, a mainstay of the tourism sector on the Dutch side of the island, was always a modest activity in Saint-Martin, since the Marigot harbour can only accommodate vessels with a modest draft. A certain category of small luxury cruise ships (100 to 360 passengers) frequented Saint-Martin, with an all-time high of around 15,000 passengers received in 2011. This activity almost completely petered out even before the Irma and Covid-19 onslaughts, as the cruise lines involved reoriented their itineraries towards Latin America (IEDOM, 2022b, p. 80). This further increases its economic dependency on its southern half-island.

Saint-Barthélemy (Saint-Barths)

History and governance

Saint-Barthélemy – commonly called Saint-Barths – is a French Collectivité, like Saint-Martin. The two jurisdictions split off from the Département d'Outre-Mer of Guadeloupe in 2008 (IEDOM, 2022a, p. 20). This gave the island a high degree of autonomy in essential areas like taxation, the admission of foreigners on a work permit, infrastructure, tourism, and the environment, policy areas previously determined by the Département of Guadeloupe. Saint-Barths further carries the distinction of being the only European Union territory to switch from Outermost Territory (OMT) status, hence subject to the EU *acquis communautaire* with a few localised exceptions, to the status of Overseas Countries and Territories (OCT). This move brought the territory out of EU jurisdiction in 2012, even though it is still part of an EU member state, and thus its inhabitants remain EU citizens (IEDOM, 2022a, p. 20). This removed EU import duties and the common market's many economic regulations, even though it also took away eligibility to tap EU structural funds. The former implications were considered decisive, due to the island's high per capita GDP that puts most EU structural funds out of reach anyway.

Economics and tourism

In the North-Eastern Caribbean, Saint-Barths developed itself into a small yet highly exclusive destination, frequented by the international jet set. It is known as the "Saint-Tropez of the Caribbean" (IEDOM, 2022a, p. 7). While the island started off as an important cruise ship destination in the 1990s, this type of vessel has largely been replaced in economic importance by luxury yachts. Of the visitors arriving by sea, the proportion of cruise passengers dropped from 50% around the turn of the century to under 10% in Covid-19 year 2021. Yacht passengers however increased to 20% of the total, the remainder arriving by ferry (from Saint-Martin/Sint Maarten) (IEDOM, 2022a, p. 79).

As a result, and in stark contrast to Saint-Martin, Saint-Barths is economically very successful, with by far the highest per capita GDP of the territories under review: more than twice that of Saint-Martin and over 50% higher than Sint Maarten. Standing at 4%, its unemployment rate is lower than that of any French department, European or overseas (IEDOM, 2022b, p. 37). Unsurprisingly however, the carrying capacity of the tiny island in terms of the harbour, electricity or sanitation is at breaking point. The island is also demographically challenged, with out-migrating local youths and in-migrating male workers, leading to an overrepresentation of men in the active age category (IEDOM, 2022a, p. 8).

Logistics

Logistically, Saint-Barths occupies a special place in the Caribbean. The Gustaf III airport possesses a 646-metre runway with a very difficult approach. It maintains connections to a few nearby islands, of which Guadeloupe, Sint Maarten, and Puerto Rico receive intercontinental flights. Between 50 and 60% of passengers arrive from Sint Maarten, followed at a distance by the other two international airports, stressing the former's importance as a regional hub (Collectivité de Saint-Barthelemy, 2023). Furthermore, Saint-Barths enjoys a reputation as a destination for top-tier yacht tourism, with between 5,000 and 6,000 yacht visits, bringing around 35,000 passengers annually over the decade up to the hurricane year 2017. This number bounced back in 2019, and surprisingly did not drop below the 50% level in the pandemic years of 2020 and 2021 (IEDOM, 2022a, p. 76). Saint-Barths however does not provide the marina facilities for long term berthing of yachts, nor the international airport to receive their high-end passengers. This is where Sint Maarten plays another role as a hub; its marinas harbour most of the yachts visiting Saint-Barths during the season, and its airport receives their passengers and crew. Many of the former arrive by private airplane, which makes reception and parking of private jets an industry in its own right. Traditionally, on the days before New Year's Eve, a flotilla of superyachts sets sail from Sint Maarten to Saint-Barths (Borden, 2019).

Apart from the extremely high-end stayover sector and the yachting industry, Saint-Barths is a significant day-trip destination for Sint Maarten as well as Saint-Martin. The island traditionally received ferry passengers in roughly the same numbers as Anguilla received day-trippers, close to 100,000. Although the available sources do not distinguish between daytrippers and stay-overs, it is safe to assume that the vast majority are excursionists. The unfavourable connection between arriving international flights in Sint Maarten and the greater distance of the ferry – compared to Anguilla – makes transferring to such a connection unattractive. For Sint Maarten, Saint-Barths is the second most popular day trip destination after Anguilla.

Anguilla

History and governance

Like all UK Overseas Territories in the Caribbean, Anguilla is typified by a small size and population. Anguilla came out of the dissolution of the West Indies Federation in 1962 as part of the associated state of Saint Kitts-Nevis-Anguilla. However, in a familiar centrifugal Caribbean pattern, the association to the relatively distant, larger, and more populous St. Kitts-Nevis archipelago was perceived as disadvantageous by the Anguillans. This led to the – comparatively peaceful – "Anguillan revolution" of 1967, which did not have independence as its goal, but rather secession from St. Kitts and a direct association with Britain. This process was concluded in 1969 with the restoration of British authority, formalised by the Anguilla Act of 1971 (Anguilla, 2023).

Economics and tourism

Anguilla followed a markedly different development path compared to its neighbour Sint Maarten. This UK Overseas Territory secured a status as a luxury, exclusive destination for stayover tourists in its small number of exclusive hotels. Additionally, Anguilla receives large numbers of day trippers from Sint Maarten/ Saint-Martin. More than half of the ferry passengers are 'excursionists': close to 100,000 in the pre-Irma/Covid-19 years. This group makes Anguilla the primary one-day excursion destination for Sint Maarten and Saint-Martin. In the stayover as well as the day tripper categories, traditionally around two thirds originate from North America (Anguilla Statistics Department, 2023c, tbl. 2.4.5-V6 and 2.4.5-T6).

Logistics

Located directly north of Sint Maarten/ Saint-Martin across a narrow sea strait, Anguilla has one airport and one ferry terminal (the Blowing Point). The ferry connections involve boat trips of just over ten kilometres, at less than half an hour. Most ferries connect to the Saint-Martin capital of Marigot, on average accounting for some 75% of passengers received (Anguilla Statistics Department, 2023c, tbl. V3; IEDOM, 2023c); others to piers in the vicinity of the Princess Juliana airport on the Dutch side. Until recently, Anguilla was fully dependent on neighbouring airports to receive non-regional tourists arriving by air, who would then change to smaller craft flying to

Anguilla's Clayton J. Lloyd International Airport with its 1,665-metre runway. In the case of Sint Maarten/ Saint-Martin, most arriving passengers rely on ferries to transport them to their eventual destination. Until recently, just under half the passengers arriving by ferry were stayover tourists, while this group constituted between 75% and 85% of annual stayovers (Anguilla Statistics Department, 2023c, tbls. V3 and V8). Therefore, only a minority of accommodation guests arrive by airplane, at least for the last leg of their trip. The impact of the pandemic was felt particularly when all ferry traffic between Saint-Martin and Anguilla was suspended from March 2020 until January 2022, making the latter fully dependent on air connections and ferries to the Dutch side (IEDOM, 2023c, p. 6). The capacity of the remaining ferries was never above 20% of the original total; numbers that illustrate the traditionally high reliance of Anguilla's stayover tourism on Sint Maarten as a hub.

Saba and Sint Eustatius (Statia)

History and governance

When the Netherlands Antilles were dissolved in October 2010, Sint Maarten opted by referendum to become a constituent country within the Kingdom of the Netherlands. Saba and Statia voted differently and became "special municipalities" of the Netherlands proper. This gave them a hybrid constitutional status, somewhere between the French 'collectivité', with its full conformity to metropolitan French laws, regulations, and social security on the one hand, and the more loosely associated UK 'Overseas Territories' model on the other. Saba and Statia (and the larger and more populous Bonaire, situated off the Venezuelan coast) are governed by a tailor-made legal framework that is significantly different from that of a European Dutch municipality, and includes tasks that in metropolitan Netherlands are executed on a provincial level (Ministry of General Affairs, 2017). Each of the three islands has its own Island Council that performs these – modified – municipal tasks, while the central tasks delegated from the different Ministries in the Netherlands are executed by the "Rijksdienst Caribisch Nederland" situated in Kralendijk, Bonaire (Ministry of the Interior and Kingdom Relations, 2023). With Bonaire situated some 1,000 km from Saba and Statia, this situation is neither ideal nor efficient.

Economics and tourism

Like the other islands in the north-eastern Caribbean archipelago, Saba and Statia are highly tourism-dependent, although a large oil transfer and storage facility is of economic importance to Statia. The latter circumstance skews the per capita GDP for this island considerably, making it almost twice as high as Saba's. The large value added of the oil terminal makes for a high level of productivity which unfortunately does not translate into a high per capita income; Statia's median household income is similar to Saba's (CBS, 2023e). As a tourist destination, Statia has a higher historical appeal, having once been a main Caribbean trade hub. Hard-to-approach Saba, on the other hand, mostly developed in splendid isolation and is currently known for its pristine natural surroundings and its unique scuba diving opportunities. Illustrating their relative dependence on tourism, pre-Irma/Covid-19 Saba received 2.5 times the number of tourists – including day trippers – per resident compared to Statia (CBS, 2023d, 2023a, 2023f).

Logistics

Both islands are logistically almost fully dependent on Sint Maarten, by sea as well as by air (both flights take less than 30 minutes). The short distance notwithstanding, until recently no regular scheduled ferry connections existed between Sint Maarten and Saba or Statia. The small passenger numbers mean only small vessels can be deployed economically. At the same time, the deep waters between the islands make for rough seas and uncomfortable crossings. This means that all tourists reach the islands via Sint Maarten and residents are fully dependent on the one airline connecting them to this island. Due to the shorter distance and its higher appeal, only Saba is a significant day trip destination for Sint Maarten visitors.

Bouncing back (or forward) from dual catastrophes? A review

Sint Maarten and the tourism archipelago

Sint Maarten is clearly the transportation and tourism centre of gravity with respect to the islands under review. It is the most intensely developed and by far the largest in volume of its tourism industry. Most or all extra-regional tourists travel to the archipelago by way of Sint Maarten. The other constituent islands serve as daytrip or otherwise secondary destinations, with Saint-Martin of course offering an entirely extra dimension to Sint Maarten as a destination. Sint Maarten's yachting industry moreover caters for visitors enjoying the surrounding islands and beyond. Furthermore, Sint Maarten is by far the largest cruise line destination in the sub-region, and in the top 5 of the Caribbean as a whole. The available data does support the notion that Sint Maarten functions as the hub in and to the North-Eastern Caribbean tourism archipelago. Indeed, there would be no such archipelagic entity without this hub function by Sint Maarten.

Sint Maarten

Sint Maarten's tourism has yet to recover beyond 2016 levels (Figure 11.3). However, of the islands under review, it is the only one with a GDP recovery over pre-crisis levels (Table 11.1). The latest estimates suggest even stronger economic

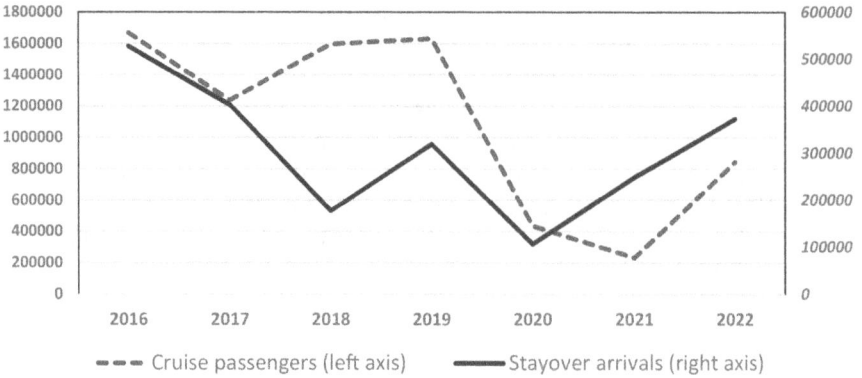

Figure 11.3 Sint Maarten tourism arrivals 2016–2022
Source: Department of Statistics Sint Maarten (2023a, 2023c).

growth than expected in 2022 and 2023 (IMF, 2023, p. 5). Hurricane damage put a number of the larger hotels out of business in 2017, and not all of these have been rebuilt as of 2023. However, pre-Covid 2019 numbers have been surpassed (IMF, 2023, p. 59). For instance, stayover recovery has been much stronger than the Caribbean average, with over 120% against around 100% of 2019 levels (IMF, 2023, p. 61). Cruise tourism, a mainstay of Sint Maarten's industry, is a different matter. The sector is severely affected, and lingering health concerns keep the number under half of the 2019 level, which was in itself an excellent post-Irma year (Figure 11.4). Sint Maarten has been in a double recovery mode since hurricane Irma. The motto of the post-Irma recovery process, strongly aided by a World Bank-managed recovery fund of US$ 553 million granted by the Netherlands, was "build back better" (NRPB, 2023; World Bank Group, 2018). This slogan however was not primarily aimed at re-thinking of Sint

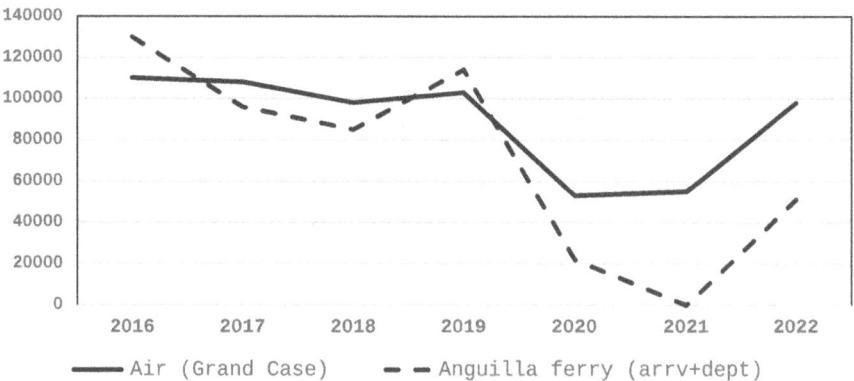

Figure 11.4 Saint-Martin tourism arrivals 2016–2022
Source: IEDOM (2023a, 2023c).

Maarten's carrying capacity-challenged tourism formula, but rather as an attempt to become more hurricane-resilient in a physical and institutional sense. The subsequent Covid-19 pandemic was economically softened by a large Dutch loan program, accompanied by a set of reform measures, again aimed at making the island more resilient against future disruptions. This program however, is mainly aimed at improving the workings of government in a financial and institutional sense, and involves no systematic reform of the economic structure, nor does it consider the carrying capacity of the small half-island at its heart (Ministry of the Interior and Kingdom Relations, 2020). In spite of a strong economic recovery, unemployment numbers are still higher than in 2016. This may be due to a decrease in tourism value added per capita, possibly due to Airbnb-type accommodation rentals and the increased share of passenger expenditure 'staying on board' cruise ships (IMF, 2023, p. 60). If persistent, this trend is the exact opposite of what would be needed to attain a more resilient and sustainable tourism formula, that is, higher value added per visitor as well as per employee.

Saint-Martin

Overall tourism numbers increased in 2021 after the worst of the Covid-19 pandemic, but still fall short of pre-hurricane Irma levels (CEROM, 2023, p. 3; IEDOM, 2023d, p. 79). Strong reconstruction investments and French financial support for businesses drive growth and employment across economic sectors; but 2014 remains the latest best year. As with the rest of the archipelago, Saint-Martin bounced back strongly after Irma/Covid-19, but not to the same extent as Sint Maarten (IEDOM, 2023d, p. 2) (Figure 11.5). Saint-Martin's economy generally moves in sync with developments on the Dutch side, but the French jurisdiction always falls deeper and never recovers quite as quickly or fully as its neighbour (Figures 11.4 and 11.5). Its physical entanglement with the Dutch side means that local arriving passenger numbers do not reflect overall visitor frequency, as most tourists arrive at the larger Dutch airport. Figure 11.5 thus shows the combined passenger data of both airports. Passage to Anguilla deserves special mention, as this connection was fully suspended for part of 2020 and the entire year 2021. All remaining ferry movement to Anguilla in that period originated from the Dutch side. Since resumption, it has not yet reached the halfway mark of the – normal – 2019 level.

During the pandemic, Saint-Martin benefited from its inclusion in the entire French social security system, complemented with a state guaranteed loan scheme for businesses (Prêts garantis par l'État, PGE) out of which local companies received €39 million up to mid-2022 (IEDOM, 2023a, p. 8). Even though GDP has shrunk compared to 2016, so has total population, resulting in stable unemployment numbers that are normal for Saint-Martin, though very high by any other standard. The dual (hurricane-coronavirus) disasters have not led to a sober re-evaluation of the territory's logistics. Also, there is

index 2016=100

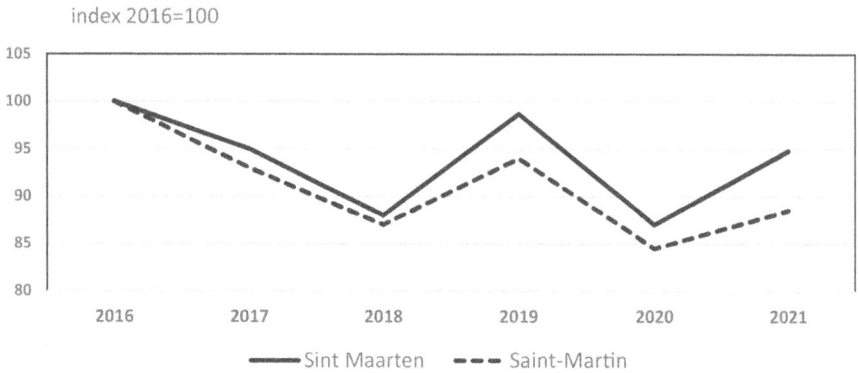

Figure 11.5 Sint Maarten/Saint-Martin comparative GDP growth 2016–2021
Source: IEDOM (2023d).

no observable reorientation of Saint-Martin's approach to tourism. In marketing terms, its positioning as the 'gourmet capital of the Caribbean' remains its unique selling point (Office de Tourisme, 2023).

Saint-Barths

In stark contrast to the other islands in the north-eastern Caribbean, tourism on Saint-Barths recovered to pre-pandemic levels in 2022 (Figure 11.6). The 2021–2022 season was even considered "excellent", with high occupancy rates, strong receipts for the hospitality sector, and record numbers of flights received by the airport during the first months of 2022, even though the main feeder airport of Sint Maarten was still in disrepair and has yet to reach pre-disaster passenger levels (IEDOM, 2022a, p. 12). Passenger arrivals by sea

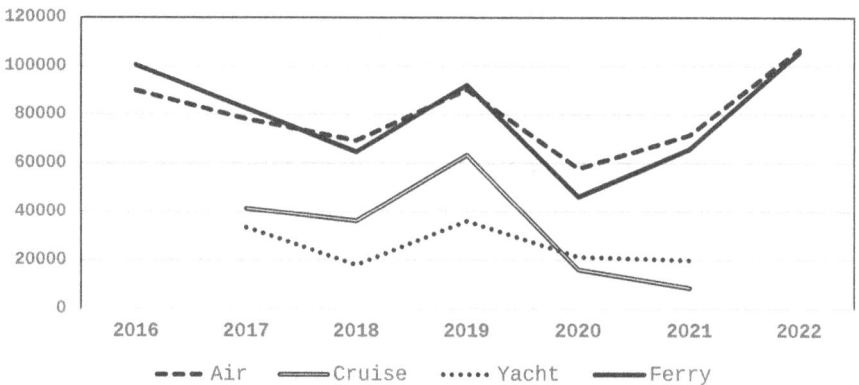

Figure 11.6 Saint-Barths tourism arrivals 2016–2022
Source: IEDOM (2022a, 2023b).

and air in 2022 exceeded those of 2016 (IEDOM, 2023b, p. 2,3). The yachting sector also recovered to its usual end-of-year splendour in 2022 (Dobson, 2022). Interestingly, even the year 2021 saw some 20,000 yacht passengers disembark on the island, against some 40,000 in the pre-disaster years (IEDOM, 2022a, p. 9). In no category and in no disaster-year did the island's tourism numbers fall much below 50% of the 2016 level. This relatively low impact and high pace of recovery is attributed to the specific niche of tourism Saint-Barths caters to: an extremely wealthy North American clientele. The very high-end facilities on the island appear to serve as a healthy financial cushion; while the ultra-wealthy actually tried to use their yachts to escape the pandemic ravaging their home countries, keeping the number of luxury vessels relatively stable (Faus, 2021; Hall, 2020). Saint-Barths' high-end positioning makes for excellent resilience. There is no impetus to reinvent the winning formula.

Anguilla

Anguilla is positioned at the high end of the tourism spectrum, though not quite as exclusive as Saint-Barths. The foundation of its tourism industry, catering for the wealthiest segments of North-American tourists, has not been shaken (Anguilla Beaches, 2023; ICCaribbean, 2022). Where its 'hard to reach' status was once part of its allure, its poor connectivity is now recognised as a weakness. Dependence on the ferry connections with Saint-Martin became glaringly obvious, and stayover numbers plummeted as a result of their total suspension. However, the bounce back is solid, fast approaching 2019 numbers (Figure 11.7).

The resilience of Anguilla's few but highly profitable resorts is beyond discussion; but its logistics are a different matter. Starting on the eve of 2022, Anguilla secured its first direct air connection to the United States by way of

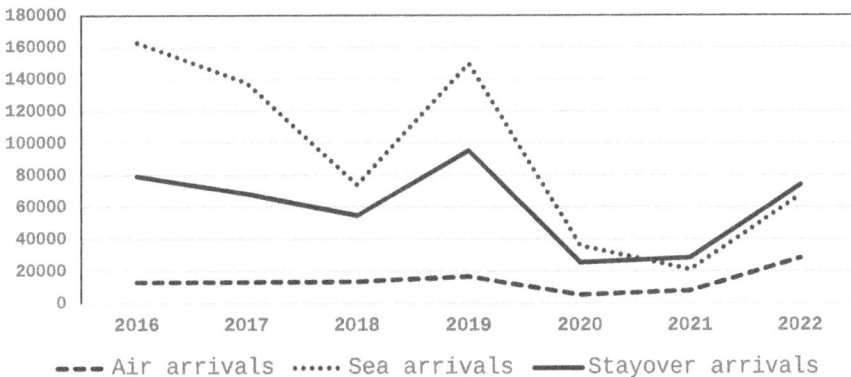

Figure 11.7 Anguilla tourism arrivals 2016–2022
Source: Anguilla Statistics Department (2023c).

American Airlines (AA) (Newsdesk, 2022). This revolutionised the island's logistical situation overnight, raising air arrivals to about 28,000 annually, more than double the pre-Covid-19/Irma levels. This is totally due to substitution in mode of transportation, since total stayover numbers have not quite recovered to the original levels (Anguilla Statistics Department, 2023c, tbl. 2.4.5-V3). In other words, this happened fully at the expense of Sint Maarten's hub function. This trend continues, as numbers for the first five months of 2023 already exceed those of the same period of 2022 by 50%, reflecting a further increase in AA's flight schedule (Major, 2022). Though Anguilla will remain an important daytrip destination for Sint Maarten/ Saint-Martin, the latter's hub position is now seriously challenged for the first time by this jurisdiction. Interestingly, this development preceded the pandemic and was a direct reaction to the damage from hurricane Irma sustained by the Blowing Point pier. With UK financial support, the airport runway was upgraded to higher safety standards while enabling it to receive more and larger airplanes, thus making the island less dependent on sea traffic ("Clayton J. Lloyd International Airport Prepares for Transatlantic Flights", 2021; "Markings Complete Airport Runway Resurfacing", 2021).

Saba and Statia

Saba and Statia show very similar impacts of both disasters, as well as parallel lines of recovery (Figure 11.8). The impact of hurricane Irma, devastating on Sint Maarten, Saint-Martin, and Anguilla, was modest on the smaller two islands, situated just far enough outside the path of the centre of the cyclone. Indirectly though, the damage to the Sint Maarten airport affected both islands deeply. The subsequent pandemic hit both islands with equal force compared to the rest of the archipelago (Figure 11.8). Recovery has been less than convincing. As with the other islands under review, however,

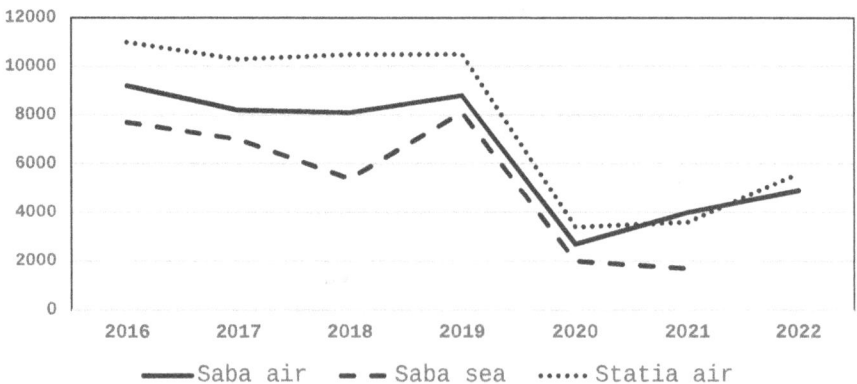

Figure 11.8 Saba and Statia, tourism arrivals 2016–2022
Source: CBS (2023d, 2023f).

no fundamental course changes have taken place. Statia is somewhat of an exception, with its important economic reliance on its oil storage and transfer terminal, and its uncertain future. This development is comparable to the phasing out over previous decades of the oil industry in the other Dutch dependencies of Aruba and Curaçao, as well as in nearby St. Croix (US Virgin Islands). Statia doubled down on its commitment to a tourism future with the opening of a high-end resort in late 2021: a major development that is banking on the island's relatively pristine status (Caribbean Journal, 2020; Hospitality Management, 2023). Saba on the other hand is unwavering in its positioning as a small-scale, laid-back, nature-loving scuba dive paradise.

After the dual disasters and due to the existing airline monopoly connecting the islands to Sint Maarten, the Dutch government initiated and subsidised a new ferry service that started operating late in 2021. It connects both Saba and Statia with Sint Maarten as well as with neighbouring St. Kitts. So far, this service is mainly used by local residents – including Kittitians – and therefore complements rather than replaces the air connections. Passenger bookings suggest that residents often combine an early flight to Sint Maarten with a return by boat, making same day purchases and transport of (bulkier) necessities possible. The ferry carries some 3,000 passengers monthly (one way), compared to a 2022 monthly average of 1,770 air passenger movements on Saba and 1,890 on Statia. The projected 2023 passenger total of 47,000 signifies an important increase compared to the previous year (26,000, with Covid-19 measures in place for the first part of the year, and the St. Kitts destination added in the last quarter). This has prompted the Government of the Netherlands, the authority responsible for the inter-island connections, to extend its subsidy into 2024 and 2025 (Public Entity Saba, Public Entity St. Eustatius, 2023a, 2023b; Slagt, 2023).

Conclusion

In the centre of its archipelago, Sint Maarten is the only one of its territories whose GDP we know is now at a higher level than in 'the last normal year' of 2016 (Table 11.1). Tourism numbers suggest that Saint-Barths' economy has grown too; but no GDP data is available after 2014.

In all instances, we see the clear pattern of two consecutive disasters hitting the North-Eastern Caribbean. Hurricane Irma in September 2017 brought all traffic to a stop for some time, and it took the islands different lengths of time to recover infrastructure and amenities, causing a visible dip in 2018 as well. The destruction of Sint Maarten's airport plays a main role for the whole archipelago. On the other hand, Saba and Statia were mostly outside the path of Irma, and although hit by subsequent storm Maria, their level of damage was far less, as was the direct impact on their travel and tourism sectors. Secondly, we clearly see the impact of the Covid-19 pandemic on tourism numbers, much greater than that of the hurricanes on all six subnational jurisdictions.

Was the manner of recovery influenced by the archipelagic nature of the region? First, there is a strong indication that Sint Maarten's hub position within the archipelago contributed to its favourable outcome compared to the other constituent islands. Second, among the other islands, and as borne out in other destinations, a higher tourism spending segment correlates with higher resilience, and a lower elasticity of demand. Saint-Barths has bounced back quickly and Anguilla as well, but to a lesser extent, since the latter has been hampered by logistics.

The future of Sint Maarten and the archipelago

What hurricanes and pandemic did not achieve was a re-evaluation of the highly concentrated, carrying capacity-exceeding formula of some SITEs, in particular Sint Maarten/ Saint-Martin. At the time of writing (summer 2023), none of the six territories under review have reconsidered restructuring, or – using Butler's term – *rejuvenating* their tourism industry. Rather, the emphasis everywhere is – so far, at least – on a return to *business as usual*. The moves that are made in reaction to the Covid-19 crisis are mostly defensive and focus on addressing connectivity with the archipelago's centre, whether in terms of improving or diversifying these, or replacing them altogether. Sea and air connections are scrutinised in Saba, Statia, Anguilla, and Saint-Martin; while Sint Maarten – in spite of substantial external aid – still struggles to get its airport hub fully operational again. The delay in the full reconstruction of the Princess Juliana Airport has indeed probably slowed down recovery in the entire archipelago.

The North-Eastern Caribbean tourism archipelago, however, still has the opportunity not to 'waste a good crisis' and re-evaluate the nature of its island tourism. The four 'outer' islands already offer a mixed bag of high-end, low key, and small-scale tourism; although, in the case of Saint-Barths, the limits of its carrying capacity are challenged nonetheless. The centre of the archipelago – Sint Maarten/Saint-Martin – would however be well served by a Butlerian reinvention of its tourism product, to bring it back within the boundaries of its carrying capacity. This would allow the two island halves to mitigate the negative consequences of overtourism, increase the quality of their tourism product, and improve the quality of life of their residents, while better exploring the synergies with the surrounding islands of their impromptu archipelago.

References

Alberts, A. J. (2016). Immigration-dependent extensive growth in small island tourism economies: The cases of Aruba and Sint Maarten. *International Development Planning Review*, 38(1), 75–93. http://dx.doi.org/10.3828/idpr.2016.4.

Alberts, A. J. and Baldacchino, G. (2017). Resilience and tourism in islands: Insights from the Caribbean. In R. W. Butler (Ed.), *Tourism and Resilience* (pp. 150–162). Wallingford, UK: CAB International.

Alberts, A. J. (2020). *Small island tourism economies and the tourism area lifecycle: Why Aruba and Sint Maarten have exceeded their carrying capacity* [PhD, Universiteit van Amsterdam]. https://hdl.handle.net/11245.1/67446c6f-950d-4738-8b5a-6f1082924523.

Anguilla. (2023). In *Wikipedia*. https://en.wikipedia.org/w/index.php?title=Anguilla&oldid=1163500361.

Anguilla Beaches. (2023). *Anguilla vacations: An introduction*. Anguilla Beaches. www.anguilla-beaches.com/anguilla-vacations.html.

Anguilla Statistics Department. (2023a). *Anguilla Statistics Department | 1.1 Population Statistics* [Excel file]. http://statistics.gov.ai/AllDocuments/1.1%20Population.xlsx.

Anguilla Statistics Department. (2023b). *Anguilla Statistics Department | 2.2 National Accounts Statistics* [Excel file]. Anguilla Statistics Department. http://statistics.gov.ai/AllDocuments/2.2%20NA%20Statistics.xls.

Anguilla Statistics Department. (2023c). *Anguilla Statistics Department | 2.4.5 Tourism* [Excel file] Tourism Statistics. Anguilla Statistics Department. http://statistics.gov.ai/AllDocuments/2.4.5%20Tourism.xlsx.

Anguilla Tourism Bureau. (2018, May 24). Visit Anguilla | The official travel guide of Anguilla. *Anguilla British Caribbean*. https://ivisitanguilla.com/.

Baldacchino, G. (2010). *Island enclaves: Offshoring strategies, creative governance, and subnational island jurisdictions*. Montreal, Canada: McGill Queens University Press.

Baldacchino, G. (Ed.) (2015). *Archipelago tourism: Policies and practices*. Farnham: Ashgate.

Borden, T. (2019, 31 December). The Caribbean is the place to be on New Year's Eve if you're a billionaire with a superyacht. Here's a look at the islands that currently have the most docked yachts. *Business Insider Nederland*. www.businessinsider.nl/how-billionaires-spend-new-years-eve-superyachts-caribbean-2019-12/.

Butler, R. W. (1980). The concept of a tourist area cycle of evolution: Implications for management of resources. *Canadian Geographer*, 24(1), 5–12.

Butler, R. W. (1996). The concept of carrying capacity for tourism destinations: Dead or merely buried? *Progress in Tourism and Hospitality Research*, 2(3), 283–293. https://doi.org/10.1002/pth.6070020309.

Butler, R. W. (2006a). The origins of the tourism area life cycle. In R. W. Butler (Ed.), *The tourism area life cycle, Applications and modifications* (Vol. 1) (pp. 13–26). New Delhi, India: Viva Books.

Butler, R. W. (2006b). *The tourism area life cycle. Conceptual and theoretical issues* (Vol. 2). Bristol: Channel View Publications.

Caribbean Journal. (2020, 16 October). Statia is getting its first luxury resort. *Caribbean Journal*. www.caribjournal.com/2020/10/16/statia-luxury-resort-new/.

CBS. (2023a). *Caribisch Nederland; bevolking, geboorteland, nationaliteit*. StatLine. https://opendata.cbs.nl/statline/#/CBS/nl/dataset/84757NED/table?ts=1689668817649.

CBS. (2023b). *Caribisch Nederland; bruto binnenlands product (bbp)*. StatLine. https://opendata.cbs.nl/statline/#/CBS/nl/dataset/84789NED/table?ts=1682511530536.

CBS. (2023c). *Caribisch Nederland; bruto binnenlands product (bbp) per inwoner*. StatLine. https://opendata.cbs.nl/statline/#/CBS/nl/dataset/85251NED/table?ts=1682511850116.

CBS. (2023d). *Caribisch Nederland, Saba; veerboot passagiers*. StatLine. https://opendata.cbs.nl/#/CBS/nl/dataset/85008NED/table?searchKeywords=shirley%20ortega%20azurduy%20caribbean.

CBS. (2023e). *Caribisch NL; inkomen van personen met inkomen in particuliere huishoudens*. StatLine. https://opendata.cbs.nl/statline/#/CBS/nl/dataset/83381NED/table?ts=1689774179976.

CBS. (2023f, 29 March). *Caribisch NL: Inkomend toerisme per vliegtuig* [Webpagina]. Centraal Bureau voor de Statistiek. www.cbs.nl/nl-nl/cijfers/detail/83104NED?dl=7B948.

CEROM. (2023). *Produit intérieur brut de Saint-Martin*. CEROM.

Clayton J. Lloyd International airport prepares for transatlantic flights. (2021, 22 November). *The Anguillian Newspaper – The Weekly Independent Paper of Anguilla*. https://theanguillian.com/2021/11/clayton-j-lloyd-international-airport-prepares-for-tra nsatlantic-flights/.

CLIA. (2023). *Norovirus on cruise ships*. Cruise Line Industry Association. http://cruis ing.org/en-gb/about-the-industry/policy-priorities/health/nororvirus-on-cruise-ships.

Collectivité de Saint-Barthelemy. (2023). *Aéroport de Saint-Barthelemy Rémy de Haenen— Statistiques 2022* (Synthese Octobre 2022; Statistiques mensuelles). www.comstbarth.fr/ in/rest/annotationSVC/Attachment/attach_cmsUpload_458e8d4e-5e68-4173-8574-43e2c b676b8d.

Creedy, K. (2015). St. Maarten, Part I: A regional hub for Caribbean island hopping. *Forbes*. www.forbes.com/sites/kathryncreedy/2015/11/21/st-maarten-part-i-a-regiona l-hub-for-caribbean-island-hopping/.

Croes, R. R. (2006). A paradigm shift to a new strategy for small island economies: Embracing demand side economics for value enhancement and long term economic stability. *Tourism Management*, 27(3), 453–465. https://doi.org/10.1016/j.tourman. 2004.12.003.

Department of Statistics Sint Maarten. (2017). *Statistical Yearbook 2017* (pp. 1–74). Department of Statistics Sint Maarten. http://stat.gov.sx/downloads/YearBook/Sta tistical_Yearbook_2017.pdf.

Department of Statistics Sint Maarten. (2023a). *Cruise arrivals 2015–2022*. http://stats. sintmaartengov.org/tables_n_charts/tourism/cruise/cruise_arrivals.xlsx.

Department of Statistics Sint Maarten. (2023b). *Population estimates and vital statis- tics 2022 and 2023* [Press release]. http://stats.sintmaartengov.org/press_release/Pop ulation/2023/Press_release_est_Population_2023.pdf.

Department of Statistics Sint Maarten. (2023c). *Stayover 2022*. http://stats.sintmaa rtengov.org/tables_n_charts/tourism/stay_over/stay_over_2022.xls.

Dobson, J. (2022, 22 December). *Billionaire superyacht showdown: Who's who in St. Barths for New Years 2023*. Forbes. www.forbes.com/sites/jimdobson/2022/12/22/bil lionaire-superyacht-showdown-whos-who-in-st-barths-for-new-years-2023/.

Faus, J. and Pons, C. (2021, 9 November). Superyachts enjoy pandemic tailwind as rich seek zen at sea. *The Sydney Morning Herald*. www.smh.com.au/world/europe/superya chts-enjoy-pandemic-tailwind-as-rich-seek-zen-at-sea-20211109-p597hh.html.

Fernandez-Stark, K. and Daly, J. (2017, 22 September). The global cruise industry: Impacts in the Caribbean countries. *CEPAL*. VI Conferencia REDLAS: Tendencias y perspectivas para la producción y el comercio de servicios en América Latina y el Caribe, Costa Rica. https://comunidades.cepal.org/redlas/es/grupos/evento/vi-confer encia-redlas-tendencias-y-perspectivas-para-la-produccion-y-el-comercio-de.

Getting Here | Saba Tourism. (2021, 19 April). www.sabatourism.com/getting-here/.

Hall, Z. D. (2020, 5 May). 'An escape from people and disease': The billionaires' flight from Covid-19 (and why a yacht might not keep you safe). *The Telegraph*. www.telegraph.co.uk/luxury/property-and-architecture/escape-people-disease-inside- flight-billionaires-covid-19-yacht/.

Hospitality Management. (2023, 31 August). Golden Rock Resort viert tweejarig bestaan en benoemt Anthony Reid tot general manager. *Hospitality Management*. www.hosp itality-management.nl/golden-rock-resort-viert-tweejarig-bestaan-en-benoemt-anthony -reid-tot-general-manager.

ICCaribbean. (2022, 15 February). Anguilla's ultra-luxury blueprint to attract super-rich tourists in 2022. *ICCaribbean*. www.iccaribbean.com/anguilla-luxury-blueprint-super-rich-attraction/.

IEDOM. (2022a). *Rapport annuel économique Saint-Barthélemy 2021*. IEDOM. www.iedom.fr/IMG/rapport_annuel_iedom_st-barthelemy_2021/.

IEDOM. (2022b). *Rapport annuel économique Saint-Martin 2021*. IEDOM. www.iedom.fr/IMG/rapport_annuel_iedom_st-martin_2021/.

IEDOM. (2023a, 28 February). *Comite consultatif et Reunion de place – Saint-Martin, 28 fevrier 2023*.

IEDOM. (2023b, 3 March). *Comite consultatif et Reunion de place – Saint-Barthélemy, le 3 mars 2023*.

IEDOM. (2023c, June). *Bilan 2022 et conjoncture 2023 à Saint-Martin*.

IEDOM. (2023d, June). *Évolution comparée des PIB à Saint-Martin et à Sint-Maarten*.

IMF. (2023). *Kingdom of the Netherlands-Curaçao and Sint Maarten: 2023 Article IV Consultation Discussions-Press Release; and Staff Report* (IMF Staff Country Report 2023/285; IMF Country Report, p. 93). International Monetary Fund. www.imf.org/en/Publications/CR/Issues/2023/07/27/Kingdom-of-the-Netherlands-Curaao-and-Sint-Maarten-2023-Article-IV-Consultation-Discussions-537067.

IMF. (2022). *Kingdom of the Netherlands-Curaçao and Sint Maarten: 2022 Article IV Consultation Discussions-Press Release; and Staff Report*. IMF. www.imf.org/en/Publications/CR/Issues/2022/08/09/Kingdom-of-the-Netherlands-Curaao-and-Sint-Maarten-2022-Article-IV-Consultation-Discussions-521977.

La Flamme, A. G. (1983). The archipelago state as a societal subtype. *Current Anthropology*. 24(3), 361–362. https://doi.org/10.1086/203006.

Major, B. (2022, 23 September). *American Airlines increasing flights to Anguilla*. Travelpulse. www.travelpulse.com/news/destinations/american-airlines-increasing-flights-to-anguilla.

Markings complete airport runway resurfacing. (2021, 30 August). *The Anguillian Newspaper – The Weekly Independent Paper of Anguilla*. https://theanguillian.com/2021/08/markings-complete-airport-runway-resurfacing/.

McElroy, J. L. (2003). Tourism development in small islands across the world. *Geografiska Annaler: Series B, Human Geography*, 85(4), 231–242. https://doi.org/10.1111/j.0435-3684.2003.00145.x.

McElroy, J. L. (2006). Small island tourist economies across the life cycle. *Asia Pacific Viewpoint*, 47(1), 61–77. https://doi.org/10.1111/j.1467-8373.2006.00303.x.

McElroy, J. L. and de Albuquerque, K. (1998). Tourism penetration index in small Caribbean islands. *Annals of Tourism Research*, 25(1), 145–168. https://doi.org/10.1016/S0160-7383(97)00068-6.

McElroy, J. L. and Hamma, P. E. (2010). SITEs revisited: Socioeconomic and demographic contours of small island tourist economies. *Asia Pacific Viewpoint*, 51(1), 36–46. https://doi.org/10.1111/j.1467-8373.2010.01412.x.

Ministry of General Affairs. (2017, 12 September). *What are the different parts of the Kingdom of the Netherlands?* [Onderwerp]. Ministerie van Algemene Zaken. www.government.nl/topics/caribbean-parts-of-the-kingdom/question-and-answer/topics/caribbean-parts-of-the-kingdom/question-and-answer/what-are-the-different-parts-of-the-kingdom-of-the-netherlands.

Ministry of the Interior and Kingdom Relations. (2020). *Country Package Sint Maarten*. Government of the Netherlands, Government of Sint Maarten. https://english.two-acs.

com/documents/besluiten/landspakket-sint-maarten/landspakket-sint-maarten/landspa kket-sint-maarten/index.

Ministry of the Interior and Kingdom Relations. (2023, 9 February). *Rijksdienst Caribisch Nederland* [Webpagina]. Ministerie van Binnenlandse Zaken en Koninkrijksrelaties. https://english.rijksdienstcn.com/.

Newsdesk. (2022, 5 January). *American Airlines to begin daily service to Anguilla.* Travel Agent Central. www.travelagentcentral.com/transportation/american-airlines-begin-daily-service-anguilla.

NRPB. (2023). Trust Fund: National Recovery Program Bureau. *National Recovery Program Bureau.* https://nrpbsxm.org/trust-fund/.

Office de Tourisme. (2023). *Tourisme à Saint-Martin aux Caraïbes.* St Martin Caraïbes. www.st-martin.org/.

Pereira, E. E. and Croes, G. G. (2018). *Tourism maturity in Aruba* (pp. 1–14). Centrale Bank van Aruba.

Peterson, R. (2020). *Whence the twain shall meet: Weathering overtourism and climate change in small island tourism economies* (pp. 1–67). Centrale Bank van Aruba. www.cbaruba.org/cba/readBlob.do?id=6088.

Premium Caribbean. (2023). Sint-Maarten/Saint-Martin. *Premium Caribbean.* www.premiumcaribbean.nl/en/category/sint-maartensaint-martin/.

Princess Juliana International Airport. (2023). In *Wikipedia.* https://en.wikipedia.org/w/index.php?title=Princess_Juliana_International_Airport&oldid=1168693159.

Public Entity Saba, Public Entity St. Eustatius. (2023a). *Evaluation of two-year Makana Ferry.* Public Entity Saba, Public Entity St. Eustatius. www.tweedekamer.nl/kamerstukken/detail?id=2023D49739&did=2023D49739.

Public Entity Saba, Public Entity St. Eustatius. (2023b). *Plan of Approach for Makana Ferry Service.* Public Entity Saba, Public Entity St. Eustatius. www.tweedekamer.nl/kamerstukken/detail?id=2023D49739&did=2023D49739.

Reid, R. (2012, 24 January). *Caribbean DIY island-hopping: St-Martin/Sint Maarten.* Melbourne, Australia: Lonely Planet. www.lonelyplanet.com/articles/caribbean-diy-island-hopping-st-martinsint-maarten.

Ridderstaat, J. (2015). *Studies on Determinants of Tourism Demand: Dynamics in a Small Island Destination. The Case of Aruba: Vol. PhD.* https://research.vu.nl/en/publications/studies-on-determinants-of-tourism-demand-dynamics-in-a-small-isl.

SHTA. (2023). Hub St Maarten / St Martin is easy to reach and explore. *Visit St. Maarten / St. Martin.* /www.visitstmaarten.com/center-stage/.

Slagt, R. (2023, 22 December). *Ferry Caribisch Nederland populair.* NRC Handelsblad, E7.

St Barts Tourism Committee. (2023). *St Barts island tourism – French west indies.* Comité Territorial de Tourisme de Saint Barthélemy. www.saintbarth-tourisme.com/en/home-of-st-barts-tourism/.

St. Maarten TouristBureau. (2023). *Island hopping.* St. Maarten. www.vacationstmaarten.com/things-to-do/island-hopping/.

Statia Tourism Bureau. (2023). How to get here. *Statia Tourism.* https://statia-tourism.com/plan-your-trip/how-to-get-here/.

TTCI. (2004). *St. Maarten carrying capacity study draft final report.* Tourism & Transport Consult International.

TTCI. (2005). *St. Maarten Tourism Master Plan: Tour map; Final report & action plan* (pp. 1–252). Tourism & Transport Consult International.

Viswanathan, G. (2023, 11 July). After years of decline, norovirus outbreaks surge on cruise ships. *CNN*. www.cnn.com/2023/07/11/health/norovirus-outbreaks-cruise-ship s-wellness/index.html.

World Bank Group. (2018). *Sint Maarten National Recovery and Resilience Plan. A roadmap to building back better.* Government of Sint Maarten, World Bank Group, Government of the Netherlands. www.sintmaartenrecovery.org/sites/sxm/files/stra tegicdocs/NRRPfinal_0.pdf.

Saint Brandon

Port Mathurin
Rodrigues

Port Louis
Mauritius

Saint-Denis
Réunion

Indian Ocean

La Fourche· **The Mascareignes Islands**

Agelega

Saint Brandon

Rodrigues
Réunion Mauritius

0   1,200   2,400 kms

0   500   1,000 kms

0   200   400 kilometres

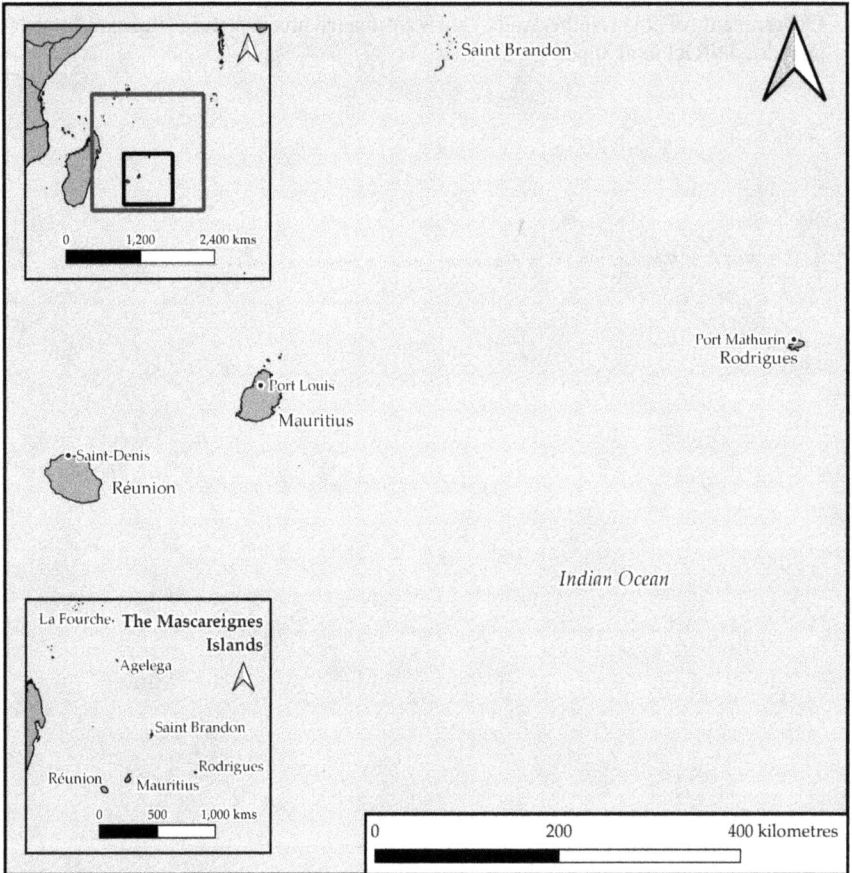

Figure 12.1 The Mascarene Islands: Mauritius, Réunion, Rodrigues

12 The complex and unequal tourism triangle of the Mascarene Islands

Mauritius, Réunion, Rodrigues

Hélène Pébarthe-Désiré

Introduction

The name of the Mascarenes archipelago is not well known. Scattered across the south-western Indian Ocean, to the east of Madagascar, these islands owe their name to the sixteenth-century Portuguese navigator and explorer Pedro Mascarenhas. It is not a very evocative name for an archipelago whose three main islands – Mauritius, Réunion, and Rodrigues – have very different types and levels of tourism development. So: does a chapter on archipelago tourism concerning these islands make sense? Yes, if only to highlight all the nuances of a complex triangle shaped by a partly shared history and a dual core-periphery relationship played out on different scales: Réunion with metropolitan France as a French overseas department; and Rodrigues as a sub-national island jurisdiction of Mauritius (with its own Regional Assembly since 2002).

The level of complexity of the inter-island links, and the diversity of the tourism offer, are among the highest in the world for an archipelago of just three destinations. The Republic of Mauritius also claims Tromelin and the Chagos Islands; and includes Saint-Brandon and Agaléga, but these are small islands with very low populations (274 inhabitants in 2022) and are not served by commercial flights and can speak of no tourism development; while both the Chagos Islands, location of a large US base, as well as Tromelin, have no permanent populations. The three Mascarene destinations recorded 1,383,500 arrivals for Mauritius in 2019, on the eve of the crisis linked to the Covid-19 pandemic, compared with just 78,000 arrivals for Rodrigues and 533,600 for Réunion. However, in relation to the population of these islands, the level of visitor numbers is all the more significant: both Réunion and Mauritius are already densely populated, with a population in 2022 of 873,100 for Réunion and 1,217,600 for Mauritius; while Rodrigues has 44,660 inhabitants. The economic and social shock was therefore particularly strong during the pandemic-induced lockdowns, which were strict and long in the case of those imposed by the Mauritian government: from 19 March to 30 May 2020 and then from 10 March to 30 April 2021. Meanwhile, the French government imposed two strict confinements in mainland France in 2020; Réunion was in lockdown from 17 March to 11 May 2020; then again from 31 July to 18 September 2021.

DOI: 10.4324/9781003451037-16

Most scientific publications on the development of tourism in the Mascarene Islands, written mainly in French and English, focus on one or other of the islands, or put just two of them into perspective: in most cases Mauritius and Réunion; or, more rarely, Mauritius and Rodrigues, with very little writings on the latter (e.g. Jauze, 1998; Wergin, 2012). Moreover, tourism in Réunion has tended to be studied in the context of the French overseas territories (Gay, 2009; 2021). In the context of my own work on the islands of the region, conducted since 1999, Mauritius has been the focus of my doctoral research (Pébarthe, 2003). This included work on air connectivity (Pébarthe-Désiré and Mondou, 2014; 2016) and the Mauritian model of hotel development and its spread to other Indian Ocean islands (Pébarthe-Désiré, 2015). This focus led me to study tourism in Réunion and Rodrigues, as well as other islands, both in the region (such as Seychelles, Maldives); as well as beyond (such as French Polynesia) (Blondy and Pébarthe-Désiré, 2018).

This chapter is based on an analysis of statistical data, tourism development plans and interviews with local tourism stakeholders, enabling us to analyse changes in the structure of tourism in the Mascarene Islands. Recent fieldwork carried out in Mauritius, during two visits in 2022 and another at the end of 2023, enabled us to understand the post-Covid-19 situation, particularly for Mauritius and Rodrigues, with major stakeholders in the tourism system from both the private sector (professional associations and training institutes in particular) and the public sector (Ministry of Tourism). Writing here on the scale of the Mascarene archipelago, and in the wake of the Covid-19 crisis, is all the more stimulating. It is even the right time, since these islands, led by Mauritius as a tourist leader and vibrant international destination, have witnessed their tourism take off again in 2023 on the basis of pre-Covid-19 visitor numbers, while at the same time questioning the islands' future and their need to develop a more sustainable form of tourism that addresses environmental and socio-cultural concerns.

The question that arises from this chapter is therefore: can we speak of archipelago tourism when referring to the Mascarene Islands, and has the Covid-19 crisis changed these islands' relationship with tourism development and the interrelationships between them?

This leads us to analyse the links between the three islands, but also with the centre: that means mainland France in the case of Réunion. The starting point is a consideration of the development trajectories taken by these three island jurisdictions and the place of tourism in these rather divergent paths. Second, the question of air access to and between these destinations, a key issue because air access is crucial for islands that are so far from their source markets, will also provide a broader understanding of the tourism system in each of these territories and in the archipelago as a whole. Last, the ambivalence of the post-Covid-19 situation will be highlighted, as these islands face the challenge of strengthening environmental and cultural concerns in their tourism offer, while at the same time seeking to revive growth in arrivals, and to do so as quickly as possible. The aim will be to see whether a strengthening of the links between the three destinations in terms of tourism practices and policies is perceptible and/or possible and likely to move in the direction of a post-Covid-19 scenario that is both prosperous and sustainable.

The Mascarene Islands: Which archipelago? Between common characteristics and individual visibility

What makes the Mascarenes an archipelago? It is a question worth asking. A number of factors bring these territories closer together: the geographical proximity between them is all the clearer given that there are fewer islands in the Indian Ocean than in the Caribbean, and they are more scattered and distant from one another.

Jauze (2009) reminds us that, as Auguste Toussaint (1972) pointed out his *Histoire des îles Mascareignes*, the term 'Mascarenes' only appeared in geographical nomenclature around 1825. Prior to that, the term *Îles françaises orientales* was used to designate this group of islands, comprising Réunion (Île Bourbon), Mauritius (Île de France), and Rodrigues, as well as the Seychelles and the Chagos, an association that would explain the error, still made today, of sometimes including the Seychelles in the Mascarenes. Apart from their names, the three main islands of the Mascarenes share a common geological history (the same hot spot, now active on Réunion, formed these volcanic territories), as well as commonalities in their climate, politics, peoples, and economies. This does not prevent their recent histories from evolving in very different ways, allowing for fruitful analyses on micro- and macro-regional scales.

Mauritius and Réunion are often referred to as 'sister islands'. They are separated by only 220 km and are very similar in size: 2,512 km² for Réunion, which is partly uninhabited due to the presence of the still active Piton de la Fournaise volcano in the south-east quarter; and 1,865 km² for Mauritius. Rodrigues is further away, 580 km east of Mauritius, and also smaller, at 109 km².

Still uninhabited when their colonisation began in the sixteenth century, Mauritius – Dutch between 1598 and 1715, then French from 1715 to 1810, and British from 1810 to 1968 – and Réunion – continuously French since 1640 – were successively populated by waves of migrants, most of whom came from neighbouring lands in the Indian Ocean. The current composition of the population bears witness to this, with similarities emerging between the two islands: Hindu communities represent just over 50% of the population in Mauritius and 25% in Réunion (Paris, 2011), and the Indo-Muslim group 17% and 5% respectively. The so-called "Creole" community in Mauritius and "Cafre" community in Réunion is estimated at 27% of the population of the former and 33% of the population of the latter, and is made up of black and mixed populations with distant African and European origins, mainly Catholics. Another less represented group, people of Chinese origin, who make up around 3% of the population in Mauritius and 5% in Réunion, nevertheless play an important economic role. The "regional" logic of colonial history largely explains the settlement of the two islands. There are, however, two main differences between their populations: the presence of a large white population on Réunion, 25% of whom are descendants of settlers and 5% of recently settled metropolitan residents, while the white Franco-Mauritian community represents only 2% of the Mauritian population but dominates major economic sectors, including tourism. Another difference, which is even more critical for understanding the political and social functioning of Mauritius and Réunion, is the fact that Réunion has a more mixed society, less divided along religious, racial, and linguistic lines, probably thanks in part to its membership of the French Republic, as well as the European Union (EU). In Mauritius, society is communitarian but also tolerant, which helps to avoid inter-community crises that would have greatly harmed the development of tourism and the image of the rainbow nation promoted by successive governments.

Rodrigues was also uninhabited at the time of the first human arrivals, which took place, for the most part, after France took possession of the island in 1725, and with the arrival, first, of a few soldiers, and then settlers and slaves, in small numbers. This attachment to Mauritius has continued ever since, with successive waves of settlement, although the island has never experienced the agricultural exploitation of sugar cane and the slave trade linked to the plantation system. After the abolition of slavery, and unlike Mauritius and Réunion, Rodrigues had no influx of Indian populations. It can also be described as an island neglected by the British colonial power, and then by the Mauritian authorities, and its fairly modest development has been based on agriculture and fishing, essentially for local consumption. The island has even regularly lost inhabitants to Mauritius in the form of work or study migration. The population of Rodrigues, which is almost entirely "Creole" in the Mauritian sense of the term, is therefore more homogeneous from a community point of view than that of Mauritius, where political power is held mainly by the majority Hindu community, and economic power is still largely held by the white descendants of the settlers. The latter have played a major role in the development of tourism in Mauritius, thanks to their control of land through sugar production. A large proportion of the sugar cane land on the estates bordered the coastline, which was gradually developed for beach resorts in line with the '3S' model – sun, sea, sand – familiar for many small tropical islands. Mauritius' accommodation offer is the most developed in the Mascarene Islands. In 2023, Mauritius has 108 hotels in operation (AHRIM, 2023), almost back to the number of 112 in 2019, while the equivalent figure for Réunion is 107 hotels (INSEE, 2023). However, Mauritius' hotel capacity is 3.5 times higher, with 13,300 rooms, the majority of which are in upscale establishments, to which should be added 8,400 rooms in guest houses and in the form of rental properties listed by the Mauritius Tourism Authority. In Rodrigues, tourist accommodation capacity amounts to 1,100 rooms at the end of 2023, but 70% of these rooms are in small accommodation units, most often in guest houses.

In all three islands, tourism was initially, and largely remains, coastal and hotel-based. However, from the 1970s–1980s, Mauritius moved towards an image (and a reality) of a high-end seaside resort destination, while Réunion gradually developed its offer in its hinterland with the development of well-marked hiking trails from the 1980s onwards, and greater access to the three volcanic cirques – formed by collapses of the volcano and ceaseless erosion – and the mountainous landscapes of Piton de la Fournaise, which sets the island apart from its lower-altitude Indian Ocean island neighbours and gives it a clear comparative advantage as a 'non-beach' tourism destination. Rodrigues has followed the Mauritian path, with its seaside hotels, of which there are seven today, with an average of around 40 rooms, built later (1991 for the first) and decked more modestly than in Mauritius, due mainly to the fact that there are fewer beaches and a very extensive but shallow lagoon, less suitable for swimming. But Rodrigues has also followed a path taken by Réunion, with the gradual and deliberate development of a form of tourism that is more closely integrated into local society and that allows visitors to discover the island's landscapes and culture, in particular its gastronomy, music, and dance. Thus, it highlights its way of life, which is closer to Réunion's intention to open up tourism to the island's interior, the discovery of its landscapes and Creole society (Wergin, 2015).

Analysing the three destinations in terms of the five attributes of archipelago tourism (Baldacchino, 2015) reveals the power imbalance between Mauritius and Rodrigues in particular, which we will come back to later, but also introduces elements of image and representation.

First, in terms of visibility, the Mascarene Islands do not communicate as an archipelago, nor do they offer any joint promotion to attract tourists. Instead, they are perceived and branded individually by their respective tourism authorities. Check out: Mauritius Tourism Promotion (2024); Rodrigues Tourism Office (2024); and La Réunion Tourisme Durable (2024).

In the case of the Mauritius logo, the 'M' in Mauritius represents the island's peaks, notably the emblematic Pieter Both, and the emphasis is placed on the four colours of the national flag, which themselves echo the red of the flamboyant trees and the symbol of the blood spilt at the time of slavery, the blue of the Indian Ocean and the sky, the yellow of the sandy beaches and the sun, and the green of the sugar cane. While Mauritian schoolchildren are all taught this variation, it is by no means certain that tourists will decipher it at first glance, and the logo, chosen in 2012, is regularly supplemented by a slogan, such as "Mauritius, it's a pleasure", now abandoned in favour of www.Mauritiusnow.com to underline the destination's dynamism on the internet, with continuous access to webcam images from around the island and frequent changes to news posts on the MTPA (Mauritius Tourism Promotion Authority) website.

Rodrigues, whose autonomy has also been strengthened in terms of promotion, is given little prominence by the MTPA, and so we do not find 'the tweaked effect'; that is, a tendency to "magnify the small islands of a whole to give the illusion of equality between the islands and of geographical proximity" (Baldacchino, 2015, p. 9). Nevertheless, the Rodrigues logo, developed by the Rodrigues Tourist Office, does feature the four colours of the national flag, but with the pirogue, the traditional hat, and the chilli pepper highlighting the island's specific cultural characteristics, the latter acting as a strong marker of the island's identity and of what tourists, particularly Mauritians, buy. The attempt to differentiate the secondary island from the main island is clear from the outline here of a distinct cultural offering, an 'authenticity', or at any rate a particular narrative.

Réunion has its own tourism authority, just like the other French regions, via its Regional Tourism Committee, as well as financial support from both the French state and the European Union (ERDF funds in particular) for its promotion and development, due to its status as an outermost region of the EU. Through its logo, the island has chosen to highlight its landscapes – sea, mountains, and volcano – a diversity that is also the strength of the tourist experience for visitors. Since the 1990s, the island has maintained the same slogan of "intense" island, to signify the strength of its landscapes and the opportunities for outdoor activities, particularly sports (Naria, 2022).

The reticular and archipelagic approach now invites us to place at the heart of the reflection, the political status, and the core-periphery links which play out in terms of two, quite distinct geopolitical and economic scales: that of the link which unites Réunion with metropolitan France; and that of the link between Rodrigues and Mauritius, the capital territory of the Republic of Mauritius. These core-periphery links influence tourism development on the three islands, the place of tourism in their economies, the way they enrich their tourism offer, and their resilience in the face of crisis, in particular the Covid-19 crisis.

Distinct development paths: One tourist centre and two tourist peripheries?

While there are obvious geographical and historical links, especially between Réunion and Mauritius on the one hand and Mauritius and Rodrigues on the other, the divergence in development is significant. In terms of political status, the Mascarenes trio comprise the three main categories of islands. Mauritius is a sovereign state which secured independence from the United Kingdom and has been an independent state since 1968; Rodrigues is an island territory of and within the Republic of Mauritius, although it has enjoyed special autonomous status since 2002 and has its own 'Regional Assembly'. Réunion has been a department of France since 1946, and is an overseas region of the French Republic which has no intentions of becoming independent, unlike other French overseas territories in the Pacific (Connell and Aldrich, 2020).

Réunion and Mauritius each play key but distinct roles in this part of the world. Réunion, whose level of development rivals that of European countries, is the main guarantor of France's presence in the Indian Ocean. Mauritius, a member of 36 multilateral agreements and very active within regional organisations (Indian Ocean Commission, Southern African Development Community, Common Market for Eastern and Southern Africa, Indian Ocean Rim) is a "dynamic hub" (Jauze, 2009). The different levels of political autonomy and links to the mainland – distant but wealthy in the case of Réunion; closer but weaker in the case of Rodrigues – go a long way to explaining their widely divergent economic development trajectories. Tourism is an excellent indicator of these dynamics: it has taken on a very different role in each case.

Like most other French overseas territories, Réunion suffers from the "Dutch Disease" (Corden and Neary, 1982, Poirine, 2007) where specific policies or leading economic sectors can have long-term negative effects on the competitiveness of other industries. Réunion's economic development is similar to that of the MIRAB model, although we cannot really speak of a TouRAB model for Tourism, Remittances, Aid and Bureaucracy (Guthunz and von Krosigk, 1996), even though tourism is gradually developing on the basis of an increasingly solid system of practices, places, and players. The economic literature argues that, in overseas France, the transfer of income from the metropole, in the form of increased salaries for civil servants or the payment of social benefits, has a long-term negative effect on the competitiveness of economic sectors open to competition, including tourism. In the case of Réunion, the predominance of public sector and administrative jobs has totally distorted the labour market. This island, a European periphery with a higher standard of living than the other territories in the region, cannot really compete successfully in tourism. Moreover, in the private sector, local elites have preferred to invest in retail business and car dealerships. As a result, while Réunion had a clear lead over Mauritius in terms of accessibility and hotel development in the 1950s and 1960s, this is no longer the case. French territories in the Caribbean and Pacific experienced a similar change of economic fortune (Gay, 2012).

Mauritius, on the other hand, has made tourism one of the main components of the diversification drive it has pursued since independence in order to develop the country in ways other than through sugar cane production alone, an economic activity that has seen its role gradually diminish as a result of the decline, from the 2000s onwards, of the comparative advantages linked to the ACP (Africa, Caribbean, Pacific) agreements with the EU, a general decline in sugar consumption in mature markets, and the rise of other sectors in the island's economy (Pébarthe, 2003, 2015; Koop, 2013). These sectors include textiles (with the establishment of a free trade zone), offshore banking, and information and communication technologies, the latter accompanied by the creation of Ebene Cyber City and tertiary activities linked to the development of software and call centres.

At the turn of the 2000s, Mauritian development corresponded to the SITE – small (warm water) island tourist economy – model proposed by

McElroy (2006), itself the result of the TouRAB process. It has succeeded in becoming doubly anchored in globalised tourism. First, from the point of view of its customer base, which is largely European but diverse (not monolithic as in Réunion, where the vast majority of customers come from mainland France), but also Asian (Indian and then Chinese) or more regional (from Réunion and South Africa). With regards to the diversity of stakeholders in the tourism sector, whether air or hotel, Mauritius has attracted a wide range of investors, and Mauritian groups are also active internationally, in the hotel industry in particular (Pébarthe-Désiré, 2015). This dynamic is reminiscent of the hotel groups of the Balearic Islands (e.g., Pons, Salamanca, and Murray, 2014).

Since the 2010s, Mauritius has been moving towards the PROFIT model (Baldacchino, 2010), which stands for People (immigration, talent, human resources), Resources (natural asset management), Overseas Management (diplomacy), FInance (banking, insurance, taxation), and Transport, highlighting its political stability, security and quality of life, and increasingly diversifying its attractiveness. Mauritius is gradually moving beyond the tourism stage, while continuing to nurture this sector, which it is using as a showcase to encourage new forms of residency, mainly connected with the purchase of luxury real estate which grants foreigners access to resident status and encourages the tax-free installation of European and South African fortunes on the island.

Rodrigues appears to be the forgotten island in this development trajectory. Tourism only began to develop there in the 1990s, but only modestly due to the difficulties in accessing the island, which has to be by air. Although in 20 years, tourist arrivals – including domestic tourists, who are the majority – have risen from 45,120 in 1999 to 78,000 in 2019, this success is a little misleading because most tourists come from Mauritius, to the detriment of potentially more lucrative international customers. This situation will be hard to dislodge, as long as Air Mauritius maintains a practical monopoly over air access (Wergin, 2015).

On the scale of the Mascarene Islands, Mauritius' role as a tourist destination is central. While Réunion may be able to avoid a form of competition with its neighbour, due to the development guarantees offered by its status as the 'privileged' periphery of a rich country, the contrast in development is particularly marked between Mauritius and Rodrigues, where the latter appears to be a periphery under dependence and whose capacity to evolve towards the TouRAB model is questionable.

The archipelagic approach will now allow us to focus on the logistics developed in terms of transport, particularly air transport, to and between these islands. All three share the characteristic of being very distant from their source markets, several thousand kilometres apart, even in the case of South Africa and India, which are nonetheless markets for Mauritius alone, and still secondary compared with even more distant Europe.

Tourism calibrated by air access and traffic flows?

Concerning air access to the Mascarene islands, the literature generally focuses on access to one of them, or compares Mauritius and Réunion (Gay, 2009; Pébarthe-Désiré and Mondou, 2014, 2016), but does not highlight the air links *between* these islands, and in particular the question of access to Rodrigues, thus often leaving aside the study of this island sometimes nicknamed the "Cinderella of the Mascarene Islands" to signify its neglect by its two big 'sisters'.

It is clear that the development of international tourism, so palpable in Mauritius, has made little headway in Réunion and even less so in Rodrigues, largely due to the structuring of air services to these islands. An analysis of air services to Mauritius and Réunion reveals very clear contrasts between these two island neighbours. It is almost a textbook case of island tourism, and remains so after Covid-19.

On Réunion, visitor numbers have indeed picked up, with 556,000 tourist arrivals in 2023 (i.e. 4% more than in 2019); but 83% of tourists in 2019 and 80% in 2023 are still from mainland France. The island is mainly served from France by Air France and Air Austral, as well as Corsair and a low-cost carrier, French Bee. Air Austral also operates regional routes, including those to Mauritius; and, in Asia, Air Austral only offers flights to Bangkok twice a week.

Mauritius, on the other hand, has had a very different strategy since independence, with the creation of Air Mauritius in 1967, and has gradually woven a web of connections to Europe (first Paris and London; then Geneva, Rome and Frankfurt in the 1990s), Asia (Hong Kong and Singapore since the 1980s, as well as Mumbai and New Delhi; then Shanghai in the 2010s) and Africa (the islands of the western Indian Ocean region, namely Réunion, Madagascar, and Seychelles, plus South Africa), Australia, and then the Middle East (Dubai). In this way, Mauritius has gradually established itself as a regional air hub by relying on its national airline, in which the State is still the majority shareholder in order to retain air sovereignty, a choice that the Seychelles and the Maldives, the real regional competitors to Mauritius, have abandoned. This Mauritian strategy has been backed up by two major commercial agreements, one with Air France and the other with Emirates. The partnership with Air France ensures easy access to customers from mainland France, Mauritius' number one tourist clients (with 25% of arrivals in 2023 and a level of visitor numbers and spending even higher than in 2019), but also to other European customers, particularly German, British, Swiss, and Italian, who together account for 25% of visitors to the island, both before and after the Covid-19 crisis, and use the Paris CDG hub extensively. The Emirates service to Mauritius, which began in 2002, is indicative of the rise of the Persian Gulf airlines, now partly relayed by Turkish Airlines, and once again demonstrates the confidence that the world's major tourism investors have in the destination. Emirates serves Mauritius twice a day with Airbus

A380s, both before and after Covid-19. This allows the island to play in the league of serious international destinations by relying on the Dubai hub from Europe and Asia.

On the main routes, Air Mauritius operates in code shares with foreign airlines, which strengthens the air capacity to the destination and has proved vital in overcoming the Covid-19 crisis. Indeed, with the stoppage of the tourism sector linked to the pandemic and the strict lockdowns decided by the Mauritian government, which led to a near-total stoppage of tourism for one year in 2020–2021 and another difficult year in 2021–2022, Air Mauritius came close to bankruptcy, made massive layoffs, and only resumed some of its routes in 2023: namely Paris then London and Geneva in Europe, and only Mumbai and Kuala Lumpur in Asia. At the end of 2023, the airline communicated more on the relays offered by its partners in the hubs of Paris, Dubai and Mumbai, and the logic of commercial success now prevails over the logic of national sovereignty asserted in the past. Other airlines serve Mauritius, such as Saudi Airlines to Jeddah and Vistara, an Indian airline, to Mumbai; while services to China have not yet resumed.

Against this backdrop of tourism recovery, regional air services have almost been restored to pre-crisis levels, both from Mauritius and Réunion, as well as between them. Air Austral and Air Mauritius are each fulfilling a regional service role, with a more complete resumption from Reunion of the destinations served in 2023. Air Austral, like Air Mauritius from Mauritius, have resumed flights to Madagascar and South Africa, as well as to the Comoros, Mayotte and the Seychelles. The Mauritius-Seychelles route is now operated by Air Seychelles but no longer by Air Mauritius, which prefers to concentrate its resources on Europe and Asia to boost tourism. However, Réunion and Mauritius are both members of the Vanilla Islands Association, which was set up in 2010 and also brings together the tourism authorities of the Seychelles, Madagascar, Mayotte, and the Comoros. Its aim is to use the image of vanilla, a product they all share, to promote combined trips between the islands. The most concrete step forward concerns air travel: the 'Vanilla Alliance', officially set up in 2015, brings together Air Austral, Air Madagascar, Air Mauritius, and Air Seychelles and aims to significantly improve air services between the islands. In practice, it is mainly Air Austral that is implementing a 'Vanilla Pass' to enable visitors to combine a stay in Réunion with another destination at lower airfares. Réunion appears to be more inclined to encourage inter-island tourism, given that combined holidays are already a reality there, with 15% of tourists visiting Réunion in 2019 (i.e. around 80,000) having visited other islands in the Indian Ocean. Réunion is also receiving European funding, for animating this cooperation, under an EU Interreg programme.

The Réunion-Mauritius service, which is known to be one of the most expensive, if not the most expensive in the world per kilometre, in order to maintain a high entry fare and the upmarket positioning of Mauritius, is shared between Air Mauritius and Air Austral, with an average of five to six flights per day in 2023. Despite this apparent reciprocity in terms of air traffic, the tourism links between Réunion and Mauritius, and between Réunion and Rodrigues, can to some extent be described as a one-way relationship, with no real reciprocity. Indeed, if we take a closer look at the nature of the flows of travellers, people living on Réunion have been Mauritius' third largest market for the last ten years or so, accounting for 11% of arrivals, on a par with the British in 2023, while Mauritians account for just under 5% of arrivals on Réunion, with their stays, moreover, being shorter and more focused on business than leisure. This distortion can be explained by the fact that prices and the value of the currency are much higher on Réunion, illustrating the lack of competitiveness mentioned above and the difficulty the island has in competing in the open tourism sector. On the other hand, it is relatively easy for Réunionese to travel to Mauritius, particularly for shopping, and also to Rodrigues, but almost exclusively via Mauritius, which reinforces the centrality of Mauritius in the archipelago and penalises Rodrigues, which remains dependent on a flow decided for and from Mauritius and has difficulties generating international flows.

Rodrigues is served four to five times a day from Mauritius in 2023 by Air Mauritius, using ATR 72 jets with just 66 seats: same as before the pandemic. This falls short of demand during the busiest months, particularly December and January, and is causing frustration among Rodrigues tourism operators. In fact, the accommodation supply listed by the association of Rodrigues tourism professionals (ATR, Association du Tourisme Réunie) has grown significantly on the island, reaching 1,100 rooms by the end of 2023, divided between seven hotels, which account for 30% of rooms, and no fewer than 179 small accommodation structures, in the form of boarding houses or rental accommodation, and good air service is necessary to support local tourism. The seat capacity / room capacity (ratio between aircraft seats and accommodation capacity) is currently strong, especially as tourist demand is very clear post-Covid-19, as an ATR manager confirmed to us in January 2024: the number of tourists welcomed in 2023 had reached 90,000, thus exceeding the figure for 2019, which stood at 78,000. Between 2015 and 2019, direct flights between Réunion and Rodrigues enabled international customers to visit the island without passing through Mauritius, but in small numbers: up to just 1,970 arrivals in 2019. The resumption of a weekly flight by Air Austral to Réunion at the end of 2023 has brought relief to Rodrigues' tourism industry. As for Air Mauritius, the purchase of a fourth ATR 72 will further boost arrivals from Mauritius, which by 2023 consist of 75% Mauritians and 25% foreigners. Work to lengthen the runway at Rodrigues airport, long overdue, should begin in 2024 and will eventually allow larger aircraft to land: the World Bank, the EU and AFD (Agence Française de Développement) will be the main financiers. Strong tourist demand, particularly from Mauritian customers after two years of strict travel restrictions during the pandemic, is therefore being met by a gradual improvement in accessibility to Rodrigues, which nevertheless remains extremely dependent on its mainland for tourism.

All in all, from the point of view of air links to and within the archipelago, it would appear that the onset of Covid-19 and then the end of the crisis have prompted each of the islands to strengthen its fundamentals, with a return to the previous level of activity, or even more intense activity, and profiles that remain those of a diversified clientele, even if mainly European in the case of Mauritius, and mainly from the respective mainland in Réunion (that is France) and Rodrigues (that is Mauritius).

A time of major upheaval for the tourist industry, the Covid-19 crisis gave way in 2023 to a clear upturn in activity in the three island contexts studied, but was also an opportunity to reposition and/or reinforce the image of these islands. In addition to the quantitative data on visitor numbers, the resumption of air services and the reopening of the accommodation sector, the aim is to provide some qualitative data on the changes taking place in demand, supply and the strategies of the tourism players. In other words, tourism is taking off again, but how? Through new ways of dealing with demand and between island players? Has the Covid-19 crisis changed the relationship between tourism and development in the Mascarene islands and the inter-relationships between these three islands?

Strengthening inter-island links in a post-Covid context of expected sobriety?

The Covid-19 crisis and its aftermath have been accompanied by a paradox that, for islands far from their markets, is coupled with a challenge: to cope with societies, particularly European ones, that are now critical of air travel, while at the same time strongly aspiring, from 2022 onwards, to travel again, which they have done, as shown by the clear upturn in tourism activity. However, the attention paid to environmental changes is now a reality and is fuelling and conditioning the demands of (at least some) tourists.

In the case of Réunion and Rodrigues, the 'sustainability dimension' is reflected in the integration of tourism practices with nature and society as a central hallmark of the destination. On Réunion, even before Covid-19, the tourism offer was already focusing on the discovery and protection of the environment. Since 2007, the island has been home to one of France's eleven National Parks, covering two-thirds of the island's surface area and offering 816 km of footpaths through a wide variety of landscapes from the sea to just over 3,000 m above sea level. The park is a continuation of public policy aimed at preserving and developing the island's highlands. Sporting events – such as the *Diagonale des Fous*, a mountain 'ultramarathon' race – also bring international recognition and acclaim to Réunion for the quality of its land-scapes. Réunion's tourism authorities have also created the *Village Créole* label, awarded to 16 localities, each with its own specific characteristics based on products such as spices, geraniums, local architecture, or, of course, natural landscapes. Rodrigues has also developed, on its own scale, a range of activities geared towards walking, fishing, or discovering the lagoon, islets, and caves (Caverne Patate). According to its tourism director, a site like the François Leguat reserve, which is reintroducing turtle species, welcomed around 36,000 visitors in 2023, half the number of visitors to the island. The local authorities are clearly committed to preserving the environment, both on land and in the sea, having also forged a number of partnerships with environmental international NGOs. These links also enable the island to free itself somehow from the control of Mauritius. The promotion of the island, decided in Rodrigues, also places local society at the centre of its message, its Creoleness – when in Mauritius this is drowned in a rainbow communication grouping communities together – and its gastronomic, music, and dance traditions. The link with Réunion is clearly emphasised, and during his promotional tour of Réunion in June 2019, the president of the ATR, interviewed on Réunionese television, opined that Rodrigues "is a destination that is reminiscent of *Réunion lontan*" in terms of nature activities and human experience. The use of the Creole word *lontan*, to mean the Réunion of yesteryear, is significant: creolity is, in Réunion as in Rodrigues, a subject of common pride which lives on through the sega music. It is present, moreover in the three islands of the Mascareignes, and widely proposed, sung, and danced, in their hotels.

In Mauritius, high-quality accommodation keeps tourists mainly on the coast. Over time, however, the destination has matured and tourist habits have diversified, both in terms of loyal customers – 20% of tourists welcomed each year on average since the 2000s are repeaters – and in terms of the expectations of a more recent clientele, such as the Chinese, which increased sharply between 2012 and 2015. Hoteliers realised that these customers were more likely than Europeans to leave their establishments to do some shopping, visit heritage sites or explore wilder coastlines. And the Covid-19 crisis seems to have led to a growth in alternative accommodation and non-hotel spending. In the first half of 2023, according to the initial results of a study financed by the World Bank, commissioned by the Mauritian government and to be published as the *10-Year Blueprint for the Mauritian tourism sector 2024–2033*, a third of tourists have not stayed in hotels, compared with 22% in 2008, and they have accounted for half of all overnight stays and 44% of accommodation revenue.

Against this backdrop, the Mauritian authorities are considering new development scenarios aimed at improving revenue rather than the volume of arrivals, including the scenario of zero net growth in tourist arrivals. This strategy would not have been seriously considered by the Ministry of Tourism, nor by hoteliers, before the Covid-19 crisis. Indeed, during the pandemic, the Mauritian government imposed an almost complete shutdown on business, and many establishments remained closed for over a year. During the lock downs, only a few of the island's hotels remained open to serve as quarantine centres where Covid-19 patients and tourists who had recently arrived in the country were isolated in a room for 14 days. The hotel industry employees, most of whom were put out of work in this way, received little assistance, unlike those in Réunion's tourism industry, who were supported by the French government. Many of them left the sector, and 4,500 of them had still not returned by the end of 2023, as the director of AHRIM told us. This shortage of staff in the context of post-Covid-19 recovery is putting a major strain on the hotel industry, but it is also forcing hoteliers to envisage scenarios such as the one mentioned above, which are at odds with the growth strategies for arrivals habitually favoured in Mauritius. It can therefore be said that the crisis has shifted the balance towards a more comprehensive consideration of the economic, social, and environmental challenges facing the destination. The main players in the tourism industry are now also taking into account the limited space still available to be developed into seaside hotels, and the social discontent that has prevented any conversion of a public beach into a tourism development project for the last ten years. A slightly tense social situation on a densely populated, multi-community island calls for caution and for environmental concerns to be given greater prominence in the development strategy. In this case, the crisis has the potential to accelerate the transformation towards a more sustainable destination, from both an environmental and a social point of view.

Conclusion

The development of tourism in the Mascarene Islands is clearly different in each island, and analysing them from the perspective of archipelago tourism has enabled us to put into perspective in an innovative way the links, similarities, and distinctions between these three contexts and development paths.

From a geographical and historical point of view, we are indeed dealing with an archipelago. But we are dealing with three distinct destinations in terms of practices and places, the way tourism operates and its place in the economy, and the political regime. Their tourism systems are more diverse than interrelated. Mauritius manages to play a more central role in international tourism; but also, vis-à-vis, Rodrigues and, to a certain extent, Réunion. Réunion Island is certainly a tourist periphery in relation to its European metropolitan mainland, but it is a redistributor on the scale of the archipelago, since it sends a proportion of its visitors to Mauritius and Rodrigues, the latter being itself a periphery dependent on Mauritius in terms of its air service and its visitors, even if the post-Covid-19 prospects are favourable in terms of air accessibility.

The Mascarenes triangle is uneven, but the signs of a post-Covid-19 recovery in 2023 are readily visible in all three constituent islands, reinforcing the need to take sustainability parameters into account. Even before the Covid-19 crisis, the emphasis on an environmentally sustainable tourism that was better 'integrated' into the lives of the locals was at the heart of the tourism strategy for both Réunion and Rodrigues. On the Mauritian side, the end of the corona crisis seems to be accompanied by a shift towards the search for equivalent tourism profits without an increase in arrivals, and a greater focus on alternatives to beach and hotel tourism. These trends could lead the Mascarene Islands towards more shared objectives, although closer tourism cooperation between the islands is not yet on the agenda.

References

AHRIM (Association des hôteliers et restaurateurs Ile Maurice) (2023). *Annual report 2022–2023*, Port-Louis. www.ahrim.mu/wp-content/uploads/2023/10/HD-WEB-AHRIM_AR-2023_SPREAD.pdf.

Baldacchino, G. and Ferreira, E. C. D. (2013). Competing notions of diversity in archipelago tourism: Transport logistics, official rhetoric and inter-island rivalry in the Azores. *Island Studies Journal*, 8(1), 84–104. https://islandstudiesjournal.org/issue/7907.

Baldacchino, G. (2014). Small island states: Vulnerable, resilient, doggedly perseverant or cleverly opportunistic? *Études Caribéennes*, 27–28, 217–238. http://dx.doi.org/10.4000/etudescaribeennes.6984.

Baldacchino, G. (Ed.). (2015). *Archipelago tourism: Policies and practices.* Farnham: Ashgate.

Barat, C. (2013). Rodrigues, de l'administration par l'Isle de France à l'autonomie dans la République de Maurice. [Rodrigues, from its administration by Réunion to autonomous status within the Republic of Mauritius.] *Études Océan Indien*, 49–50. https://doi.org/10.4000/oceanindien.1996.

Bertram, I. G. (2006). Introduction: The MIRAB model in the twenty-first century. *Asia Pacific Viewpoint*, 47(1), 1–13. https://doi.org/10.1111/j.1467-8373.2006.00296.x.

Blondy, C. and Pébarthe-Désiré, H. (2018). Les îles tropicales, lieux de l'extraordinaire? Construction et maturation touristiques en Polynésie Française et à l'Île Maurice. [Tropical islands, extraordinary places? Construction and maturity of tourism in French Polynesia and Mauritius.] *Bulletin de l'Association de Géographes Français*, 95(4), 468–491. http://journals.openedition.org/bagf/3931.

Connell, J. and Aldrich, R. (2020). *The ends of empire.* Singapore: Springer.

Corden, W. M. and Neary, P. (1982). Booming sector and de-industrialisation in a small open economy. *The Economic Journal*, 92(368), 825–848. https://doi.org/10.2307/2232670.

Dehoorne, O. (2014). Les petits territoires insulaires: Positionnement et stratégies de développement. [The small island territories: Positioning and strategies of development.] *Études Caribéennes*, 27–28. https://doi.org/10.4000/etudescaribeennes.7250.

Durbarry, R. (2002). The economic contribution of tourism in Mauritius. *Annals of Tourism Research*, 29(3), 862–865. http://dx.doi.org/10.1016/S0160-7383(02)00008-7.

Gay, J.-C. (2009). *Les cocotiers de la France: Tourismes en outre-mer.* [France's coconut trees: Tourism in the overseas territories.] Paris: Belin.

Gay, J.-C. (2012). Why is tourism doing poorly in Overseas France? *Annals of Tourism Research*, 39(3), 1634–1652. https://doi.org/10.1016/j.annals.2011.08.008.

Gay, J.-C. (2021). La France d'outre-mer: Le rôle fondamental de la relation à l'hexagone. [Overseas France: The fundamental role of the relationship with continental France.] In P. Violier (Ed.), *Le tourisme en France II: Approche régionale* (pp. 171–199). Paris: Iste éditions.

Guthunz, U. and Von Krosigk, F. (1996). Tourism development in small island states: From MIRAB to TouRAB? In L. Briguglio, B. Archer, J. Jafari, and G. Wall (Eds), *Sustainable tourism in islands and small states: Issues and policies* (pp. 17–35). London: Pinter.

Houbert, J. (2001). Mauritius, an island of success: A retrospective study, 1960–1993. *Africa: Journal of the International African Institute*, 71(2), 333–337.

Ile de la RéunionTourisme. (2023). *Fréquentation touristique 2022*. https://observa toire.reunion.fr/frequentation/destination-et-ocean-indien/frequentation-touristique-an nee-2022.

INSEE (Institut National de la Statistique et des Etudes Economiques) (2023). *Tourisme 2023*, Département de la Réunion. www.insee.fr.

Jaffur, Z.-K., Tandrayen-Ragoobur, V., Boopen, S., and Gopy-Ramdhany, N. (2022). Impact of COVID-19 on a tourist dependent economy and policy responses: The case of Mauritius. *Journal of Policy Research in Tourism, Leisure and Events*. http s://doi.org/10.1080/19407963.2022.2113090.

Jauze, J.-M. (2009). Coup d'œil sur les Mascareignes: Avant-propos. [A glance at the Mascarenes: Prologue.] *Les Cahiers d'Outre-Mer*, 245, 3–6. http://journals.openedi tion.org/com/5492.

Jauze, J.-M. (2008). *L'île Maurice: Face à ses nouveaux défis*. [Mauritius: Facing new challenges.] Paris: L'Harmattan.

Jauze, J.-M. (1998). *Rodrigues: La troisième île des Mascareignes*. [Rodrigues: Third island of the Mascarenes.] Paris: Université de la Réunion, L'Harmattan.

Koop, K. (2013). La trajectoire d'émergence de l'île Maurice: Rattrapage puis ajuste- ment à la globalisation. [Development trajectory for Mauritius: Capture and adjustment to globalisation.] In A. Piveteau, E. Rougier, and D. Nicet-Chenaf (Eds), *Émergences capitalistes aux Suds* (pp. 169–184). Paris: Karthala.

La RéunionTourisme Durable (2024) *Images*. www.tourisme-durable.org/images/stor ies/flexicontent/item_1140_field_90/l_irt_alt_vec.jpg.

Mauritius Tourism Promotion (2024) *Mauritius Tourism logo*. https://doodleworld wide.com/portfolio/mauritius-tourism-promotion-authority/.

McElroy, J. L. (2006). Small island tourist economies across the life cycle. *Asia Pacific Viewpoint*, 47(1), 61–77. https://doi.org/10.1111/j.1467-8373.2006.00303.x.

McSorley, K. and McElroy, J. L. (2007). Small island economic strategies: Aid-remit- tance versus tourism dependence. *e-Review of Tourism Research*, 5(6), 140–148. http://ertr.tamu.edu/pdfs/a-140.pdf.

Mondou, V. and Pébarthe-Désiré, H. (2016). Dépendance économique au tourisme et accessibilité aérienne des espaces insulaires: Exemples des Maldives, des Seychelles, de l'île Maurice et de la Réunion. [Economic dependence on tourism and air access to island spaces: Examples of Maldives, Seychelles, Mauritius and Réunion.] In J.-F. Hoarau (Ed.), *Spécialisation touristique et vulnérabilité: Réalités et enjeux pour le dével- oppement soutenable des petits territoires insulaires* (pp. 351–364). Paris: L'Harmattan.

Naria, O. (2022). Sports practices and governance of local actors in the islands: What are the challenges for sustainable development and territorial dynamics in the Indian Ocean? In C. Sobry and K. Hozhabri (Eds), *International perspectives on sport for sustainable development* (pp. 299–322). New York: Springer International.

Paris, F., (2011). La Réunion et l'île Maurice: Jumelles, sœurs ou cousines de l'océan Indien? [Réunion and Mauritius: Twins, sisters or cousins in the Indian Ocean?] *Population et Avenir*, 704, 16–19. www.cairn.info/revue-population-et-avenir-2011-4-pa ge-16.htm.

Pébarthe, H. (2003). *Le tourisme, moteur du développement de la République de Maurice? Un secteur à ménager, des lieux à intégrer.* [Tourism: Engine of development in the Republic of Mauritius? A sector to manage, sites to integrate.] PhD thesis. University of Paris IV – Sorbonne. www.theses.fr/2003PA040260.

Pébarthe-Désiré, H. and Mondou, V. (2014). The island and the plane: From the technical constraint to the economic choices of the tourist islands of the Indian Ocean: Reunion, Mauritius, Seychelles, Maldives. *Géotransports*, 3(3), 39–56. https://univ-a ngers.hal.science/hal-02384127.

Pébarthe-Désiré, H. (2015). Les acteurs touristiques mauriciens à la conquête de nouveaux Suds. [New Mauritian tourism actors out to conquer the new South.] *Autrepart*, 76, 161–181. www.cairn.info/revue-autrepart-2015-4-page-161.htm.

Poirine, B. (2007). *Eloignement, insularité et compétitivité dans les petites économies d'outre-mer,* [Development, insularity and competitiveness in the small overseas economies.] AFD, Document de travail, No. 52. https://hal.science/hal-00974440/.

Pons, A., Salamanca, O. R., and Murray, I. (2014). Tourism capitalism and island urbanization: tourist accommodation diffusion in the Balearics, 1936–2010. *Island Studies Journal*, 9(2), 239–258. https://doi.org/10.24043/001c.81625.

Rodrigues Tourism Office (2024). *Logo on Facebook page.* www.facebook.com/photo/?fbid=623105592954767&set=a.623105566288103.

Statistics Mauritius. (2023). *Survey of inbound tourism, 1st Semester 2023.* Port Louis: Ministry of Finance and Economic Development.

Wergin, C. (2012). Trumping the ethnic card: How tourism entrepreneurs on Rodrigues tackled the 2008 financial crisis. *Island Studies Journal*, 7(1), 119–134. https://doi.org/10.24043/isj.265.

Wergin, C. (2015). Travelling the Mascarenes: Creoleness in tourism policies and practices. In G. Baldacchino (Ed.), *Archipelago tourism: Policies and practices* (pp. 227–240). Farnham: Ashgate.

Conclusion

Unsettling cartographic imaginations

Godfrey Baldacchino

Introduction: Can the centre hold?

The added focus on core-periphery dynamics in this volume may have thrust a political economy dimension to what would otherwise might have been a more sedate discussion of tourism geographies. The 12 case studies have proved sensitive, to various degrees, to the jostling and positioning that pits island against mainland, island against island, and capital city against the rest.

The pecking order can be strongly determined and driven from the centre. In Seychelles, exclusivity and cost of accommodation are measured by non-modernity and flying/sailing distance from Mahé, the main(is)land, where over 90% of the population lives anyway. Madeira island decides the tourism signature for Porto Santo. Grenada's politics offers limited room for manoeuvre to the Minister responsible for Carriacou and Petite Martinique; and Mauritius maintains its airline monopoly for travel to and from Rodrigues.

Elsewhere, the power differential is less obvious. Romblon capital town jockeys with (larger) Tablas island for pre-eminence. In all respects, Mallorca rises head and shoulders above the other trio of Balearic islands: in size, tourist revenue, political clout. But Ibiza has its hands full with its own mass tourist visitations, and consequences; while Minorca and Formentera have grown wary of being recipients of similar 'golden hordes' and prefer to craft a different tourism development trajectory. Formentera does not even have an airport; and may want to keep it that way.

Relationships and dependencies can and do change with time, reminding us that power can be fluid, and destinations can be reconstituted, in line with government policy, marketing campaigns and transportation infrastructure revisions. An Atlantic Bubble – a part-time, four-province, five-month 'archipelago' of sorts – came conveniently and smoothly into being and then just as smoothly and conveniently melted away, literally overnight. Stringent Covid-19 lockdowns revived traditional practices in the outer islands of Vanuatu; and these may be the authentic customs that *some* tourists may want to sample, as the residents of such far-flung islands may discover. Lingering communal traditional practices in Miyako have also met tourism approval and custom. With the connivance of the central government in Manila,

DOI: 10.4324/9781003451037-17

Romblon province and its residents envisage – not necessarily happily – a future with hardly any developments in transport connectivity, so its sense of exclusivity stands better prospects of surviving. Residents on small Italian islands complain about amnesia and lethargy amongst policy makers, whether in Palermo (Sicily's regional capital) or Rome. Ferry schedules and flight availability will also constrain and shape market flows. Some places may be 'out' one year; 'in' the next; and 'shut down' the year after. In spite of international borders, various jurisdictions in the North-East Caribbean have come to depend on the tourist handling capacity of the Dutch half-island of Sint Maarten, which acts as their 'mainland' for transiting their tourists. Bridged near islands (like Pag, in Croatia; and Prince Edward Island, in Canada) are practically assured 24/7 connectivity to mainlands.

Random observations

The spatial is political. How we configure space is a direct function of forces that direct our understanding of that same space. Materiality and geography do not come with their own, inbuilt 'logic', but are socially, politically, and economically constructed (Elden, 2007; Lefebvre, 2014). As a 'tourism destination', the archipelago offers a distinct yet complex set of affordances and assemblages that deserve recognition (e.g. Briassoulis, 2017). Amongst these, questions such as: why and how is a particular (island) space deemed a 'core' and another is a 'periphery'? And how do island size, population, jurisdictional clout, air and sea transport infrastructure, and inter-island or island-mainland distance configure these labels, at any point in time?

The 12 case studies showcased in this book, and investigated using an archipelagic imagination, afford multiple, sober observations. A number of these appear below in point form. They foreground how geography is invariably interlaid with politics and economics to explain the signature of tourism.

- There is not much evidence of a "rejuvenation" of tourism industries (in Butler's sense), whereby destinations take the crisis situation exacerbated by Covid-19 – and possibly accompanied by environmental disasters, such as one or more hurricanes in the Caribbean; or a cyclone or typhoon in the Pacific – as an opportunity to re-evaluate and reimagine their tourism infrastructure (Butler, 1980). The short-term, windfall profits resulting from revenge tourism have all but sidetracked exhortations towards some critical reassessment of the industry. Any "strategic repositioning" of island tourism (Campbell and Connell, 2021, p. 13) has so far been glaringly absent.
- Tourism and other mobilities come into being, change, and decline over time; in the process, they construct and deconstruct places via processes of inclusion and exclusion. Geographically peripheral islands – occupying positions of double or triple insularity – typically continue to suffer a marginalisation compounded by administrative neglect, logistic nightmares

and connectivity bottlenecks. In the special case of Atlantic Canada, the bubble was transient: it emerged and evaporated as the urgency of Covid-19 lockdowns rose and eased respectively.

- Small island tourism comes with many bottlenecks and monopolistic arrangements. The operation of a suitable ferry and/or an airline, with their schedules, regularity, and price options, can make or break a tourism industry (and more besides). For small islands at the mercy of such supply side constraints, their best hope is to build on characteristics that render their tourism offer complementary to, and not in competition with, their main logistic hub, thus potentially luring tourists to sample their idiosyncratic fare. Consider volcanic landscapes in Réunion; Creoleness in Rodrigues and Petite Martinique; caves, geoparks, and the unique ecology in Minorca; and robust communitarianism and social solidarity in Miyako. Note that beaches are exceptionally absent here.

- What the length of an island's only airport runway can do to its economic prosperity is quite surprising. Island geography – a rugged, volcanic landscape in particular – and wind flows – such as windshear patterns – can dictate whether an airport is at all feasible; and, if yes, how long a runway can be. This in turn will have consequences on the type of airplanes that can navigate the take-off and landing conditions, and so also determine the number and purchasing power of arriving and departing visitors.

- Claiming authenticity, rusticity, seclusion, and tranquillity can be easier to implement in uninhabited or sparsely populated islands (e.g. Šolta, in Croatia; Alicudi, in Italy's Aeolian islands; or the outer islands of Seychelles). The claim will however be challenged by local residents' desire for a decent quality of life in more populated islands (e.g. Philippines).

- Islands, and islanders 'at the end of the world' may find that their geographical remoteness and marginalisation is politically and economically enhanced, even legitimated – by others, or by themselves – in order to be rendered and represented as more exotic, as more traditional, as more community-oriented, as more authentic, as more cut off, spared, and insulated from the burdens of modernity … anyway, as more *different*. They thus appeal to a narrower but more up-scale tourism market segment. This narrative however can also be used to justify these islands' underdevelopment and they being short-changed for infrastructure and investment.

- Islands with well-developed air and sea connectivity, including suitable port infrastructure, will continue to reap the lion's share of tourism visitations, and receipts; but they also face challenges relating to their carrying capacity and overtourism. Centralised institutions of economic and political power are likely to protect such arrangements. The politico-economic power imbalance is likely to be reflected, if not reproduced, geographically: the main island, and the capital city on that main island, is likely to dictate much of what goes on, even on outer islands. This applies to tourism as much as to other industries.

- There is no love lost between neighbouring islands. They may be obliged to present a common front and practise co-opetition for certain marketing campaigns, and to build robust regional, multi-island brands. But they can also compete vigorously with each other, seeking to stand out from and against their island neighbours. All talk of island hopping and inter-island travel would still conjure up nervous concerns about revenue allocation and distribution.

Sustainable tourism for archipelagos? Five issues

In this book and in the 2015 archipelago tourism volume, five attributes of archipelago tourism are proposed. In their seminal contribution to comparative archipelago tourism, Bardolet and Sheldon (2008) also identify five issues that archipelagos confront as they seek to develop sustainable tourism and which differ from those faced by single islands. It is pertinent to revisit and expand on them here, now that so much material and experiences have been critically examined in the preceding pages of this text.

First, *complex governance*: power in archipelagos is often shared between and across many levels. Tourism (and other) policies must negotiate the complexity of hierarchical governmental structures and decision-making (mainland, region, archipelago, island), potentially leading to conflict and communication and political difficulties not faced in individual islands (e.g. Trousdale, 1999). The ensuing strategic games can have quite dramatic results; and 'zero sum' scenarios would result in winners and losers. The level of complexity gets worse when dealing across political borders, including "impeded archipelagos": neighbouring islands, like French Corsica and Italian Sardinia, that belong to different polities, resulting in stunted cross-island tourism and trade (Farinelli, 2021).

Second, *cultural diversity*: different cultural and community interests on each island will challenge stakeholder involvement in any centralised tourism planning (Sheehan and Ritchie, 2005). Balancing the views of each island's community as the archipelago develops tourism is ideal but elusive. Nor is one to essentialise islands and assume that each island's community can easily come together and speak as one voice.

Third, *differential development*: the different islands in an archipelago are likely to find themselves at different stages of the tourism lifecycle, requiring distinct product and market policies. As one island reaches maturity, visitors may shift to another island, leaving the first to redefine and renovate its image. Inter-island competition can cause market confusion and/or political tension among the various island communities. The rational way forward is to promote a 'win-win' complementarity of the tourism offer, even if this is contrived through savvy marketing and branding (e.g. Baldacchino and Ferreira, 2015).

Fourth, *transport infrastructure*. Travel between the islands in an archipelago is critical to their tourism development and to their inhabitants' quality of life. Distances between islands, and the character of transportation linking the islands, are a vital part of an archipelago's tourism policies. If most visitors stay on one island, then transport logistics – passenger safety; and frequency, cost, regularity and choice in modes of travel – will be partly responsible (e.g. Karampela, Kizos and Spilanis, 2014). Connectivity concerns highlight the crucial status of islands as open economies and societies, whose survival depends on what comes from or on the sea, for better or for worse. This, by the way, is a beckoning area for comparative, quantitative island tourism research.

Fifth and lastly, *data availability*. The standardisation of tourism statistical and economic data collection is more challenging given different governmental jurisdictions on the various islands. Without some standardisation of data, centralised planning and policy-making for any archipelago become more difficult. All things being equal, it is more likely that good quality data exists and is collected regularly when one is dealing with island *jurisdictions* (sovereign states or sub-national territories with some level of autonomy) (Baldacchino, 2004).

Conclusion: Reordering how we look at the world

Archipelagos remain a less examined metageography. Thinking *with* the archipelago can change how we think about the world and our place in it (Pugh, 2013). Such a 'turn' foregrounds more fluid tropes of assemblages, mobilities, and multiplicities associated with island-island movements (Tsai, 2003). Admittedly, the plurality of an archipelago can be elusive; it may not easily lend itself to control and profiling; it may not settle submissively into tight historical, cultural, or discursive compartments; it could defy coordination and organisation; and it would tend to express itself via a cacophony of voices, aspirations, identities, and histories that clash with the 'official', smart logo, brand, identity, and history of a pluri-island group (Baldacchino and Ferreira, 2013). "Each island, however small, tends to have a distinct history, certain unique cultural characteristics, and often its own language or dialect" (Hamilton-Jones, 1992, p. 200); one can only imagine how many more differences than these tend to lurk and linger in *island-island* relations (La Flamme, 1983; Stratford et al., 2011).

During Covid-19, many have just been impatiently waiting for the revival of a familiar economy that worked to provide food on the table (Connell and Campbell, 2021, p. 520). Lockdowns, deserted hotels, and disrupted supply chains have resurrected and foregrounded traditional practices and informal economy pursuits – artisanal fishing, subsistence agriculture, barter, petty self-employment – which offered solace to many while Covid-19 was rampant. In the post-Covid-epoch, the same life-saving 'traditional practices' may offer niche tourism opportunities, especially to peripheral islands which are not likely to benefit, or suffer, from mass tourism numbers.

Post Covid-19, there remain no compelling reasons why islands and archipelagos need to consider, let alone choose, new development trajectories. The status quo is strong, and is back with a vengeance. At the same time, however, it *does* help to usurp the 'territorial trap' (Agnew, 1994) and resist the raw and awesome power of the political map in shaping our views of the world (Sidaway, 2007). There are new insights and epistemologies awaiting, making us better able to see our geographical environment as not just consisting of the 'islands of the world' but of 'our world of islands' (Baldacchino, 2007). If we see islands at all.

Here lies the opportunity to reorder how we look at the world (Lewis and Wigen, 1997, p. ix); and how to unsettle and recentre cartographic imaginations.

References

Agnew, J. (1994). The territorial trap: The geographical assumptions of international relations theory. *Review of International Political Economy*, 1(1), 53–80. https://doi.org/10.1080/09692299408434268.

Baldacchino, G. (2004). The coming of age of island studies. *Tijdschrift voor Economische en Sociale Geografie*, 95(3), 272–283. https://doi.org/10.1111/j.1467-9663.2004.00307.x.

Baldacchino, G. (2007). Introducing a world of islands. In G. Baldacchino (Ed.), *A world of islands: An island studies reader* (pp. 1–29). Malta and Canada, Agenda Academic and Institute of Island Studies, University of Prince Edward Island.

Baldacchino, G. and Ferreira, E. C. D. (2013). Competing notions of diversity in archipelago tourism: Transport logistics, official rhetoric and inter-island rivalry in the Azores. *Island Studies Journal*, 8(1), 84–104. https://doi.org/10.24043/isj.278.

Baldacchino, G. and Ferreira, E. C. D. (2015). Contrived complementarity: Transport logistics, official rhetoric and inter-island rivalry in the Azorean archipelago. In G. Baldacchino (Ed.), *Archipelago tourism: Policies and practices* (pp. 85–102). Farnham: Ashgate.

Bardolet, E. and Sheldon, P. J. (2008). Tourism in archipelagos: Hawai'i and the Balearics, *Annals of Tourism Research*, 35(4), 900–923. https://doi.org/10.1016/j.annals.2008.07.005.

Briassoulis, H. (2017). Tourism destinations as multiplicities: The view from assemblage thinking. *International Journal of Tourism Research*, 19(3), 304–317. https://doi.org/10.1002/jtr.2113.

Butler, R. W. (1980). The concept of a tourist area cycle of evolution: Implications for management of resources. *The Canadian Geographer/ Le Geographe Canadien*, 24(1), 5–12.

Campbell, Y. and Connell, J. (Eds) (2021). *COVID in the islands: A comparative perspective on the Caribbean and the Pacific*. New York: Palgrave Macmillan.

Connell, J. and Campbell, Y. (2021). Aftermath: Towards the 'new normal'? In Y. Campbell and J. Connell (Eds), *COVID in the islands: A comparative perspective on the Caribbean and the Pacific* (pp. 517–528). New York: Palgrave Macmillan.

Elden, S. (2007). There is a politics of space because space is political. *Radical Philosophy Review*, 10(2), 101–116. https://doi.org/10.5840/radphilrev20071022.

Farinelli, M. A. (2021). The impeded archipelago of Corsica and Sardinia. *Island Studies Journal*, 16(1), 325–342. https://doi.org/10.24043/isj.142.

Hamilton-Jones, D. (1992). Problems of inter-island shipping in archipelagic small island countries: Fiji and the Cook Islands. In H. M. Hintjens and M. D. D. Newitt

(Eds), *The political economy of small tropical islands: The importance of being small* (pp. 200–222). Exeter: University of Exeter Press.

Karampela, S., Kizos, T., and Spilanis, I. (2014). Accessibility of islands: towards a new geography based on transportation modes and choices. *Island Studies Journal*, 9(2). 293–306. https://doi.org/10.24043/isj.307.

La Flamme, A. G. (1983). The archipelago state as a societal subtype. *Current Anthropology*, 24(3), 361–362.

Lefebvre, H. (2014). The production of space (1991). In J. J. Gieseking, W. Mangold, C. Katz, S. Low, and S. Saegert (Eds), *The people, place, and space reader* (pp. 289–293). London: Routledge.

Lewis, M. W. and Wigen, K. E. (1997). *The myth of continents: A critique of metageography*. Berkeley CA: University of California Press.

Pugh, J. (2013). Island movements: Thinking with the archipelago. *Island Studies Journal*, 8(1), 9–24. https://doi.org/10.24043/isj.273.

Sheehan, L. and Ritchie, B. (2005). Destination stakeholders: Exploring identity and salience. *Annals of Tourism Research*, 32(3), 711–734. https://doi.org/10.1016/j.annals.2004.10.013.

Sidaway, J. D. (2007). Enclave space: A new metageography of development? *Area*, 39(3), 331–339. https://doi.org/10.1111/j.1475-4762.2007.00757.x.

Stratford, E., Baldacchino, G., McMahon, E., Farbotko, C., and Harwood, A. (2011). Envisioning the archipelago. *Island Studies Journal*, 6(2), 113–130. https://doi.org/10.24043/isj.253.

Trousdale, W. J. (1999). Governance in context: Boracay island, Philippines. *Annals of Tourism Research*, 26(4), 840–867. https://doi.org/10.1016/S0160-7383(99)00036-5.

Tsai, H.-M. (2003). Island biocultural assemblages: The case of Kinmen island. *Geografiska Annaler B*, 85(4), 209–218. https://doi.org/10.1111/j.0435-3684.2003.00143.x.

Index

For Product Safety Concerns and Information please contact our EU
representative GPSR@taylorandfrancis.com
Taylor & Francis Verlag GmbH, Kaufingerstraße 24, 80331 München, Germany

www.ingramcontent.com/pod-product-compliance
Lightning Source LLC
Chambersburg PA
CBHW060237220326
41598CB00027B/3958

9 781032 586786